Atlas of Economic Mineral Deposits

Atlas of

Economic Mineral Deposits

COLIN J. DIXON

Senior Lecturer in Mining Geology, Imperial College, London

Cornell University Press

Ithaca, New York

© 1979 Colin J. Dixon

First published 1979 by Cornell University Press

Printed in Great Britain by Fletcher & Son, Ltd., Norwich

Library of Congress Cataloging in Publication Data
(For library cataloging purposes only)

Dixon, Colin J.
 Atlas of economic mineral deposits.

 Bibliography: p.
 Includes index.
 1. Mines and mineral resources.   2. Geology, Economic.   I. Title.
 TN263.D57   553   78-65360
 ISBN 0-8014-1231-5

Contents

Introduction

The Scope, Purpose and Layout of the Book

Air, sea, surface water and soil support life, from which comes our food; the fossil remains of life, that is: coal, oil and gas, together with solar and terrestrial radiation provide energy; but almost all the artifacts of human civilization are made from substances taken from the earth's crust. This book deals specifically with the geology of economically workable deposits of the solid, non-combustible mineral resources of the earth. Its purpose is to describe the geology of a carefully selected group of mineral deposits; to show where they are, what they consist of, how big they are and how they are related to the geological environment in which they occur. It is also intended to show what place each example has in the scientific study of deposits, and its importance as part of the natural resources available to mankind.

I am convinced that, to a large extent, geologists think and communicate best in pictures. The basis of almost all geological work is the making of maps, plans, sections and the like. The graphical form is the most efficient way of displaying and explaining geological observations and ideas. I have, therefore, decided to present this work in the form of, and to call it, an Atlas.

Choice of Material. Forty-eight mineral deposits or groups of deposits are described with world distribution maps of five selected groups of commodities. The former are intended to exemplify the principal types of deposits known to geologists; the latter to show how the major deposits are related to the general geology of the earth's crust. It is not easy to define an individual 'economic mineral deposit'. It is an accumulation of a mineral or minerals at a particular place in the earth's crust and is thus a geological object; but, by the word 'economic' is implied that the mineral accumulation has some value to mankind. Because the needs of mankind change from time to time and place to place, it is never possible to define the set of geological objects called mineral deposits in purely geological terms.

A single body of rock that is in some way valuable to man is called an 'ore body'. (Leaving aside for a moment the arguments about the usage of the word 'ore'), ore bodies that are geologically similar often occur in groups at the same locality. Also, we often find groups of ore bodies that are quite different occurring together, but with features that make us conclude that they have a common mode of formation. Where a single ore body is characteristic of a group, I have chosen to describe it on its own. In other cases, I have chosen to describe groups; either 'mineral fields' (generally in the range 1–10 km in size), 'mineral districts' (10–100 km) or in a few cases a 'mineral area' (100 km and over).

The choice of the forty-eight examples has not been easy, and is necessarily a personal one. In making the choice I have been guided by certain principles. I have tried to include the widest possible range of types from a geological point of view, but also to cover the widest possible range of commodities. Deposits of metalliferous ores figure rather more than the so-called 'industrial' and non-metallic minerals. There are several reasons for this, the most important being the availability of information. I have not included any examples of deposits from the U.S.S.R., China and Eastern Europe, and although information on deposits in these countries is now generally available, much of it is in the form of scientific discourses rather than primary descriptive material. Furthermore, I would not want to add to the excellent work of Smirnov, whose book on the mineral deposits of the U.S.S.R. was recently published in English. I have tried to include as many as possible of the large and famous deposits and those that may be regarded as type examples, confining the choice to those that are well documented in freely available scientific publications, and those that can be readily displayed in graphical form. Even with all these limitations, choice was still difficult and the final selection, based on my experience as a teacher, is in the form of an introductory syllabus for students of the subject.

Readership. The Atlas is intended to be a basic work of reference for any reader who wishes to study economic mineral deposits. I have in mind that it could be the basic descriptive part of a university course on the subject. Many teachers of economic and mining geology prefer to lecture on the formative geological processes and origin of mineral deposits, and most of the existing textbooks do likewise. The Atlas is intended to be a compendium of descriptive material on which a more analytical series of lectures, or course of reading, could be based.

The book is also intended for those many students of geology who do not have the opportunity to study mineral deposits as part of their formal education, but who wish to know more; either for reasons of general interest, or because they want to enter the mining and exploration industries. The Atlas should be useful to students who study geology as a subsidiary subject, particularly mining engineering. I also hope that it will give an understanding of the size, scale and distribution of mineral resources to those who study geography, economics, planning and the environmental sciences. There is a growing interest in mineral resources from a wide section of the public, and the highly illustrated form of this book is, in part, intended to make the subject comprehensible to the non-geologist.

Sources. As far as possible I have taken the main material from published sources that are widely available. All the maps and diagrams have been simplified and modified after the originals to emphasize the key features within the limitations of scale. A very large proportion of the material has come from the maps and publications of national geological surveys, and from the many excellent review papers presented to international congresses. Unfortunately, the learned journals in economic geology are not as rich as they might be in graphical material. It is a common experience to read a fascinating account of some famous mineral deposit but to be frustrated by the lack of illustration showing where it is and what it looks like. In a few cases I have had to use material from university doctoral theses which, although available in the libraries of the universities to which they were presented, are not published as such.

Format. The description of each mineral deposit is presented as a page of diagrams, and a text that is intended to be a set of notes on the diagrams. In each case I have included a location map, a map showing the regional geological setting of the deposit, and maps and sections of the deposit itself, together with stratigraphic columns and other appropriate diagrams. Although in many cases very simplified and generalized, the maps and diagrams are for the most part real ones, not idealized sketches.

All maps and sections have been drawn on simple scales because this helps one of the most essential parts of geological education; the appreciation of the scale of geological phenomena. The colour scheme is common throughout the book. The objects of economic interest are printed in red, the immediate host rocks or rocks most closely related to the deposit, are printed in a pale red tint. The rest is printed in black using, as far as possible, the conventional and self-explanatory symbols. A grey tint has been used in most cases to indicate the rocks which form the 'basement' to the units in which the deposit occurs.

The explanatory notes follow a common pattern. I have given a precise statement of the location of each deposit because my experience of much of the scientific literature is that it is very weak on this point. Notes are included on the geographical setting, history and mining of each of the deposits because they will interest many readers and act as an aid to memory by placing it in a context. The geological notes are divided into four parts: background, detail of the deposit, some ideas of size and richness, and a separate section in which the main ideas that have been advanced for the origin of the deposit are summarized. A full genetic discussion is not included because there are many other texts that do this, but some introductory pages are included placing these short sections on interpretations into a wider context. The Atlas should be studied in conjunction with other texts that go more deeply into theories of origin, and a list of the major works is included at the end of this introduction. There is also a selection of references for 'further reading'. These are not intended to be full bibliographies, but they do include references to the source material. Full bibliographies are in many cases unnecessary because of the monumental work of Ridge (Ridge, 1972 and 1976).

Terminology. This is a persistent problem in geology. What I have tried to do is use a consistent, and internationally acceptable set of terms, making as much use as possible of the recent attempts by international organizations to standardize geological terminology. In many cases this has meant changing the words used by original authors. Much of the source material for the Atlas was published before attempts were made to establish a more rigorous lexicon, and changes have to be made; but it is difficult to do this without making some mistakes, and impossible without risk of offending some authors. International standardization of terminology has been successful in mineralogy, partially successful in stratigraphy, but rather less so in petrology.

In stratigraphy I have eliminated words like 'series' or 'system' unless they are correctly used as chrono-stratigraphic terms. It is not always easy to find the correct term with which to replace them, particularly when dealing with Pre-Cambrian litho-stratigraphy. The definitions of stratigraphic boundaries are constantly being revised and in many cases older literature does not contain enough information to enable a redrawing of boundaries to be made. To help the reader a table of the chrono-stratigraphic terms used in the book is included (p. 138). Where I have made important changes in stratigraphic or petrological terms they are explained in footnotes.

I have not burdened the text with any explanation of the names of minerals. Instead, there is a glossary of mineral names; but only the barest outline of information is given and the readers will have to refer to the available texts on mineralogy for more information.

Units of measurement used are those of the 'Système Internationale'. I have used several units that are acceptable to, rather than being an integral part of the S.I. where I feel that they would be better understood. (Thus, I have used the tonne (t) in preference to the megagram.) For the benefit of readers who are not accustomed to using units of the S.I., there is a table of conversion factors (p. 137).

Mineral Deposits, their Discovery and Exploitation

In this book the definition used of mineral deposits will be 'bodies of rock of actual or potential value to mankind'. The use of the term 'mineral' here is in the sense used by mineralogists, that is to say, a naturally occurring substance with an atomic structure that approaches a defined ideal and a chemical composition that varies within a range consistent with the atomic structure. Rocks, being aggregates of one or more minerals, can be of value if they can be used for some purpose as a whole, or if they contain one or more constituent minerals that can be used. There are four kinds of rocks that are useful.

The first kind includes those that contain one or more minerals from which metals can be extracted and are usually referred to as 'metallic ores', or simply 'ores'. The term ore is sometimes used in other senses, but the meaning quoted will be the only one used in this book. The second kind of valuable rock includes those that contain minerals from which the non-metallic elements may be extracted or from which compounds of the non-metallic elements may be derived. These may be referred to as 'non-metallic ores', but some people do not use this term. Since most of these minerals are used as raw materials in the chemical industry one can refer to them as such. The third kind of valuable rocks are those consisting of, or containing, the so-called industrial minerals. These are minerals that are used as such, because of their properties rather than as sources of an element or compound. Some industrial minerals are used in the form in which they occur, others are feed-stocks to pyro-chemical or other industrial processes. The fourth kind includes rocks that are used in bulk, primarily for the construction industry, and in which it is the properties of the rock *en masse* that are important rather than those of the individual mineral constituents.

The difference between these four kinds of useful rock can be explained by examples of minerals that belong to more than one kind. Chromite for instance may be smelted to produce ferro-chrome, and thus one can call a chromite-bearing rock a chrome ore. But the same mineral can be roasted with soda-ash to produce sodium dichromate for use as an oxidizing agent in the chemical industry. Chromite may also be mixed with magnesium oxide and formed into a brick that will withstand high temperatures, in which case it is being used as an industrial mineral. A block of marble may be cut into slabs to cover a building and thus is a bulk constructional material, but the dust from the cutting may be used as a filler in plastic floor tiles, in which case it would be regarded as an industrial mineral.

The phrase 'value to mankind' needs some explanation. As recently as 1963 a book on mineral deposits (Park and MacDiarmid) appeared which defined them as rocks that can be recovered at a profit. The word 'profit' has come to mean many things and, as a result of the political idealogies derived from the Industrial Revolution, the word tends to be emotive. However, whatever the economic or political structure of the part of the world in which a mineral deposit is worked, there will always be some criteria by which the effort put in to exploiting it may be compared with the incentive for doing so. A mass of rock becomes an economic mineral deposit if the incentive compares favourably with the effort needed to work it.

The Discovery of Mineral Deposits

The art of discovering mineral deposits is called prospecting* and geology is the science that has greatest application to that art. A mineral deposit is found when someone sees a piece of it and recognizes it for what it is. There are some deposits that outcrop, but there are many more that only outcrop in a modified form or are obscured by superficial materials. The surface expressions of mineral deposits are a study in their own right and vital to the prospector. Deposits containing metal sulphide minerals, for example, are frequently oxidized at outcrop. Massive or diffuse, zones of iron oxide minerals (with others) occur at the surface, known respectively as 'gossans' or 'leached outcrops'. The significance of large, well-developed gossans has been known for several millennia, but we are still learning how to recognize the more subtle forms of leached outcrop which are often difficult to distinguish from 'rusty' looking weathered rocks unrelated to mineral deposits. Outcrops of deposits that do not contain sulphide minerals may also be modified, but forms that occur are numerous and defy generalization.

Many mineral deposits are obscured by superficial materials. Soil is the most obvious of these; but large areas of the world are covered by unconsolidated materials such as sand, alluvium, glacial till, loess, shallow water and ice. Chemical processes taking place at the surface often produce a superficial layer that is cemented by iron oxides, silica or calcium carbonate (known respectively as ferricrete, silicrete and calcrete) that obscures the appearance of rocks at outcrop.

Mineral deposits which do not outcrop have to be found by indirect methods. The foremost indirect method of prospecting is geological inference. Certain types of deposit occur in specific types of rock or in association with certain kinds of geological structure. The fundamental method of prospecting is the search of the land (or bed of the sea) for surface expressions and for geological features from which may be inferred the possible presence of a mineral deposit. The record of such a search is in the form of maps, the making of which can be helped by photographs and images of the surface taken from air- or space-craft. But there are a number of aids to prospecting that depend on one of two things: either the effect that some minerals have on physical force fields, or the dispersion of minerals or of their chemical constituents from, or associated with, their sub-outcrops.

In the first case, geophysical prospecting, a number of force fields can be measured at or above the surface. These are the magnetic, gravitational and electrical fields; radiation flux and the respose of the rocks to electrical or electro-magnetic induction. There are a large number of geophysical aids to prospecting that are of two kinds; methods which aim to detect anomalies caused by accumulations of minerals with certain properties (e.g. ferromagnetism, high density) and methods which aim to detect anomalous fields due to rocks or structures that may be related to deposits.

The second case, geochemical prospecting, is carried out by sampling and chemical analysis of surface sediment, soil, water, outcropping rock or the interstitial air in soil or sediment. The results sometimes show anomalies that indicate the presence of concentrations of certain minerals in the bedrock. Somewhat similar to geochemical prospecting is the direct search for mineral grains in superficial materials which survive the environment at the surface. This includes the old traditional method of prospecting for heavy minerals with a pan, and the more sophisticated versions of the same method that have been developed for certain types of mineral deposit.

Geophysical and geochemical methods have had their successes, but nonetheless, a great deal of the world's mineral resources have been found just by looking over the surface and quite a few deposits have been found by chance, following the digging of a hole in the ground for some other purpose, such as railway or road cuttings. No real discovery happens until the prospector actually sees a piece of the valuable rock. From the earliest times, indications of mineral deposits have been tested by digging pits, trenches and the like. But the greatest aid to prospecting to be invented in the modern era is the drill. There are various forms of drill; percussion, rotary, helical augers, etc. It is the diamond-core drill that is the greatest single prospecting tool. This device can penetrate the earth's crust well beyond the reach of mining techniques, bringing back samples of rock that can be studied and tested almost as if they were outcrops. In several areas of the world it has been shown on various occasions that the more you drill, the more mineral discoveries are made. But drills have an even more important use because they enable mineral discoveries to be explored and evaluated.

The Exploration of Mineral Deposits

The purpose of exploring a mineral deposit is to find out whether it is worthwile to exploit it. What must be shown is that, with a chosen level of confidence (correspondingly to an acceptable level of risk) the deposit contains enough of what is wanted, and of sufficient quality, to repay the effort of developing and exploiting it. The quantity of rock that can be mined must be enough to sustain a certain rate of production for a definite period of time (usually between 10 and 30 years) and the rate of production must be great enough to profit from economies of scale, but not so great as to cause overproduction in comparison to demand.

There are three essential aspects of this process. The first is to establish how big is the deposit. This is a matter of investigation (by means of drilling or mine workings) of the boundaries of the deposit; in other words, subsurface geological interpretation. Some idea of the relationship that the deposit has to its host rocks frequently aids this part of the process. The second aspect is to determine all the features of the deposit that will enable a method of mining to be designed. The shape and size of the deposit is all-important here, as well as its position, depth from the surface, the mechanical properties of the enclosing rocks and the distribution of water in the rocks. The third aspect is the quality of the material itself. In a few cases, it is the whole rock that is wanted but, generally, the important factors are the amount and distribution of one or more valuable minerals that are contained in it.

The quality of the material is estimated by sampling (by drilling or other methods) and carrying out mineralogical, chemical or other analyses of the rock. Care must be taken to ensure that the values indicated by the samples are not biased and there are a variety of methods, collectively called geostatistical, that are usually capable of giving unbiased estimates of the quality, or grade, of the desired substance that is present, and an estimate of the uncertainty of these estimates. Large samples of the deposit are usually taken for tests that enable a design to be worked out for mining and processing of the material into usable form.

Mining

This term is used here to cover all methods of getting rock out of the ground (in spite of the common tendency to restrict the term to underground methods). Methods of mining are determined by size, shape and position of the deposit and a number of other factors. Methods differ from one another in: being open to the sky or underground; in what is done with the hole left after the ore is taken out; and the sequence of stages in which it is carried out. Surface workings, open-pits, open-cast workings, quarries, etc. are applicable to ore bodies near the surface. In the special case of unconsolidated materials the working can often be by dredges working under water. Otherwise the pit has to be drained, both for access and to increase the bulk strength of the walls of the working. Open-pit working is sometimes carried out by making a conical hole in steps or benches and leaving it open. This depends for its success on maintaining the slope of the sides flat enough to prevent collapses. Strip-mining is the name generally given to open-pit mining where the deposit is worked in strips side by side, using the waste rock covering each strip to fill in the hole left by the previous. Loosely consolidated materials can sometimes be worked by high pressure water (hydraulic mining), softer materials can be dug with mechanical digging machinery, but harder rocks have to be drilled and blasted with explosives.

Underground workings have to be used when the deposit is deep, or if there is some reason for not disturbing the surface, such as the preservation of farmland. Means of access are provided by shafts, adits or declines, the size and number of which usually limits the rate of production. Underground, the stopes (or places from which the ore is taken) can be left open, but usually they either have to be supported or allowed to collapse in a controlled way. Support is provided by leaving pillars, which limits the amount of material that can be mined, by timbering or by back-filling with waste rock, or similar material. Underground, rock is broken by drilling and explosives, by special digging machinery if the rock is soft, or by allowing the rock to crush under its own weight.

Mineral Processing

Most valuable rocks have to be processed in some way to produce a usable product. In the case of metallic ores and a number of other types of deposit, this usually means a process known as 'benificiation' that produces a concentrate of the wanted minerals that is less bulky to transport.

A beneficiation process is usually in two parts: comminution and concentration. Comminution is a matter of crushing and grinding the rock to a size range such that, as far as possible, the grains of the different minerals are liberated from one another. In many mineral workings this is the most difficult and expensive part of the process. On average it takes about 15 kW h of energy to liberate the minerals in a tonne of rock, and this may account for 80% of all the energy used in producing the final product. Not infrequently comminution must reduce all the rock to the size of fine sand, silt or mud.

Concentration processes depend on the properties of the minerals that are present. Various physical or physico-chemical properties of minerals enable machinery to be designed that will separate one set from another. The properties include density, magnetic susceptibility, electrical conductivity, surface chemical properties, colour, reflectance and several others. One important aspect of geology applied to mining is the investigation of the properties and style of intergrowth of the minerals, upon which the design of benificiation processes depends.

Apart from benificiation processes there are a number of others that have to be used in the preparation of the raw rock for use. Many mineral products have to be graded into sizes by the use of screens or gravity-separation processes. In some cases the grain size to which the material must be reduced for libration and concentration is smaller than that acceptable for subsequent use, and the concentrate has to be pelletized, agglomerated or bricketted. Many industrial and bulk-minerals are feed-stocks to pyrochemical processes such as cement-, refractory-, ceramic- or brick-making processes which involve burning or heating the material in a kiln. The

* I use this term deliberately as it is the traditional one. I cannot understand why some people want to substitute for it the word 'exploration'. A person who prospects may be called a 'prospector'. One cannot use the word 'explorer', since this has another meaning, which leads to the use of the ghastly and unnecessary word 'explorationist'.

efficiency of most of these processes depends on careful control of grain size, and the chemical composition of the material.

Metallurgical and Extractive Processes

Raw rock or mineral concentrates are often used to produce a metal, non-metallic element or compound by some process. Broadly these fall into three classes, pyro-, hydro- and electro-chemical. Smelting of metallic ores or concentrates is an example of a pyro-metallurgical process. This involves heating the material in a furnace so that minerals such as sulphides may be oxidized and oxides reduced to molten metal. One of the best known hydrometallurgical processes is that used to extract precious metals from their ores. The pulped rock is mixed with sodium cyanide solution, the solution cleaned and the gold, silver, platinum, etc. precipitated. Hydro-metallurgical processes are used to recover uranium, copper, and in the extraction of alumina from bauxite. Electro-chemical processes are frequently used to refine raw smelter products because they can be very selective. Pyro-chemical processes are used to extract many non-metals such as sulphur; hydrochemical ores are widely used, for instance, to extract phosphoric acid from phosphatic rock and, in a few cases, electro-chemical processes are used (for example to produce chlorine from salt).

With all these extractive processes there are two important aspects. The first is the main purpose, that of extracting the desired product. The second is the removal of unwanted impurities that would otherwise alter the properties and value of the product. This second aspect is particularly important for geologists because many of the impurities found in the final product are in the original deposit, the minor components of which must be studied and understood. In quite a few instances rock may contain what seems to be a workable grade of a valuable substance, but is rendered valueless because an economic treatment or refining process cannot be designed to remove unwanted impurities.

The Economics of Mineral Working

At a basic level, mineral working is encouraged by rising demand and discouraged by lack of it. For long periods in history the incentives or disincentives to mineral working were set by the establishment of a price for the product in a market. But at various times, not least at the present day, minerals have not been produced in a free market situation. Mineral producers try to monopolize the market for particular commodities, or they band together into cartels to maintain high prices. Mineral consumers do the same for the opposite reasons. Governments intervene to maintain levels of employment or to maximize foreign currency earnings. They also intervene to discourage mining by imposition of taxes or quotas, to curtail the effects of pollution, despoliation of land, or to conserve natural resources for the future. Countries with centrally planned economies often try to set production quotas according to expected demand, but are not usually more successful at getting the equation right than is the case in free market economies.

A mineral deposit becomes economic if the reward exceeds the effort of exploiting it. It is convenient to translate these terms into those that apply to the capitalist and mixed economies, and talk of cost rather than effort, and selling prices rather than reward; while keeping in mind that there are other terms that must be used in some places and at other times. ('A slave miner works because the whip of the overseer is the incentive'.)

The cost of a mining operation is in two parts. The first is the cost of the provision of equipment, the construction of plant and execution of the preliminary works necessary to begin, expand or maintain the necessary rate of production. This, for the sake of simplicity, we will call the investment cost. The second part is the cost of actual production. The size of the investment is roughly proportional to the required productive capacity and depends on the location of the deposit, being high in remote places, or where mining is disruptive to a developed region. The cost of production tends to be lower per unit produced as the rate of production increases; the economy of scale. An essential part of the cost of production and to some extent also of the investment, is the cost of transporting the product to the user, particularly if the location is remote.

A general feature of mineral industry is that investment is high compared with annual turnover, often of the order of three times in monetary terms. (In many manufacturing industries the annual turnover is usually similar to the invested capital of the enterprise.) Also the time taken to execute the investment works is long compared with manufacturing industry. It can often be five or more years from the time a deposit is discovered to the time production begins, whereas a factory can often be built in a year. The result of this is that the initiation or expansion of a mineral operation, while it may be encouraged by rising demand, depends very much on the ability of society to generate the surplus wealth necessary for the required investment. It sometimes happens that investment begins because of demand which can be falling by the time production begins. Another consequence of the capital-intensive nature of mining is that the added cost of failing to produce to the full designed capacity is great. These features, that are characteristic of, but not unique to, the mineral industries, make, as a consequence, a very erratic industry that oscillates between over- and under-production.

A further characteristic of the industry is the distribution of resources. Centres of population and industry, the users of mineral raw materials, have tended to grow alongside sources of energy and other natural resources. Since the Industrial Revolution the range of mineral products used has become very great, and sources of mineral raw materials are frequently remote from the centres where the products are used. It follows that transport is a vital factor in the industry, but perhaps more important, political and social changes have a marked effect on supply.

The Study of Mineral Deposits

This is probably the oldest branch of geology and predates the development of the science and the invention of the name. The nomenclature of the science is complicated and confusing; there is no name that can be universally applied to the study of mineral deposits. (This may be no bad thing, because names seem to foster separatism in branches of science.) Mining Geology is the branch of geology applied to the art of mining, but it is normally taken to cover only the applications to developing and running mines. The best-known scientific publication that deals with mineral deposits is called Economic Geology, but that name can be taken to cover other things besides the study of mineral deposits. There is a term 'metallogeny', the study of the origin of metals in the earth's crust. But there is more to the study than the discovery of origins, and nobody has managed to invent an acceptable term which includes non-metallic mineral deposits. Some years ago the term *'gîtologie'* was invented in France to cover the descriptive part of the study, a kind of 'physiology' of deposits separate from *'métallogenie'*. If this book had been written in French it would doubtless be called 'l'Atlas de Gîtologie'. But the word does not fit happily into the English language, and is not widely used.

This section of the introduction is an outline of the various aspects of the scientific study of economic mineral deposits. The study can be thought of as providing the answers to five questions for each deposit; where is it? what does it look like? of what does it consist? how is it related to its geological surroundings?, and how did it get there? The answer to the first question is a matter of geography rather than geology, but should be included because it is often forgotten. It is surprising how many learned works on a mineral deposit do not tell you were it is (on occasions there may be commercial reasons for keeping it a secret). It is mandatory on a physicist or chemist to describe his experiments in such a way that they may be repeated by others, in order to check the results. It is likewise mandatory on a geologist to describe the locality of his observations with sufficient precision to enable others to go there and look for themselves.

The Morphology of Mineral Deposits

The earliest classifications of mineral deposits were based on form, probably because this is the principal factor affecting the way they can be mined. A large number of terms for describing what mineral deposits look like have come into use, but few are purely morphological as they also imply relationships with surrounding rocks. A 'vein' is a deposit of tabular form, but the term also implies that it cuts across the predominating structure of its wall-rocks. The term 'bed' also describes a tabular form, but one that has a conformable relationship to its host rocks; moreover, it is only applied to sedimentary rocks. Some terms are more or less purely morphological; such as 'lens', 'pipe', 'pocket', 'pod' and are for the most part self-explanatory.

There are a number of important terms used to describe form and relationship. Some deposits are described as 'stratiform' that is to say, having the form of a stratum. It might be thought to be synonymous with 'bed', but it is generally applied to deposits that are members of a squence of strata and which display internal stratification, lamination or similar features generally conformable to those of the host rocks. It is unfortunate that some geologists are hesitant to use this term because they fear that its use may imply a sedimentary origin. Others refuse to use it if a part of the body of mineral shows disconformable relationships. This is quite wrong. We must accept some terms that are morpho-relational, but we should not accept any confusion with genetic ones. 'Strata-bound' is another useful term that suffers from the same confusion. It means that the mineral deposit is in a series of strata, but is confined within it rather than having that form. The difference can be readily appreciated by referring respectively to the Sullivan and Laisvall deposits in Part II.

Veins are common forms. They may be single or may occur in groups. A group of veins that has some preferred orientation or organization is called a 'vein system'. Small veins that are too small to be mined individually are called veinlets; one can also find veinlet systems. Where a group of veinlets has little or no organization it is called a 'stockwork'. The nomenclature of mining contains a number of words such as lode, reef, ledge, etc., but most of these are ill-defined and of little value as scientific terms.

The Mineralogy and Petrology of Mineral Deposits

Under this heading we can deal in part with the answer to the question 'of what does it consist?' The study of the minerals for which mankind finds a use is sometimes called 'economic mineralogy', but is is not in reality a different subject from mineralogy in general. However, there are certain features of minerals that are of special relevance in the study of mineral deposits and quite a few minerals, commonly encountered in mineral deposits, are rather unusual in ordinary rocks. We find a use for just over a hundred minerals, and perhaps another hundred are encountered during the study of economic mineral deposits, in all about a tenth of the total number recognized by mineralogists. For a proper understanding of mineral deposits it is necessary to understand certain factors that control their composition, structure and form.

Most minerals are substances in the solid, crystalline state that can be regarded as being a more or less symmetrical arrangement in space of atoms, or groups of atoms of a certain kind, between which are 'sites' that can be occupied by various others. To explain this somewhat simplified view of crystal structure it is best to take an example. The mineral chromite has the ideal composition, $FeCr_2O_4$ and can be regarded as an isometric lattice of oxygen atoms between which there are two kinds of site, the difference between which is in the number of oxygen atoms that are in close proximity to the atom that occupies the site. In the ideal chromite crystal, one set of sites is occupied by iron and the other by chromium. But magnesium atoms can equally well occupy the sites normally occupied by iron, and aluminium and iron can replace chromium. The composition of the mineral can thus be represented as $(Fe, Mg).(Cr,Al,Fe)_2O_4$. Chromite is said to be a member of an isomorphous group of minerals (known as the 'spinels') which has a very wide range of possible composition. Clearly if this mineral is to be used as an ore of chromium, this potential for variation in composition is of some economic importance.

Some isomorphous groups of minerals, many of economic importance, display another and related phenomenon, that is the range of possible composition changes with physical conditions of formation. The atoms in a crystal lattice are to some extent free to spin, vibrate and to move from one lattice site to another. The rate at which they do these things is related directly to what we call temperature. Where a source of heat is available there is a tendency for atoms, of varying sizes and natures to be included in the same lattice. If the crystal is cooler there is a tendency towards a more ordered lattice, and initially heterogeneous crystals will tend to separate into two or more distinct ones. This process of 'unmixing' or 'ex-solution' occurs in many groups of minerals, one of the best known of which is the feldspar group which can end up at surface temperatures as very complex intergrowths called perthites, or anti-perthites. Unmixing is very common in several groups of economic minerals; it commonly occurs among the metal sulphides for instance. If the scale of the unmixed intergrowth is small, the separate phases cannot be liberated and separated in a benificiation process, and it then has to be treated as a single mineral which may carry impurities through to the extractive process. To some extent the study of these unmixed intergrowths can help in measuring the physical conditions of formation of the minerals. The optical microscope and, more recently, the electron microscope and the electron-probe microanalyser, have provided a great deal of information about this phenomenon.

There are some cases where minerals form with a composition outside the normal range at a given temperature. These are principally ones formed by an organic process, or in the presence of organic materials. Once removed from the effects of the organic environment, they revert to mixtures of mineral phases of the expected composition, but in the process their form and texture will have changed.

Minerals, like all substances in the crystalline state, depart from their ideal atomic structures and it is on these 'lattice imperfections' that depend many of the properties of minerals. Imperfections depend on the way the crystal grew, on irregularities in composition and on subsequent events that affect it, such as damage by radiation or mechanical stress. Crystal lattice disorders affect colour, magnetic susceptibility, mechanical strength and may affect the kinetics of chemical reactions in which the mineral takes part whether man-made or natural. In general, crystal structure is a more significant feature of the industrial minerals than of the ore minerals.

The behaviour and properties of minerals not only depends on composition and structure, but also on the exterior form of the crystals. The size, shape, fabric of intergrowth of mineral crystals results from their mode of formation. All minerals form from fluids and we can identify several modes of formation depending on the nature and physical conditions of the parent fluid. Some crystallize from melts such as silicate magmas or molten sulphides, oxides, carbonates, etc. Other minerals crystallize from simple ionic solutions such as sea water and ground water. Around volcanoes in particular, minerals form as sublimates directly from gases, but deeper in the earth's crust the distinction can no longer be drawn between the liquid and gaseous state; the range of temperatures and pressures encountered in the crust covering the critical points of some of the common solvents like water and carbon dioxide. Several kinds of mineral deposit are thought to form as a result of transitions that take place between what may be regarded as solvent-rich melts and solutions, i.e. the unmixing of phases in the liquid state akin to boiling. Quite a number of minerals display what are known as colloform and framboidal textures, commonly interpreted as indicating formation from colloidal solutions. Several substances precipitate, in the first instance, as gels which then dehydrate and crystallize. Some do this directly from ionic solutions, but others do so because of the influence of organic matter. A number of minerals have separate names for the macrocrystalline and colloform varieties; names that were given before X-ray crystallography and precise chemical analysis showed that they were essentially the same: compare for example, specularite and kidney iron ore, forms of hematite.

After the initial formation of a mineral, changes may take place which are of great importance in the study of mineral deposits. Minerals often recrystallize in the solid state, albeit aided by interstitial fluids. There is a tendency for more ordered crystals to form. Fine-grained colloform-textured minerals are often found with overgrowths of larger crystals. Fine-grained mineral precipitates, and mechanical agglomerations of minerals, tend to be annealed in much the same way as crystals in a metal during heat treatment. Aided by interstitial fluid, the minerals in many sediments gradually recrystallize as they turn to consolidated rocks, 'diagenetic processes'. Changes in the fabric of crystal intergrowths can be induced by stress. Some of these post-depositional effects are important in the designs of industrial processes that use minerals. They are of fundamental importance in the interpretations of the modes of formation and origin of minerals, because the texture or fabric of a mineral asemblage will tend to reveal the latest geological event that affected it, and it may not retain a detectable record of its earlier history.

The Distribution of Minerals in Mineral Deposits

The other important aspect of the constitution of a potentially valuable mass of rock is the amount of the usable minerals that it contains, and the way in which they are distributed within the rock mass. Some mineral deposits are almost solid economic mineral, whereas, at the other end of the scale, some deposits of precious metal can be worked even though the metal content is as low as 100 mg/t. The amount of the desired mineral in the rock necessary for successful working, or workable grade, depends on the availability of a technology for extracting it with an effort commensurate with the reward for so doing. Not only the amount, but the distribution of the minerals in the deposit affects the design of the mining and extraction processes. The spatial distribution of the minerals also has an effect on the ease with which we can make reliable estimates of the recoverable grade.

There are a number of, more or less, self-explanatory terms used to describe the distribution of minerals in rocks such as massive, disseminated; many of which are the same as those used to describe the morphology of deposits. In order to estimate grade and grade distribution, we have to take samples of the material and subject them to analysis. The uncertainty of these estimates is controlled by the amount of sampling we do, and by the relationship between the geometry of the sampling scheme and the geometrical aspects of the grade distribution.

Many kinds of mineral deposits, particularly those containing a low content of some highly valuable mineral, consist in part of rock containing less than the average content with a few rich concentrations in the form of zones, pockets, veins, etc. When such a deposit is sampled, some of the measurements will show rich values, but many more will contain far less, and the sample will have a statistical distribution that is known as an asymmetrical or 'skewed' distribution. Provided the sample distribution parallels that of the whole deposit, all will be well; but differences, often very difficult to detect, will lead to bias. An unfavourable interaction between the geometry of the sampling scheme and that of the mineral distribution can be one of the causes of bias. For example, if a deposit consisting of a mass of rock containing a stockwork of near-vertical mineralized veinlets is explored by a series of vertical drill-holes, a reliable estimate will only be produced if the chance intersections of drill-holes with veinlets is in the same proportion to intersections with non-mineralized rock, as the proportion of veinlets to the whole rock-mass. In cases like this bias is common, and the exploration has to be augmented by driving horizontal sampling tunnels through the deposit.

Many mineral deposits have boundaries which cannot be seen or mapped and must be fixed by sampling and analysis, so the uncertainty of grade estimation affects the estimation of the size of the deposit as well. In some cases it is possible to mine a large mass of rock and send only selected portions for treatment, dumping the remainder as waste. In such cases the discrimination between valuable rock and waste depends on sampling and analysis. Thus the study of the configuration of the valuable minerals inside the deposit is vital to the control of the efficient estimations of grade, size and the overall efficiency of the mining operation. In so far as this configuration results from the processes of formation of the deposit, it is also a factor in understanding the origin of the deposit.

The Relationship of Mineral Deposits to Host Rocks

A great deal can be learned from the study of the nature of the rocks associated with a mineral deposit and the relationships between them. These features may be studied on a variety of scales, some things can be learnt from geological relationships at the contact between the deposit and its host rocks on a scale of a few millimetres; other things from a study of the geological setting of the deposit in regions of hundreds of kilometres extent. It is not possible to separate the purely descriptive aspects of such a study from interpretations, and the terminology usually refers to both. Even our primary subdivision of rocks into igneous, sedimentary and metamorphic is genetic as well as descriptive. Relationships with host rocks are fundamental to any typology or classification of mineral deposits.

Dealing first with the local relationships between a mineral deposit and its host rocks, one can note that these can be 'conformable' or 'disconformable' in relationship to the predominating structure or fabric of the host rock. Examples of conformable deposits are those which are beds in a sequence of sediments, or layers in a layered igneous body. Disconformable deposits are veins, fracture fillings and the like. Many essentially conformable deposits have locally developed disconformable features; some display the two kinds of relationship without one being more important than the other. For instance, the Kuroko-type sulphide deposits (see Kosaka) commonly have a basal part that is a decidedly disconformable stockwork, but at the top they are stratiform and conformable with the enclosing sediments and pyroclastic rocks. The importance of these features is in the interpretation of the time relationships between the formation of the mineralization and the host rocks. A disconformable relationship means that the deposit formed after the host rock. The converse, however, is not true. Some conformable deposits form at the same time as the host rocks, but confirmation of this must come from other evidence than the conformity itself. Many disconformable features associated with essentially conformable deposits are post-depositional modifications.

Two terms are in common use for describing the time relationships between mineral deposits and host rocks. A 'syngenetic' deposit is one formed at the same place and time as its host rocks; the converse is an 'epigenetic' deposit. Unfortunately, these two terms are difficult to use because their application depends on scale. For example, if a quartz-cassiterite vein is seen cutting a granite outcrop, then it would seem to be epigenetic. But if one considers the granite pluton as a whole, the tin vein may have formed as a phase of the cooling history of the pluton, and thus could be regarded as syngenetic. The two terms are particularly difficult to apply in the case of minerals in sedimentary rocks, in which the relationship of the minerals to the host rock may be more influenced by diagenesis and metamorphism than the original sedimentary process. It is probably reasonable to say that the majority of disconformable ones are epigenetic, the majority of conformable ones are syngenetic, and that in each case exceptions occur that can only be revealed by other evidence.

Many disconformable mineral deposits are associated with rocks that have been modified. Zones of alteration surround or accompany deposits although it is not always easy to show that alteration and mineralization are connected. In some instances, the results of laboratory experiments have shown that a process that could have formed a certain assemblage of economic minerals could also have caused the associated alteration. The nomenclature of alteration is usually self-explanatory, being based on the name of the predominating new mineral formed, or the one destroyed. Many alteration processes involve the change of silicate minerals to hydrous forms, such as the change of feldspar to mica or clay minerals. (Weathering also effects the same kind of changes, but rather more slowly and less extensively than in the deeper, hotter parts of the crust.) These two features of alteration are, perhaps, the best justification for the use of the term 'hydrothermal' for the formative processes of many mineral deposits. A great deal of evidence of formative processes comes from the study of alteration.

On a larger scale, we can frequently obtain useful information both for scientific interpretations and for prospecting from a study of the rocks and structures in the area or region around the deposit. It is often easier to understand the formative processes of the ordinary rocks in association with a deposit, and thus deduce what was the general geological environment in which the deposit was formed. Some kinds of mineral deposit are only found in certain specific rock types; chromite is rarely found except in ultrabasic igneous rocks. But others occur in a great variety of rocks. For example, massive pyritic deposits occur in basalts, andesites, rhyolite, in mixtures of them, both in lavas and in pyroclastic rocks, but almost always in an assemblage of volcanic rocks that is characteristic of oceanic island arcs of mid-oceanic ridges. The study of the wider geological context in which deposits are found is often more revealing than a study of the local. These features will be discussed in more detail in the introductory sections to Sections I, II, III and IV of the book and the more global aspects in Section V.

The Origin of Mineral Deposits

The majority of the known mineral deposits in the world have been discovered without much idea of their geological origin, but it is probably true that we would be better able to find more if we understood how they were formed. The search for origins is a preoccupation of the geologists who study mineral deposits, and their conclusions can help the prospector. However, prospectors are continually making chance discoveries of new deposits, new objects for the scientists to study. A definitive treatise on the subject cannot be written because, as fast as conclusions are reached, so will changing economic factors stimulate the discovery of different objects to study. This is a very important factor that is often not fully appreciated. It is possible to account for much of the change in opinion about the origin of deposits as resulting from changes in the types of deposit being mined and studied. Couple this with the sheer size and complexity of the earth's crust, and it is not difficult to understand why the subject is probably second only to cosmology for provoking violent controversy. Like most sciences, present opinion is affected by the debates of the past, and it is helpful to know a little of the history of the subject.

The History of Ideas

Mineral deposits have been worked for well over 7000 years, but there is remarkably little recorded about the ideas or beliefs that people may have had about the origin of metals and minerals. From the dawn of the Bronze Age to the end of the Middle Ages no literate caste seems ever to have been involved in mining and metallurgy. There are a few references in Greek writings and Egyptian inscriptions which usually refer only to the places from where metals came. The Bible describes and explains the origin of the world, but hardly mentions minerals and metals at all. There are even fewer references in the Koran. Roman writers, described the technology of mining and metallurgy in some detail, but there are few philosophical musings on the origin of the substances themselves. But there is evidence in mythologies of a secret caste of metalsmiths who passed on their knowledge from father to son, and it seems probable that ideas about minerals and metals remained in an oral tradition until the Renaissance. By the time George Bauer (Georgius Agricola) published his famous treatise, *De Re Metallica* in 1556, there was clearly a mature body of knowledge and ideas, not only on the technology of mining, but also on methods of prospecting and theories of origin. By Bauer's day there was a well-established religious tradition among the mining communities of Central Europe, centred around their patron, Saint Barbara. Mining communities figured prominently in the Reformation; Martin Luther was the son of a miner.

The natural philosophers of the seventeenth century had a good deal to say about mineral deposits and it is then that we see, for the first time in print, the dichotomy of approach to the origins of minerals that persists to the present day. In the work of René Descartes, mineral deposits are seen as associated with the firey, sulphurous and generally hellish things that go on in the depths of the earth. The poet John Milton put such ideas more succinctly (see *Paradise Lost*, Book I, l.663, et seq.). However, contemporaries of Descartes who knew more about minerals in the field, notably the English scientist, John Woodward, tended rather to develop the ideas of Bauer, that mineral veins were filled by circulating groundwaters. In an age of religious fervour, it is perhaps not surprising that some of these writers looked upon the Deluge as the primary ore-forming fluid; an act of God rather than Descartes' devilish magmas.

More than a century later, when the scientific study of mineral deposits became established the same dichotomy can be observed in the famous rivalry between the 'neptunist' ideas of Abraham Werner in Freiburg, and the 'plutonist' ideas of the Scotsman, John Hutton. This clash of ideas, coming as it did at the beginning of the Industrial Revolution and at the time of the first major European war in which natural resources seriously affected strategy, stimulated a great deal of new investigation. With new observation came new ideas; but travel was difficult and theories tended to be based on the limited field experience of their proponents. By the middle of the last century the mineral wealth of colonial territories began to be developed, and travel became more common. Many of the writers of the day were colonialists, like Sir Henry de la Beche, the founder of the Royal School of Mines, in London. It is the Frenchman, Elie de Beaumont, who stands out as the most influential writer of this period. In 1847, he published a comprehensive synthesis of the ideas of the time. He held that mineral veins, and other forms of deposit, were in the main formed by thermal waters associated with the igneous and volcanic processes that acompany the development of mountain chains. Werner's surface-water theory was abandoned as were the stricter forms of Huttonian magmatic injection in favour of a magmatic and volcanic 'hydrothermal' theory. (The term was first properly defined by Bunsen of gas burner fame.)

The latter half of the nineteenth century saw much development of these ideas, and experimental work on hydrothermal processes was begun by Daubrée in France, and supporting field evidence came from the much travelled Englishman, Thomas Belt. But the period also saw a reaction. In Germany, Bishoff revived the ideas of Werner and gained support from Sandberger, Steery Hunt and later the American workers Emmons and Van Hise. From these writers emerged the idea that mineral veins were filled by groundwaters that leached material from the surrounding rocks; the so-called 'lateral secretion' theory. This theory's greatest advocate, Van Hise, widened its scope by suggesting that hydrothermal solutions were meteoric waters that descended and became heated in the crust, so increasing their ability to leach and transport the substances that later became deposited in veins. There were a few stricter adherents to Werner's descending waters, but these ideas were largely confounded when it was realized that sulphides in veins could form from descending solutions (the process of secondary enrichment) but that this could only take place if a primary or 'hypogene' mineral assemblage was already present.

At the end of the century the 'decensionists' and 'lateral secretionists' came under vigorous attack, particularly from the Czech, Pošepný. His experience was in the mineral fields of Bohemia where mineral veins have a close relationship to granite plutons, very different from the lead-zinc and fluorite deposits of northern England and the American Mid-West, upon which the ideas of Steery Hunt, Emmons and Van Hise were based. But it was an age of monistic conviction; there could only be one theory of the origin of mineral deposits. Pošepný's attack on the 'lateral secretionists' at a meeting in America in 1893, co-inciding as it did with the publication in France of a book by de Beaumont's devout follower, de Launay, heralded the heroic age of magmatic hydrothermal theory. Waldemar Lindgren, in America, proposed a comprehensive classification of deposits based on hydrothermal theory, which he eventually embodied into his book *Mineral Deposits* (published 1933), the most influential piece of scholarship ever written on the subject. Lindgren did describe some deposits as sedimentary, and allowed some flexibility for the ideas of Emmons on the lead-zinc deposits of the American Mid-West, but the core of his work was a grouping of deposits according to physical conditions of formation and proximity to their magmatic sources. His hypo-, meso- and epithermal types could be recognized by a combination of mineral paragenesis, and style of alteration. Such ideas were applied by his followers with great zeal. Experimental work developed to test and confirm the theory, and detailed studies began to be made of mineral assemblages, by means of the newly developed methods of reflected light microscopy and X-ray crystallography. By the 1930s Bateman felt able to claim that the newly discovered copper deposits in Zambia were mesothermal, because they contained the same assemblage of minerals and

range of textures as deposits established by Lindgren as mesothermal. But this time the magmatic hydrothermalists had over-reached themselves.

During the first half of this century there were great developments in mining technology; the steam shovel and large open-pit mining, the selective flotation concentrator and in geophysical prospecting. New deposits were discovered of hitherto unknown scale and type, and all over the world geologists were being faced with deposits that did not seem to fit classical theory. The development of the petroleum industry yielded a great deal of information on the nature of sedimentary basins and the composition and movement of fluids in the crust. By the middle of the century, air travel became available to many more people; the era of conference field excursions had arrived and geologists were able to make comparative studies.

The reaction to classical theory came from Southern Africa, Australia and Europe. Garlick argued forcefully for a sedimentary origin of the Zambian copper deposits; Haddon King a meta-sedimentary one for Broken Hill. Geologists rallied to the banners of 'syngenetic' and 'epigenetic', and there ensued a battle that was fought with something approaching the quasi-religious fervour of the 'neptunist–plutonist' battle of 150 years before.

These two terms, syngenetic and epigenetic, are very difficult to apply in practice, and it is not surprising that the debate became acrimonious and was carried beyond the bounds of scientific reasonableness. But it did stimulate new research. While all this was going on, the lateral secretion theory was being revived, at first by Boyle, in order to account for the geochemical features of the Yellowknife gold veins in Canada. The discovery of the large bodies of zinc-lead ore across the Lake from Yellowknife at Pine Point became critical to a revival of the ideas of Emmons. From this discovery came the idea that sulphide ore could be precipitated by reaction between a metal-bearing, saline groundwater and hydrogen sulphide generated by the reduction of sulphate by hydrocarbons. It is curious that this idea came so late. Every schoolboy knows that if you bubble H_2S into lead-chloride solution you get a black precipitate of lead sulphide. But for years geologists brought up on classical theory sought only a single solution that could carry all the constituents of the mineral formed, disregarding the solubility of lead in chloride brine and the known fact that generations of miners had been driven out of the stopes of the lead mines of the Tri-State district by foul and poisonous concentrations of hydrogen sulphide.

Another development of the period after the Second World War was the renewal of interest in the role of vulcanism. Quite early in the century, Japanese geologists recognized that certain sulphide deposits most probably were formed on the sea floor as part of a volcanic event. Despite the emphasis that Lindgren gave to the evidence of thermal springs, he made no distinction between those that were associated with extrusive as opposed to intrusive events, and overlooked the fact that magmatic hydrothermal mineralization could be conformable and syngenetic. The Japanese work remained untranslated and unknown until Watanabe, and later Sato, began to be known in the rest of the world. On the other side of the world Schneiderhohn was coming to similar conclusions, and coined the name 'submarine exhalative'. Oftedall applied the term of the pyritic deposits of the Norwegian Caledonides, many of which seemed conformable with altered lavas on one wall, and sediments on the other. The magmatic hydrothermalists had interpreted massive sulphide deposits as replacements, but they never satisfactorily explained what happened to the millions of tonnes of rock that were replaced, or the fact that the sediments in contact with them are usually unaltered. The simplicity of the submarine exhalative theory was that even if volcanic rocks were replaced during the deposition of the sulphides, the products of rock decomposition could be easily dispersed in the sea.

The beginning of the second half of the twentieth century was a time of reappraisal, but was marred by people taking entrenched positions, and some unfortunate nationalistic tendencies emerged. For a time it seemed that classical hydrothermalism was to become established doctrine for the Americans and for the Russians, whose work now began to become well known in the West. Europe, Africa and Australia became the territory of heretics. Gratton, McKinstry and Sales wrote essays attacking the syngeneticists and remaining loyal to Lindgren. It was, above all, John Ridge whose monumental works of synthesis and bibliographic works, unsurpassed of their kind (Ridge, 1972 and 1976) contained a series of critical essays and introductions that were too dogmatic for the Europeans. Pierre Routhier, in a critique of Ridge's work published in 1969, bitterly complained of the way in which he chose to ignore European work, particularly that of Schneiderhohn, Ramdorh, Bernard, Maucher, Nicolini and Amstutz. To this list one must add at least the names of Routhier himself, and the admirable ideas man and essayist, Charles Davidson, and the rebels from the Southern Hemisphere, Garlick, Haddon King and Stanton. All these workers were, in their own often very different ways, appealing for greater variety and flexibility of genetic interpretation. Amstutz and Routhier proposed classifications of mineral deposits based on observable time and space relationships with host rocks. The word 'gîtologie' was coined, and the concept of a 'metallotect' was invented for a mappable object associated with a mineral deposit divorced from genetic connotations. This amounted to an appeal for a return to objective description, and a move away from that subjective application of received doctrine which was in danger of reducing scientific debate to empty polemic. But the new thinkers had in their turn their Achilles Heel.

Workers like Garlick had shown, fairly convincingly to all but the rigid magmatic hydrothermalists, that metal sulphides could be incorporated into sediments at the time of deposition. What they failed to show was where the metals came from. If they came from the sea, or from sedimentary basins, then an effective concentration process must be found, and this was largely lacking in their explanations. It was all too easy to 'lower the problem', and propose some hidden magmatic source in depth. Those who favoured compromise pointed to volcanic sources. The discovery of the metal-rich brines in the Red Sea and the Salton Sea lent credance to this idea. For a time, everyone was scanning sedimentary sequences in mineralized areas for bands of tuff that might show the presence of volcanic activity at the time of deposition; and this approach met with some success, at least in the case of the large stratiform and strata-bound sulphide deposits.

The 1960s was not only a time of debate and re-appraisal about mineral deposits, it was also a time of rapid development of more fundamental ideas about the nature and origin of the earth's crust as a whole.

Geophysical measurements, particularly over the oceans, had provided new support for the theory of Continental Drift, long believed in by a number of European and African geologists, and it developed into the theory of plate tectonics. New insight was gained into volcanic processes, petrogenesis and the formation, deformation and metamorphism of large piles of sediment. A better understanding of processes that take place in the upper mantle of the earth was obtained, and to some extent confirmed, by the exploration of the moon. At the same time, from the older continental areas came sufficient radiometric ages for an interpretation of crustal growth by the stabilization of tectonically mobile zones. A much more developed concept of the architecture of the crust was emerging to which mineral deposits could be related.

Parallel with this, the study of minerals had advanced through a period of detailed chemical and crystallographic studies to the study of the stable isotopes of certain elements. It became possible to discriminate between sulphur of primordial origin, and that which originated from sulphate in sea water. In the present decade the ratios of the isotopes of hydrogen and oxygen have begun to show that much of the water involved in hydrothermal processes is meteoric, the idea championed by Van Hise in 1901. Pluto was in retreat on almost all fronts; hydrothermal solutions did exist, but it was Neptune that provided the 'hydro', and Vulcan the 'thermal'. Many geologists were moving towards a pluralist view of the origin of mineralizing solutions. Their sources were not just intrusive-magmatic, but volcanic, metamorphic, tectonic, basinal or other origins. But, in geology, pluralist views rarely remain unchallenged for long, and soon the rumblings of Pluto from the depths began afresh.

The theory of plate tectonics included a new view of rift systems. A firmer base was found for ideas that go back several decades when geologists had noted that mineral deposits of various kinds lay along lines or in belts (lineaments) the orientation of which often has little relationship to the structure of the rocks seen at the surface. An important concept in this connection is the so-called 'principal of inheritance'. This envisages that large, deep-seated discontinuities are built into bodies of crust during their formation, which affect all subsequent geological processes taking place in the rocks laid on top. The view began to emerge that, whereas the detailed modes of formation of mineral deposits at a local scale may be plural, deeper down there is a more fundamental source for the mineralizing agency, perhaps coming from the mantle via deep structures.

The debates on the origin of mineral deposits today, and for the next few years, will almost certainly centre around this difference of approach. There will be those who think that the origins are best understood within the context of rocks observed to be in association with the deposit, and who will maintain a pluralist view of sources; and those who prefer to pursue the quest for a more unified and fundamental origin deep in the earth. Most of this book is written on the assumption that the former will be the more productive approach, particularly for the prospector.

Genetic Interpretation

The observational information that comes from the study of mineral deposits is in two parts; that which comes from a study of the material of the deposit itself, and that which comes from a study of the enclosing rocks. From this information a number of deductions can be made on the mode of formation and origin of the deposit, at least in favourable circumstances. The important kinds of deduction are as follows:

Physical conditions of formation. The temperature and pressure under which the minerals in a deposit formed, or were subjected to, are significant. Temperature can be measured by the study of fluid inclusions in certain minerals, and to a limited extent by analogy between observed mineral parageneses and the results of laboratory phase-equilibrium experiments. Again, but to a limited extent, pressure can be estimated in a similar way, but the soundest method is based on the measurement of stratigraphic sections, and the deduction of the depth of burial at the time of formation. Unfortunately, this can only be done in rare cases.

Chemical conditions of formation. These can be deduced more readily from mineral assemblages, from the composition of fluid inclusions in minerals and from the nature of alteration associated with the deposit. Analogies based on laboratory experiments are indispensable for such interpretations.

Indirect inference of physico-chemical conditions. So long as the time relationship between the minerals in the deposit and those in the associated rocks is known, then we can infer the conditions of formation of the deposit from knowledge of those of the host rocks, which are often easier to understand. Sediments, in which deposits are found, frequently have preserved in them evidence of the environmental conditions in which they were laid down, e.g. climate, water depth, salinity. Climatic and environmental conditions can be inferred from palaeo-geographic interpretations that come from a study of stratigraphy and palaeo-magnetic latitudes.

Age of formation. Very few mineral deposits can be dated directly. Quite a few can be dated indirectly from the age of the associated rocks that existed before the mineral deposit was formed, and another that was formed later. The gap in time between these limiting dates is sometimes long, but in many cases is short enough to give a good idea of the deposit's age. The best estimates of age are those based on fossiliferous rocks that can be correlated with others with well-established radiometric ages; or cases in which the mineral deposit was formed during a period of formation of minerals containing enough of one of the radioactive isotopes that can be used for dating. Age determination might be thought of as a method of resolving differences of opinion between syngenetic and epigenetic origins, but in practice the situation is the other way round because the meaning of age of measurements depends on the interpretation of the relative chronology inferred from relationships.

Space for the minerals. It is usually important to investigate how space was made available for the minerals in the deposit (or the deposit as a whole) to occupy. (Residual deposits, by definition, occupy less space than their parent rocks so the question is irrelevant.) Deposits formed at the surface of the lithosphere or 'exogenic' deposits occupy free space at the expense only of water or air. In practice, it is not simple to show that a deposit is exogenic, and the usual line of argument that one has to resort to is something like this. If the deposit has relationships to other rocks, indicating that it could have formed at a rock, air/water interface, and shows no evidence of a space-creation process, then it is probably exogenic. The crux of the matter, therefore, is the recognition of the effects of processes that create space in rocks. Broadly these are either mechanical or chemical. The obvious mechanical process is the fracturing of rocks. When a rock-mass is deformed, there will be some part subjected to tension; mechanical failure will allow extension and the formation of a space. Numerous examples are known, the formation of openings on undulating fault planes, on the crests of folds and many others. Opening of existing fractures in rocks may occur, such as the opening of a sedimentary rock along bedding planes or laminations during folding, or by the action of groundwater under great pressure.

The creation of space by chemical processes takes place in several ways. One mineral in a rock may be replaced by another by reaction with a solution and, if the newly formed mineral occupies less volume than the original one, space is made. The two best known examples of this are the dolomitization of limestones, and the serpentinization of olivine-rich rocks. During certain sorts of hydrothermal alteration, particularly the feldspar-destructive alteration of igneous rocks, the rock is often rendered porous. (One should note that some alteration processes reduce porosity.) Space may be created by the wholesale dissolution of minerals such as happens during the formation of caves in limestones. In all cases of chemical alteration and solution, one must be able to account for the removal of the displaced material. The evidence of these processes usually comes from the study of rocks that are in a half-way stage. The critical evidence often comes from the contacts and marginal zones of a deposit (often regrettably but necessarily avoided by miners).

Transport and deposition of the material. In the case of deposits in sediments, one can often deduce from the study of the texture of the host rock that the minerals were carried by water, air or ice. The study of grain size, grain shape and sedimentary structures will reveal a great deal about the nature of the medium of transport and mode of deposition. In some cases the evidence is fragmentary and has to be assembled from observations made over wide areas. For example, only after decades of careful observation has the true picture of the sedimentary environment of the Witwatersrand gold-bearing conglomerates begun to emerge.

A not dissimilar process accounts for the accumulation of minerals in differentiated igneous rocks. Some concentration takes place when minerals settle out under the force of gravity from a magma, the rate of settling being affected by the density difference between crystals and the magma, the viscosity of the magma and any movement or turbulence.

In the case of endogenic deposits in which the minerals are emplaced into spaces in, or created in, the host rock, one must usually try to find what kind of solution or fluid effected the transport and the reason for deposition at the particular place. Fluid inclusions are generally regarded as samples of the parent fluid of a mineral, and their composition can tell a great deal. In many kinds of deposit, fluid inclusions are found to contain chloride-rich brines. Laboratory experiments have shown that the constituents of some oxide minerals may travel as volatile chlorides or fluorides, and those of sulphides can move as poly-sulphide, hydro-sulphide or thiosulphate complexes. One could say that there are two sorts of mineralizing fluid, one that carries all the constituents necessary for formation of the mineral, and the other that carries only a part, and therefore can only form the mineral if it encounters another substance that provides the remaining constituents.

Precipitation of minerals can take place by a change in physical conditions that affects either the solubility of the mineral's constituents or the kinetics of the precipitation. Minerals precipitate because of drop in temperature, or pressure, or by dilution with more of the solvent, and the configuration of minerals in a deposit sometimes yields evidence of the probable mechanism. Zoning in deposits is usually held to be caused by one of these mechanisms. Precipitation by reaction is known to take place in some circumstances. Barite is simply precipitated by the mixing of chloride brines containing the metal (which are known) with sulphate-rich groundwater (which is quite common). Galena and sphalerite may form from the mixing of lead and zinc-bearing chloride brines with sulphide ions in the form of sour gas, or sulphide-rich groundwaters. Uranium, which is readily transported (in its oxidized form) in sulphate-rich groundwaters, is readily precipitated by organic matter or hydrocarbons; iron and manganese are both transported in various kinds of solution (so long as it contains ions that buffer the effect of oxidizing agents) but are precipitated by contact with air or strongly oxygenated water.

Sources. The parent materials from which many sedimentary deposits are formed originate from the process of erosion, and if such a mode of formation is proposed, a provenance area for the material must be found to complete the interpretation, and this generally can come only from the study of palaeo-geography. Mineralizing solutions can originate either from magmatic differentiation or from differential dissolution, otherwise known as leaching. The many circumstances in which this can take place will be discussed in the introductions of the succeeding sections of this book.

The Classification of Mineral Deposits

The classification of the objects in the natural world has always been an integral part of the natural sciences. Several attempts have been made to construct classifications of mineral deposits, and it is of interest to study on what these are based and the purpose that they serve. If a classification is to be of any value it must be capable of including all the mineral deposits known. The purpose of classification is to provide a framework for thought and discussions and to define a terminology. One could add that any classification should also be useful to the prospector and exploration geologist.

Geological classifications of mineral deposits have been based on commodity, morphology, observed geological relationships with rocks and structures, on geological environment or mode of occurrence, and on mode of formation and origin. Classifications based on commodity and morphology may have uses to economists and engineers, but are too limited for geologists. There is a difference of approach among geologists between those who prefer classifications to be as far as possible non-controversial, objective and based on observable information, and those who find that classifications based on origin are the only ones that give sufficient aid to thought and debate. The main problem facing the proposer of a new genetic classification at the present time is the already discussed incompatibility between the monistic and pluralistic approaches to the origin of mineralizing fluids and solutions.

The most influential classification ever proposed was that of Lindgren. Many workers still accept and use this scheme today, particularly in America, and many more are prepared to use it in a form modified to recognize that hydrothermal solutions may not necessarily originate from intrusive igneous activity. The one unsatisfactory feature of Lindgren's classification is that we find deposits that are included under the same heading look very different in the field. It is of very limited use to prospectors who must deal with deposits as they occur, rather than as they originate.

Amstutz and Routhier have proposed classification based on time and space relationships. They class deposits into syngenetic and epigenetic (as did Lindgren) intersected with a division into those containing minerals that are indigenous to the host rock as opposed to those that are exotic.* It has been suggested that conformable and disconformable would be better than syngenetic and epigenetic, so avoiding the difficulty of applying these terms in practice. Moreover, the four-fold grouping is difficult to extend to a more detailed classification. And in any case, the two terms *familier* and *étranger* are as difficult to apply in practice as the two time relationship terms, since they too depend on the scale of one's view. Siderite, for example, could be held to be exotic to a sandstone, but indigenous to certain sorts of marine sedimentary sequence that include sandstones among other rocks.

Some classifications instead of aiding thought and discussion, tend rather to provoke controversy, but can have the positive result of stimulating new research. Geologists who dislike becoming embroiled in acrimonious debate often prefer 'typology' as an alternative to genetic classification. A 'type' of mineral deposit may be taken here as either a group of known deposits with significant numbers of common features, or a group with sufficient features in common with a famous and well understood 'type example'. There are several commonly recognized types of deposit. In general terms, everyone understands what is meant by an 'Algoma-type' iron deposit, or a 'porphyry copper'. Where (as in both these cases) there are many examples of the type, each yielding fragments of evidence that fit a general picture, we may find that deposits of a certain type have a common origin in addition to common features. This approach is in many ways more fruitful than the traditional sort of genetic classification. The approach is akin to that of the taxonomist and phylogenist in biology, and is perhaps more in keeping with geology as a natural science. We would probably arrive at the most generally useful classification of deposits if we could fit all of them into types of common origin. To some extent this book is intended as a move towards a general typology of mineral deposits, and seeks to avoid commitment to the traditional form of classification.

In order to organize types of deposit into larger groupings it seems obvious to base them, as far as possible, on observable features. This is the reason why this book is divided up according to the general environment of occurrences. Four categories have been chosen. The first includes deposits which have simple relationships to rocks that have the features resulting from processes taking place at the earth's surface. The second category are deposits that are stratiform or stratabound, in sedimentary sequences, but where the relationship to the enclosing rocks is not simple. The third and fourth categories are those associated in some way with igneous rocks. The separation into the two is rather arbitrary, but is justified because the basic and ultrabasic igneous rocks have special features which distinguish them from all other igneous rocks. Each section is preceded by a separate introduction defining the grouping more precisely.

Further Reading

SMIRNOV, V. I. (1976) (English edition), *Geology of Mineral Deposits.* Trans. of Geologija Poleznykh Iskopaemykh, M.I.R. Moscow, 1962. [This is the most modern book on the subject in English and one of the most comprehensive ever to be published. The majority, but by no means all, of the material is drawn from examples of deposits in the U.S.S.R. It is a very good book and is well-illustrated, but is rather better on magmatic and hydrothermal deposits than those in sediments.]

* There is a problem of translation here. The original terms *familier* and *étranger* can be translated in various ways and the words I have chosen are not the same as some other authors have used. The precise intentions behind the use of terms can only be appreciated by reading the original French texts. See, for example, Routhier, 1967, *Chronique des Mines et de la Récherche Minière*, **363**, 177–90.

RIDGE, J. D. (1976), *Annotated Bibliographies of Mineral Deposits in Africa, Asia (except U.S.S.R.) and Australia.* Pergamon, Oxford, 546p.

RIDGE, J. D. (1972), Annotated Bibliographies of Mineral Deposits in the Western Hemisphere. *Geol. Soc. Am. Mem.* **313**, 681p. [These two excellent volumes are indispensable for anyone who wishes to know more about mineral deposits. The bibliographies are very comprehensive and the annotations are in the form of critical essays, summarizing the works of the major authors on each deposit, with liberal doses of Ridge's own rather strict Lindgrenian ideas. One can regret that the two volumes are not published by the same house, because they usually end up at opposite ends of your library.]

WOLF, K. H., ed. (1976), *Handbook of Strata-bound and Stratiform Ore Deposits.* 7 volumes, Elsevier, Amsterdam. [This is a volume of papers by many different authors that seeks to compliment the rather lean treatment given to such deposits in other texts. It contains a lot of very useful information but it could have been put more briefly.]

STANTON, R. L. (1972), *Ore Petrology.* McGraw-Hill, New York, 731p. [Probably the best modern text covering the methodology of investigation of mineral deposits from the pure scientific point of view. It also covers the geological environment of occurrence of quite a range of examples, and gives a very balanced treatment of genetic theories.]

NICOLINI, P. (1970), *Gîtologie des Concentrations Minérales Stratiforme.* Gauthur-Villars. [The great statement of position of the syngenetic school. Not as wide a coverage as Wolf, but much more compact. Good compliment to texts that do not deal properly with deposits in sediments. Many of the examples covered would not be called stratiform by English writers. He includes many which would be better called strata-bound.]

PARK, C. F. and MacDIARMID, R. A. (1964), *Ore Deposits.* Freeman, San Franciso, 522p. [Quite a good introductory text for students, but it has a heavy emphasis on American examples, and on classical magmatic hydrothermal interpretations.]

ROUTHIER, P. (1963), *Les Gîsements Métallifères.* Masson, Paris, 2 vols, 867p and 408p. [This mammoth work is comprehensive and gives a very balanced view of genetic theory; well worth more attention than it normally receives from readers outside French-speaking countries. In addition to the main sections on the geology of deposits, it also covers the technological aspects of mining geology, but it is not so good on this aspect.]

BATES, R. L. (1960), *Geology of the Industrial Rocks and Minerals.* Harper, New York, 441p. [Almost the only book on the other half of the subject, an indispensable complement to the many texts that emphasize metalliferous deposits.]

BATEMAN, A. M. (1942), *Economic Mineral Deposits.* Wiley, New York, 916p. [For many years a standard textbook for students and still interesting, with lots of background information. Somewhat confusing to geologists at times because the material is arranged substance by substance. It also deals with coal, oil and gas, but the treatment is poor compared with metalliferous deposits.]

LINDGREN, W. (1913), (2nd edition 1933) *Mineral Deposits.* McGraw-Hill, New York, 930p. [The most famous and influential book ever written on the subject, still useful for reference and has lasted better than most texts.]

CROOK, T. (1933), *History of the Theory of Ore Deposits.* Murby, London, 163p. [Fascinating little book that traces the origin of the classical hydrothermal theory from ancient times; badly needs updating.]

Section One: Deposits in Geological Environments at the Earth's Surface

Introduction

Environments at the Earth's Surface

Looking at the surface of the earth as it is today one finds that about 34% consists of continents about a tenth of which are covered by relatively shallow water (the continental shelves), the remainder being ocean. At the poles the surface is covered by permanent ice, and a large area is covered by ice and snow for half of the year. Between the Tropics there is a region where it is hot and the climate is fairly predictable. Between the Tropics and polar regions is a zone subject to much more extreme seasonal variations and unpredictable weather. In this zone are the regions that suffer the monsoon, hurricanes and cyclonic weather. The present disposition of the continents and oceans affects climate, causing some parts to be humid and others dry, some parts along west-facing coastlines are subject to little annual variation in climate, while others in the interiors suffer extremes of climate between summer and winter. Land areas vary in topography. There are mountainous belts, hilly land and flat lands, the latter either low lying near sea level or elevated.

The solid earth is enveloped in part by water. In the oceans, this contains about 3·4% of dissolved material, principally the ions Na^+, K^+, Mg^{2+}, Ca^{2+}, Cl^-, SO_4^{2-} and HCO^- and the gases oxygen and nitrogen more or less in equilibrium with their partial pressure in the atmosphere. Ocean water is constantly stirred by the effect of prevailing winds, tidal flow and by temperature differences that cause the downward flow of melt water from the ice-caps.

On land, water is usually much fresher and much more variable in composition. It is present as rivers, lakes, and as groundwater in permeable rocks. Above this so-called 'hydrosphere' is the 'atmosphere', a gas mixture containing principally nitrogen, oxygen, water vapour and carbon dioxide. The atmosphere is constantly stirred and is fairly constant in composition except for its content of water vapour that is dependent on temperature.

Life abounds on earth. The upper part of the ocean is teeming with life and all but the coldest, dryest and most mountainous parts of the land are covered by some form of plant-life and a fauna dependent on it. Organisms have developed a means of protecting themselves against the strongly oxidizing condition of the atmosphere but once life stops, their substance, which depends on compounds that contain a reduced form of carbon, decompose. Where life is so profuse that the formation of organic matter is too rapid for oxidation to be completed, reducing carbonaceous materials accumulate. Thus peat bogs, marshes and organic-rich sediments can form. On land, there is an important interface between rock and air, namely soil; a complex material, the formation of which is controlled by the interaction of the chemistry of the rocks and that of the atmosphere, over which organisms have a vital influence.

Of the geological processes that take place on the land, weathering and erosion are the most important. Weathering is an essentially chemical process by which the mineral constituents of the rocks are changed to other forms, usually with the removal of some material in solution. Water is essential for weathering and the process is accelerated by higher temperature and by organisms (or the products of their partial decomposition). Minerals vary in their reactivity during weathering, some like the chlorides dissolve very readily. The carbonates dissolve, generating so-called 'hard' water on the one hand, and the peculiar topography known as 'karst' on the other. Sulphides oxidize readily to soluble sulphates and even silicates will dissolve in groundwater, although much more slowly. Some minerals resist weathering for very long periods of time, almost indefinitely in a few cases. Among these 'resistate' minerals are the noble metals, diamond and certain oxides and silicates like cassiterite, ilmenite, rutile and zircon. The last named will resist several cycles of weathering over millions of years.

Weathering begins by mineralogical change along grain boundaries, tending to make the rock softer and more friable by the liberation of the grains from one another. As the process continues, rock-forming silicates tend to alter to phylosilicates, particularly clays and other hydrous minerals. In addition to the resistate minerals which remain stable, some of the newly-formed minerals remain stable in the weathering environment and are collectively known as the 'hydrolisate' minerals. In particular these include oxides and hydrated oxides of iron and aluminium. Provided weathering can continue for a long time and the products are not removed by erosion, the end result is often laterite, a heterogeneous mixture of hydrolisate minerals generally with a small content of resistates inherited from the parent rock. Laterite covers many of the plateau areas of the world that have suffered tropical (or subtropical) humid climate for some time, and is found covering large parts of central Africa, South America, India, etc.

Under special conditions parts of a laterite profile can be a mineral deposit, depending on the chemical composition of the bedrock and on the details of the weathering process. Thus, where aluminous bedrocks are found or where the chemistry of weathering is influenced by organic materials to the extent that iron becomes mobile, bauxite may form. Bauxite may also form by a similar process from the clay residue known as 'terra rosa' that accumulates as a result of the removal in solution of the carbonate in limestones.

Iron-ore may be lateritic in origin. This occurs occasionally where iron-rich igneous rocks weather (although these are sometimes marred in quality by trace quantities of other elements inherited from the parent rock). More frequently, lateritic iron-ores form from other iron-ores of a different origin. In this case the weathering process produces a modification in texture rather than a concentration of iron. Large amounts of iron-ore are produced in this way, such as the so-called 'powder ores' or 'dust ores' and 'canga'; friable, cheaply mined materials generally associated with a laterite profile.

Manganese and nickel are two further metals which form ores in the laterite profile. Both occur where the laterite forms over a bedrock containing the metal in question. A few resistate minerals can be worked from laterite-type materials because they have been liberated from their rock matrix and are thus more easily recovered. Minerals like chromite and the aluminium silicate minerals, andalusite and kyanite are examples. It is probable that the resistate minerals such as ilmenite, rutile, zircon, diamond and precious metals, though rarely worked from laterite profiles themselves, are in this way liberated into surface sediment for eventual concentration by other processes. Because these various types of deposit are, more often than not, a residuum left behind after the removal by weathering of other materials they are collectively known as 'residual deposits'.

Although bound up with weathering erosion is, by contrast, largely a mechanical process. Rocks become broken up into fragments by the action of wind, water, ice and thermal changes. Mechanical fracture does not necessarily break rocks along grain boundaries and tends to produce heterogeneous fragments. Erosion exposes mineral deposits, as is evident from the fact that they are more frequently found in mountains or deeply dissected terrains; but it also destroys them. It does not form mineral deposits, but it initiates the movement of rock fragments that become available for concentration by other means.

Minerals in bedrock, which have been wholly or partially liberated by weathering, begin to move down slopes by processes such as hill-creep, landsliding and the formation of mudflows. This yields the type of deposit known as 'eluvial', in which no concentration from bedrock proportions takes place, but nonetheless, the minerals are more easily worked than from unweathered bedrock. Much cassiterite and some chromite occurs in this way. Streams and rivers move eluvial material after which it is generally referred to as alluvial; and most of the processes that take place in rivers are similar to those that occur in other watery environments such as the shoreline, in lakes, estuaries, deltas and in the sea.

Moving water can carry solid mineral particles in various ways, but by whatever method, there is one underlying factor and that is that the faster the water flows the greater the mass of the particle that can be transported. Two particles that move in the same way and to the same extent by water moving with a certain velocity are said to be 'hydraulically equivalent'.

Generally, this means that they have the same effective mass with respect to the density of water. Thus smaller particles of a dense mineral may be hydraulically equivalent to larger ones of lesser density.

However, minerals do not begin their passage through the sedimentary cycle with arbitrary grain sizes. Grain size distribution tends rather to reflect that in the bedrock, because of the predominantly grain boundary nature of weathering processes, that liberate them from their matrix.

Hence we find that, although the load carried by a river becomes finer-grained from source to estuary, in special parts particles of dense and less dense minerals of similar grain size co-exist and in such circumstances any velocity gradient in the water will induce sorting and consequent concentration of the more (or less) dense mineral grains. This is the essence of the process of alluvial and littoral mineral-concentration. In rivers, bends, obstructions, confluences, changes in slope, all cause velocity gradients and hence mineral concentration; but in the simple case this is a small-scale process yielding only local concentrations known as 'placers'. However, the courses of rivers do not remain stable, nick-points are cut back, obstructions eroded away, meanders abandoned and the base levels of rivers change, causing the burial or 'perching' of the river bed. In these ways considerable accumulations of alluvium may be formed and with them 'alluvial deposits'.

When the load of a river reaches the sea, depending on the water-sediment ratio, the form of the sea-bed and the energy of wave action, either a braided estuary or a delta is formed. Here small changes in the rate of sedimentation and the distribution of plant life can modify the way in which the sediment moves and further opportunities for local concentration of minerals may occur. Perhaps more important, new sediment is swept along the coast and all but the finest material constantly returned to the shore by wave action. The rapidly moving water of an advancing wave carries with it sediment; as the wave breaks the water looses its kinetic energy and the sediment is deposited; as the water recedes the lighter particles are preferentially removed tending to leave the heavier ones behind. This tends to put boulders and cobbles on the top of the 'berm' and the finer material down the slope towards the low-tide mark. But if, for reasons connected with the provenance of the sediment, particles of similar size but of different density co-exist in the load, density sorting will occur and in this way are formed the beach-sand deposits which contain certain dense resistate minerals like ilmenite, zircon, rutile and diamond.

The intensity of all these processes is influenced by environmental factors, the most important of which is climate, which leads to differences in the rainfall, wind-speed and the magnitude of seasonal variations. In general, one can say that the more turbulent and changeable the climate, the more likely is the formation of such mineral deposits. The contrast here with the residual deposits is interesting. The large alluvial and shoreline deposits, recently formed, may well owe their origin, in part, to the changeable climate that accompanied the glaciation of much of the earth during the Quaternary. Today, the important beach-sand deposits occur in the zone of trade winds and cyclonic storms and many alluvial deposits occur in the monsoon region.

Because weathering and erosion are destructive, it is not surprising that deposits formed by these two processes are to a large extent transient; but, there are a few important exceptions to this general rule. For instance, there are fossil bauxite deposits, but they are rarely found *in situ* and are locally re-sedimented and preserved below younger sediments. Fossil alluvial deposits are also rare, except for those buried beneath lava flows. There is, however, another important exception and this one on a grad scale.

In some of the oldest sequences of sediments known on earth, such as those of the Witwatersrand and Blind River, exist gold-uranium bearing conglomerates that resemble fossil alluvial deposits but have peculiarities that distinguish them from others found in later rocks. The mode of formation of these deposits is now beginning to be understood; they were an early form of esturine or deltaic sediment formed under climatic conditions that cannot be observed today. The association with glacial sediments may be an important indicator to the environmental conditions that exist at the time. These deposits contain carbonaceous material, apparently of organic origin, and the heavy particles of gold and uranium minerals were probably trapped in the sediments by pebbles and an early form of plant life that has since become carbonized.

The products of erosion and weathering are rock or mineral particles and the material carried in solution by natural water. Certain mineral deposits are formed as a result of chemical processes that affect both these types of material. The volume of the hydrosphere only changes slowly by the addition of water from deep in the mantle through volcanic and igneous activity; and a balance is maintained between surface water and water vapour in the atmosphere. Most water is in the oceans, which have a relatively constant composition of dissolved ions and gas molecules. However, whenever the mixing action of the tides, the wind and thermal changes associated with freezing and melting are restricted, such as in partially land-locked gulf and where the hinterland of such a gulf is arid and surface run-off small, the inevitable result is that evaporation will increase the content of dissolved ions in the water and, because this is usually accompanied by a higher temperature, a decrease in the dissolved gas content. It is under these circumstances that the precipitation of minerals from sea water occurs to form a group of rocks known as evaporites; and since some of these minerals are valuable, also of 'evaporite deposits'. The same phenomenon may occur in lakes but the result is smaller deposits of rather specialized composition, due to the much more variable composition of surface waters which are subject to a far wider variety of influences than is the sea. From sea water, calcium and magnesium carbonates, calcium sulphate, sodium chloride and a variety of complex minerals contain magnesium, potassium, chloride and sulphate, as well as traces of all else found, may be precipitated, either at sea, or in porous rocks along the shores. Such environments are often rich in life and much organic debris both in the form of organic materials and the carbonate or silicious skeletons of organisms, are incorporated into the sediments. The reduced oxygen content of the water causes much of this organic matter to remain only partially decomposed. Organic matter contains elements like sulphur and phosphorus and the carbonaceous part will react with sulphates to produce sulphides and sulphur. Small wonder that there is close association between salt, oil and deposits of sulphur. And sulphur has a great propensity to precipitate heavy metals. In a later section we will return to this association and postulate that the evaporite environment is one of the earth's most important generators of mineralizing solutions.

Certain elements undergo chemical changes in sediments. Of these perhaps the most important is iron which can exist as divalent and trivalent ions under normal conditions. Reducing conditions (principally generated by accumulations of organic matter) cause iron to dissolve in its divalated state and to be precipitated in the presence of certain other elements or ions, such as sulphur, carbonate or silicate. In its ferric state it is soluble in the presence of large concentrations of strong anions such as are found only near volcanic vents or in the local environment of rapidly oxidizing sulphides. Thus in any surface, or near-surface environment, iron may pass into solution or be precipitated according to local variations in conditions. The sedimentary iron deposits are interpreted as being formed by diagenetic processes in soft sediment, probably organic-rich, by local solution and precipitation, the iron having its origin in the detritus brought to the sea from the land. However, this does not account for the fact that some sediments contain more iron than others, sometimes enough to be mined as iron-ores. The prerequisite for this is probably an anomalous source of iron from erosion and the necessary palaeogeographic and climatic conditions to enable organisms to be active at the sediment–water interface.

Anomalous sources of iron could be volcanic, or due to the erosion of previously formed iron deposits, but it could in some cases be due to the erosion of a laterite-covered plateau, particularly one on which laterites had developed from flood basalts. These remarks are intended to apply generally but need some qualification in the case of the large iron-rich sedimentary basins that are known in sequences of Proterozoic rocks. These are of such immense size that the interpretations made for iron deposits in Phanerozic rocks need some modification. The large Proterozoic iron basins were low-energy environments, quite probably semi-landlocked seas in which were formed silicious, dolomitic, ferruginous chemical sediments. They normally contain evidence of life, stromatolites in associated carbonates and in cherts, of what may have been blue-green algae. All this evidence points to an environment in which iron was quite mobile and abundant, a circumstance that could only be produced if oxidation was much less severe than under present conditions. The most popular explanation of this is the hypothesis that the oxygen content of the atmosphere and hydrosphere has changed with plant evolution, and that it was much lower during the early Proterozoic. There is a second, less plausible, explanation based on the observation that the deeper parts of the sea are at present oxygenated by cool, oxygen-rich currents of melt-water from the polar ice-caps, and that anoxygenic conditions would exist in the deep parts of the sea if there were no ice-caps.

If the atmospheric oxygen theory is true, it would mean that before a certain stage in the earth's history, the ferrous ion could have been an important constituent of sea water and thus it would be possible to view the iron-rich sediments as a form of evaporite. The mid-Proterozoic sea would have to have been more alkaline to enable silica to be more soluble.

At the earth's surface there are a number of environments in which valuable mineral deposits may be formed. Of these, the important ones are laterite profiles, the alluvial and related coastal ones, the evaporite and the environment of chemical sedimentation, primarily in shallow seas. What follows is a selection of examples of economic mineral deposits that were formed in these environments.

13

The Bauxite Deposits of Jamaica

The small island state of Jamaica possesses the largest deposits of the so-called 'karst' or 'terra rosa' type of bauxite. They are of very high quality and are found in pockets and hollows on the weathered surface of the Cretaceous White Limestone Formation that outcrops over a large part of the island.

Location

The bauxite deposits are found in an area about 100 km by 50 km on either side of the Central Mountain Range (latitude 18° 10′ N, longitude 77° 20′ W) at elevations of between 60 m and 600 m. Most of the producing areas are in the parishes of Manchester, St. Elizabeth and St. Ann.

Geographical Setting

Jamaica is 225 km long and 80 km wide, lying astride latitude 18° N. The topography is varied; there is a narrow coastal plain round the island and an interior mostly elevated above 400 m, much of which is karst topography with many caves, sink-holes and small inland drainage basins. There are two ranges of mountains, the Central Range and, in the east, the Blue Mountains, which rise up to a peak of 2256 m above sea level. The climate is sub-tropical and strongly influenced by the easterly and north-easterly trade winds.

Rainfall is high; 2 m per year and more in the mountains. It is a lush and fairly fertile island that produces sugar and several other cash crops. Tourism is one of its prime industries and the 'import' of Americans to Montego Bay rivals the export of bauxite in economic importance. The island has a rail network connecting the bauxite mines and sugar plantations to the main ports.

History

The red earths of the limestone hill areas of Jamaica were noted by early British geologists (notably Sir Henry de la Beche) but their high aluminium content was not recognized until the work of Barrington-Brown, during the first full government geological survey in 1869. The potential of these materials as an aluminium ore was realized following the work of the agricultural chemist, R. F. Innes, who in 1938 conducted a soil survey of the island. A prominent landowner, Sir Alfred D'Costa began to promote the idea of mining during the early years of the Second World War. At this time the Governor of the island declared bauxite to be Crown property and allowed a number of mining concessions, and trial shipments of raw bauxite were made in 1943. In the later years of the war and the years following, having solved certain problems of extracting aluminium from the Jamaican bauxite, several open-pit mines were opened and alumina refineries built.

Geological Background

Jamaica is one of the islands of the Caribbean Arc and is composed of rocks of Cretaceous to Recent age. There are essentially three rock units on the island; the Blue Mountain Group, the lower Tertiary limestones with which the bauxite is associated, and coastal plan sediments. The Blue Mountain Group is exposed in the mountains of the same name and along the Central Range. Their present disposition is probably due to thrusting associated with the arc development and their stratigraphic base is not seen. The series is terminated by the Wagwater Conglomerate, a red conglomerate and shale formation with gypsum beds that marks the base of the Tertiary. It also represents the beginning of the tectonic events that emplaced the Blue Mountain rocks in their present position. Above the Wagwater is a formation of carbonaceous and calcarous shales with black limestones, and a sequence of tuffs and tuffites that represent a major episode of vulcanism in the Arc. Associated with the Blue Mountain rocks are several small plugs (and one substantial pluton) of granodiorite that do not cut the Wagwater conglomerate, and a series of porphyries and lavas that are probably the same age as the Middle Eocene tuffs.

On top of all these rocks is a 500 m thick sequence of shelf carbonates. It begins with the Yellow Limestone Formation composed of impure limestone, marl, sandy beds, silty beds and some thin lignite seams. With a fairly sharp break follows the White Limestone Formation, much thicker, with a dolomite at the base but for the most part composed of fairly pure limestones varying between nodular, chalky varieties to massive well-bedded ones. The limestone sequence ranges in age from Lower Eocene to Lower Miocene.

From the Lower Miocene onwards a series of tectonic events took place and are still taking place (earthquakes occur occasionally on the island). These events folded the limestones and faulted the whole island extensively and thrust the Cretaceous rocks against the Tertiary in places. During the mid to late Tertiary a series of upward and downward movements took place and sediment was deposited along the coastal areas and in the major drainage basins. The White Limestone Formation was exposed throughout most of this period and was extensively weathered, with the formation of karst surfaces. There are inland and underground drainage systems, caves and thick accumulations of terra rosa in many places.

Bauxite, associated with the terra rosa deposits, constitutes the principal mineral wealth of the island; besides which there is only gypsum, bulk constructional materials, and a few showings of metallic and non-metallic minerals in the Blue Mountains.

Geology of the Bauxite Deposits

Bauxite occurs widely in the central part of the island, confined to the outcrop areas of the White Limestone Formation. It contains gibbsite and boehmite in a ratio of about 3 : 1, with minor amounts of kaolinite, hematite, goethite and traces of resistate minerals such as magnetite, zircon and quartz.

Chemically the material exploited contains 46 to 50% Al_2O_3, some of which (being present as kaolinite) is not recoverable, 3·5–4% SiO_2, 17–22% Fe_2O_3, 2·4–2·6% TiO_2 and 0·3–2·8% P_2O_5. The mineralogical form in which the last two elements exist is not fully understood. The material is often highly porous; bulk densities average about 1·4 t/m³ when dried (1·7 t/m³ wet) as against the value of 2·8 t/m³ for the solid material. Much of the material is soft and friable, unbedded and displaying a variety of textures such as pisolitic and nodular. Compositional variations exist in individual masses of bauxite. Iron oxide can be evenly distributed or concentrated in nodules, and silica is often lower in the centre and more concentrated at the margins of the individual bodies. There are in fact many pockets of bauxite. They occur as fillings of the karst surface, generally with rather flat top surfaces and very irregular bases with many limestone pinnacles sticking up into the material and deep pockets descending below. There is normally a thin cover of organic-rich soil, and the lower contact with the limestone is almost always sharp. Some deposits occur on high ground up to 1000 m above sea level, while others are situated on lower ground. The former tend to be smaller, to some extent eroded and exposed on the sides of hills. It is noticeable that the greatest concentrations of deposits are found in those areas where the underlying limestone is more faulted, presumably because fractured rock weathers more rapidly.

Size and Grade

Reserves of bauxite depend very much on the limits on impurity content that the aluminium refineries are prepared to set, which in turn depends on the value of alumina as an aluminium-smelter feedstock. Pockety deposits in karst zones are also somewhat difficult to evaluate. One published figure shows reserves of 315 Mt containing 46% or more of Al_2O_3, but it may well be that the total reserves of the island exceed this by a large margin.

Geological Interpretations

There are two aspects to the formation of karst-type bauxite deposits: the origin of the terra rosa and the actual formation of the bauxite. As regards the latter, there is general agreement. An original aluminous and ferruginous clay material, when subjected to humid tropical (or sub-tropical) weathering, breaks down chemically and alkalies, alkaline earth elements, and silica are removed in solution. The actual geometrical configuration of the deposit tends to determine how groundwater moves and hence the low and high silica zones. Hydrated aluminium and iron minerals remain, since they are insoluble under these conditions, as are a number of other elements or minerals that remain as traces.

The problem is the exact origin of the primary material from which the bauxite is formed. Terra rosa has traditionally been interpreted as an insoluble residue left after the dissolution of the carbonate by groundwater. The problem in Jamaica is that the total insoluble content of the White Limestone is only 0·2%, and for the requisite amount of terra rosa to accumulate, a very great thickness of limestone would have to have been dissolved away. Although the stratigraphic top of the White Limestone is not seen, the requisite thickness is difficult to envisage as ever having existed. Yet the composition (particularly of trace elements) or parent material is consistent with an origin from the limestones. However, the main uplift period of the limestone in the mid-Miocene was also a period of vulcanism, not so much on the island (although there are some mid-Tertiary lavas) but along the main Caribbean Arc. It has been suggested that the material in the karst was considerably augmented with volcanic ash that fell on the island and was washed into the hollows.

Mining

Being soft and friable, there are few problems with mining the Jamaican bauxite, which is exploited from open pits at a large number of localities in the central part of the island. There are several companies operating, all joint ventures between the Government and the major international aluminium companies. Up to 15 Mt per year is produced, of which about half is refined on the island at several plants; the remainder is exported raw. Refining of bauxite takes place by a modified Bayer process in which the raw material is dissolved in hot sodium aluminate solution under pressure, the solution filtered to remove iron oxide and the alumina precipitated by reduction of the pressure and temperature. Alumina is exported from Jamaica to other parts of the world for smelting, principally because this requires large quantities of cheap electrical power, not available on the island.

Further Reading

HOSE, H. R. (1950), The Geology and Mineral Resources of Jamaica. *Colonial Geol. Min. Res.*, **6**, (1), 11–36. [A good description of the geological background with maps, etc.]

ZANS, V. A. (1952), Bauxite Resources of Jamaica and their Development. *Colonial Geol. Min. Res.*, (4), 307–33. [The major work on these deposits, also published as: Geol. Survey Dept., Jamaica, Pub. 12, 1954.]

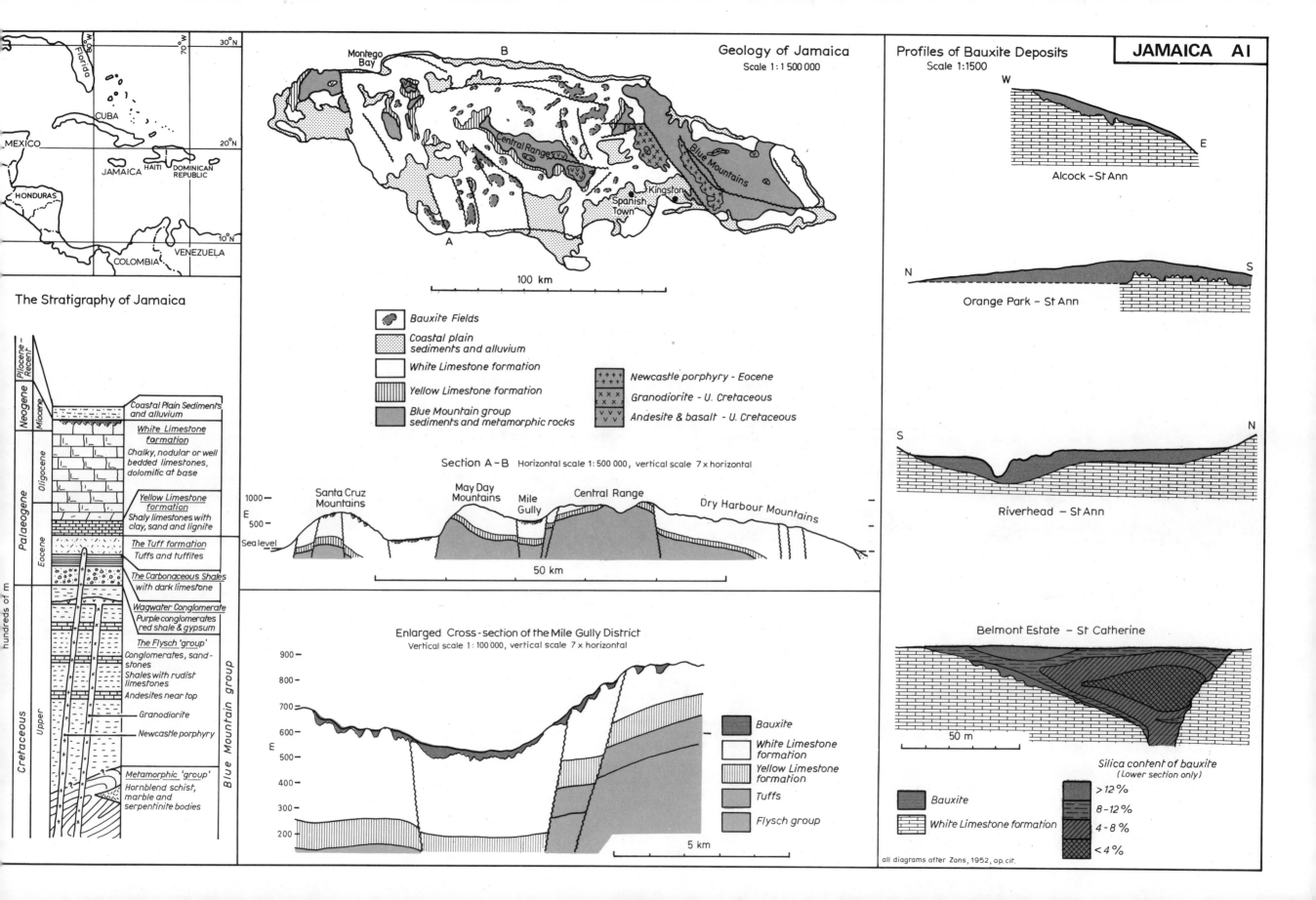

Geology of Jamaica
Scale 1 : 1 500 000

Profiles of Bauxite Deposits
Scale 1:1500

Montego Bay
B
Central Range
Blue Mountains
Kingston
Spanish Town
A

100 km

W
E
Alcock – St Ann

N
S
Orange Park – St Ann

The Stratigraphy of Jamaica

Bauxite Fields

Coastal plain
sediments and alluvium

White Limestone formation

Yellow Limestone formation

Blue Mountain group
sediments and metamorphic rocks

Newcastle porphyry - Eocene

Granodiorite - U. Cretaceous

Andesite & basalt - U. Cretaceous

S
N
Riverhead – St Ann

Section A–B Horizontal scale 1:500 000, vertical scale 7 x horizontal

Pliocene - Recent	
Miocene	Coastal Plain Sediments and alluvium
Oligocene	White Limestone formation — Chalky, nodular or well bedded limestones, dolomitic at base
Eocene	Yellow Limestone formation — Shaly limestones with clay, sand and lignite
	The Tuff formation — Tuffs and tuffites
	The Carbonaceous Shales with dark limestone
	Wagwater Conglomerate — Purple conglomerates, red shale & gypsum
Upper Cretaceous	The Flysch 'group' — Conglomerates, sandstones — Shales with rudist limestones — Andesites near top
	Granodiorite
	Newcastle porphyry
	Metamorphic 'group' — Hornblend schist, marble and serpentinite bodies

Santa Cruz Mountains
May Day Mountains
Mile Gully
Central Range
Dry Harbour Mountains

1000
500
Sea level

50 km

Belmont Estate – St Catherine

Enlarged Cross-section of the Mile Gully District
Vertical scale 1 : 100 000, vertical scale 7 x horizontal

900
800
700
600
500
400
300
200

Bauxite

White Limestone formation

Yellow Limestone formation

Tuffs

Flysch group

5 km

50 m

Silica content of bauxite
(Lower section only)

>12 %

8-12 %

4-8 %

<4 %

Bauxite

White Limestone formation

all diagrams after Zans, 1952, op. cit.

The Onverdacht Bauxite Deposit – Surinam

This is perhaps the best example of the many bauxite deposits that occur along the Guyana Coast of South America. These are residual deposits, related to the formation of laterites, but in this case (as in many) the bauxite itself is not in a typical laterite profile.

Location

Onverdacht is situated at latitude 5° 38′ N, longitude 55° 09′ W and 3 m above sea level, 23 km south of the capital city of Paramaribo.

Geographical Setting

Onverdacht is in a tropical swamp not far from the banks of the Surinam River. The Guyana Coast may be divided into a narrow coastal belt of swamp and tropical rain forest and an inland plateau-land that is relatively elevated. For 30 km from the coast in Surinam the land hardly rises more than a few meters above sea level. One hundred or more kilometeres from the coast there are substantial escarpments. Much of the escarpment and plateau regions are thickly forested, but savanna is to be found in the south of the country.

The climate on the coast is tropical with long and short wet seasons (April–July, December–January) and between them a pair of dry seasons. However rain-free days are few, and there is not much sunshine. The rainfull exceeds 2·1 m in a year (169 mm in a day has been recorded) and even in the dryest months 90 mm may fall. Surinam is sparsely populated, particularly in the interior. On the coast crops grow well; sugar cane and coffee being the main cash crops. Bauxite mining and processing is a major indutsry and is developed around two centres; south of the capital, and further east at Moengo.

History

Explorers and pioneers of the Guyana Coast did not recognize the clay-like outcrops along the lower reaches of the rivers as anything special. The First World War stimulated more interest in bauxite and by 1920 several investigations had been made along the Coast by colonial authorities and interested mining companies.

Onverdacht was discovered from small outcrops and lumps of bauxite in the local streams and exploration by drilling revealed a large ore body relatively shallow but in water-logged swamp. It was not until 1941 that production became possible.

Geological Setting

The Guyana Coast region is made up of two main geological units, an extensive area of Pre-Cambrian rocks (the Guyana Shield) and a coastal sedimentary plain that ranges in age from the late Mesozoic to Recent. In the lower coastal zone the shield rocks become progressively covered by sediment. Over wide areas of the region there is a persistent white sand formation, and the younger coastal sediments are clays, silts, sands with lignite.

The Cainozoic sequence is 250 m thick in the Berbice Basin, but in the region of the bauxite deposits is rarely more than 50 m to 100 m.

Over the inland plateau and the intermediate plateau regions are found many areas of laterite, particularly over basic igneous rocks. Some of these laterites have zones, below a ferricrete capping, high enough alumina content to make them potential bauxites. However, the present-day commercial bauxites are, for the most part, those which are buried below the younger coastal plain sediments and over most of the Guyana Coastal region they seem to occur at a level corresponding to the Eocene–Oligocene boundary.

Geology of the Onverdacht Deposits

The main Onverdacht deposit is a single lens of bauxite over 5 km long and 2 km wide averaging 6 m thickness that dips at a low angle to the north. There are, in addition, several other lenses in the area. Twenty to thirty metres below the deposit is a deeply weathered basement of schist and granite, above which is a variable thickness of poorly sorted, coarse to medium sand containing some residual feldspar. These sands are overlain by the kaolinitic clays that form the footwall of the bauxite deposit. Above the hard bauxite lens are more sands interspersed with kaolinitic clays. These are followed by mottled clays and zones rich in organic matter, which are Quaternary in age as opposed to the Tertiary age of the sediments below.

The majority of the bauxite is fairly soft, has a concretionary texture and contains up to 60% Al_2O_3 (almost entirely as gibbsite) with small amounts of impurities, iron (in the form of limonite minerals) from 1–3% (as Fe_2O_3) and between 1 and 2% SiO_2 (in the form of colloform silica or kaolinite). At the top of the lens is a zone, up to 1 m thick, with a very irregular texture and composition. Patches almost entirely composed of iron oxides occur. Between these ferricrete-like patches are zones of massive gibbsite-rich material cut by numerous veinlets of kaolinite.

Near the base of the lens the texture changes. Laminated bauxite becomes the prevailing type with pockets of a massive and of a cellular variety full of cavities (formed by a network of septa of bauxitic material). This zone is very mixed with the underlying kaolinite clay. The contacts are sharp but quite large masses of kaolinitic clay (up to 1 m across) are found above the base of the bauxite. This lower zone *en masse* tends to be silica-rich and is not mined if the silica content exceeds 15%. Associated with the laminated bauxite are thin lenses of material containing kaolinite and a small, but significant, amount of heavy resistate minerals. In this zone titanium can exceed the normal 1% by several times.

The bauxite has a loss-on-ignition of between 14 and 30%. Soft varieties have a bulk density of 1·3 t/m³ and the harder, more compact ones, between 1·6 t/m³ and 1·9 t/m³.

Size and Grade

The Onverdacht deposit itself contains over 60 Mt of bauxite capable of being produced to a specification of 5% SIO_2, 4% Fe_2O_3 and 3% moisture. The recoverable alumina from this is about 50%. Surinam has reserves of over 200 Mt of bauxite producible to present-day specifications and a further 350 Mt of lower grade material that may be worked in the future. The total resources of bauxite and potentially workable aluminious laterite along the Guyana Coast is probably of the order of 1 Gt.

Geological Interpretations

The formation of bauxite in a laterite profile is fairly well understood in general terms. In a tropical climate with alternating wet and dry seasons silicate rocks break down into aluminium hydroxide minerals and silica. At the same time any iron-bearing minerals in the original rock are broken down to iron hydroxides or oxides. During the wet season the soluble material such as the alkalies and alkaline earths are removed in solution very readily, and normal groundwater (being slightly alkaline) will slowly remove silica. Where there is an organic-rich soil cover (forest floor debris or swamp) reducing conditions are generated that will permit some iron to go into solution. During the dry season, evaporation causes upward migration of pore-solutions, and any iron in solution will be precipitated near the surface. In this way it is possible to develop a profile consisting of a ferricrete capping, over relatively pure bauxite, beneath which is a clay-rich layer and a zone of weathered rock. The process depends on the action of a tropical climate for a long time in stable crustal conditions.

The Onverdacht and other bauxite deposits of the coastal belt are not typical laterites. The texture of the bauxite, its content of resistate minerals, and the lithology of the early Tertiary sediments, suggest that the bauxite was derived from arkosic sediments, but it is by no means certain that the bauxite seen today is derived directly from the sediments that underlie them. Rather it seems that the products of weathering were transported for some distance, some ending up as kaolinitic swamp clays. The bauxite was formed from the kaolinitic clays during a long period of relative stability, probably throughout much of the Eocene and part of the Oligocene (a period of between 5 Ma and 15 Ma).

Sediments were deposited on top of the bauxite after submergence during the Oligocene and the bauxite was preserved from erosion partly by the hard ferricrete capping, and partly because the new sedimentary environment was a low-energy one. However, the same process of solution of silica in warm, alkaline, reduced groundwater continued, but this time caused downward leaching of silica into the bauxite layer and the resilification of its top and the removal of some of the iron.

It is clear that the whole assemblage of laterites and bauxites along the Guyana Coast lies on a series of early Tertiary erosion surfaces that step down from the interior plateau to the sea. The same situation is found along the West African coast and one may conjecture that these erosion surfaces formed in response to discontinuous crustal uplift as the Atlantic Ocean was opened.

Mining

The Onverdacht deposit is worked by N.V. Billiton Maatschapij Suriname at the rate of between 2 Mt/a and 3 Mt/a. The overburden is removed by cutter/suction dredges (working in poinds), by bucket-weel excavators and by draglines. The bauxite is partly sent by train to a drying plant 7 km east on the Surinam River from where it is shipped; or is sent by truck to an alumina refining plant at Paranam which also processes the material from the neighbouring mines of the Surinam Aluminium Company.

Further Reading

ALEVA, G. J. J. (1965), The Buried Bauxite Deposit of Onverdacht, Surinam, S.A. *Geol. Mijubouw*, **44**, 45–58. [The principal descriptive work on the deposit.]

MONTAGNE, D. G. (1964), New Facts on the Geology of the Young Unconsolidated Sediments of Northern Surinam. *Geol. Mijubouw*, **43**, 499–515. [Contains a detailed account of the stratigraphy and lithology of the bauxite-bearing sequence.]

MOSES, J. M. and MICHELL, W. D. (1963), Bauxite Deposits of British Guyana and Surinam in Relation to Underlying Unconsolidated Sediments Suggesting Two-Step Origin. *Econ. Geol.*, **58**, 250–62 and disc. 1002 and 1160. [An interesting genetic essay largely based on work at Moengo.]

KROOK, L. (1969), The Origin of Bauxite on the Coastal Plain of Surinam and Guyana. *Geol. Mijubouw* **20**, 173–80. [A more recent and more comprehensive genetic discussion.]

BLEACKLEY, D. (1969), Bauxites and Laterites of British Guyana. *Geol. Surv. British Guyana, Bull.* **34**, 156p. [This very comprehensive and detailed account is a useful complement to the sparse literature on the Surinam deposits.]

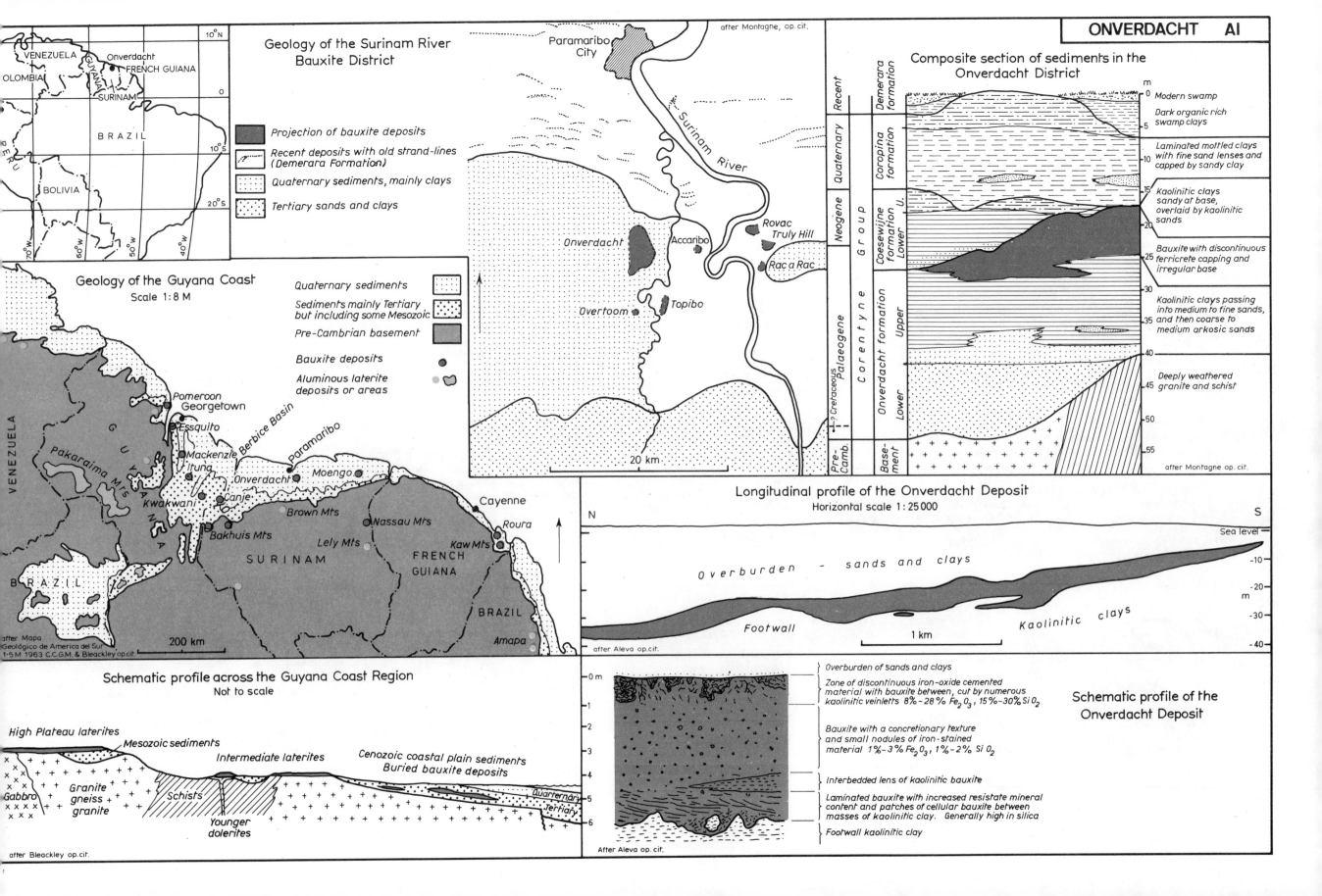

Geology of the Surinam River Bauxite District

after Montagne, op. cit.

Paramaribo City

Surinam River

Onverdacht
Accaribo
Rovac Truly Hill
Rac a Rac
Overtoom
Topibo

Projection of bauxite deposits

Recent deposits with old strand-lines (Demerara Formation)

Quaternary sediments, mainly clays

Tertiary sands and clays

20 km

Geology of the Guyana Coast
Scale 1:8 M

Quaternary sediments

Sediments mainly Tertiary but including some Mesozoic

Pre-Cambrian basement

Bauxite deposits

Aluminous laterite deposits or areas

VENEZUELA
Onverdacht
COLOMBIA
GUYANA
FRENCH GUIANA
SURINAM
BRAZIL
PERU
BOLIVIA
10° N
0
10° S
20° S
70° W 60° W 50° W 40° W

Pomeroon
Georgetown
Essquito
Mackenzie
Ituna
Onverdacht
Canje
Moengo
Paramaribo
Kwakwani
Bakhuis Mts
Brown Mts
Lely Mts
Nassau Mts
Kaw Mts
Cayenne
Roura
SURINAM
FRENCH GUIANA
BRAZIL
Pakaraima Mts
Berbice Basin
VENEZUELA
GUYANA
BRAZIL
Amapa

after Mapa
Geológico de America del Sur
1·5 M 1963 C.C.G.M. & Bleackley op.cit.

200 km

Composite section of sediments in the Onverdacht District

Recent		Demerara formation	
Quaternary	Neogene	Coropina formation	
		Coesewijne formation Lower U.	
Palaeogene	Corentyne Group	Onverdacht formation Upper / Lower	
? Cretaceous	Pre-Camb.	Base-ment	

m
0   Modern swamp
    Dark organic rich swamp clays
10 Laminated mottled clays with fine sand lenses and capped by sandy clay
15 Kaolinitic clays sandy at base, overlaid by kaolinitic sands
20
25 Bauxite with discontinuous ferricrete capping and irregular base
30 Kaolinitic clays passing into medium to fine sands, and then coarse to medium arkosic sands
35
40
45 Deeply weathered granite and schist
55

after Montagne op. cit.

Longitudinal profile of the Onverdacht Deposit
Horizontal scale 1:25 000

N   S

Sea level

Overburden - sands and clays

Footwall   Kaolinitic clays   1 km

0
-10
-20
-30
-40 m

after Aleva op.cit.

Schematic profile across the Guyana Coast Region
Not to scale

High Plateau laterites
Mesozoic sediments
Intermediate laterites
Cenozoic coastal plain sediments
Buried bauxite deposits
Gabbro
Granite gneiss granite
Schists
Younger dolerites
Quaternary
Tertiary

after Bleackley op.cit.

Schematic profile of the Onverdacht Deposit

0 m
1
2
3
4
5
6

Overburden of sands and clays

Zone of discontinuous iron-oxide cemented material with bauxite between, cut by numerous kaolinitic veinletts 8%–28% Fe_2O_3, 15%–30% SiO_2

Bauxite with a concretionary texture and small nodules of iron-stained material 1%–3% Fe_2O_3, 1%–2% SiO_2

Interbedded lens of kaolinitic bauxite

Laminated bauxite with increased resistate mineral content and patches of cellular bauxite between masses of kaolinitic clay. Generally high in silica

Footwall kaolinitic clay

After Aleva op.cit.

The Nickel Deposits of New Caledonia

These were the first major nickel deposits to be worked in the world. The deposits are residual accumulations of nickel minerals formed during the development of laterite profiles over masses of serpentinized ultrabasic rock, a type of nickel deposits now know to occur in several parts of the world, and probably representing the most important source of the metal for the future.

Location

New Caledonia is an island in the South Pacific lying between latitudes 20° S and 22° S and longitudes 164° E and 168° E, roughly half-way between the coast of Queensland and Fiji.

Geographical Setting

New Caledonia is some 500 km long, 60 km wide and rises in the centre to altitudes of over 1600 m. The climate is tropical; the island receives about 3 m of rain during a year and it lies in the belt of south-easterly winds throughout most of the year. The vegetation is tropical forest and bush, although in parts of the interior the soils are poor and the vegetation cover not dense. Mining dominates the economy of the island although fishing and agriculture are well developed. Apart from Noumea there are several towns along the less rugged south-western side of the island.

History

A French Government engineer Garnier made a report on the mineral potential of the island in 1867 during which he noted outcrops of a greenish nickel-bearing rock. A thorough study was made from 1874 to 1876 under the direction of Emile Heurtean who reported, from the result of test pits and trenches, that large reserves of nickel ore could be mined and open-cut operations began. The operations were so successful that almost at once New Caledonia became the world's most important nickel-mining area. Mining by convict labour continued for twenty years, after which the mines began to rely on immigrant labour from Indo-China, Indonesia, and from the other French-administered Pacific islands.

The Geology of New Caledonia

There are four distinct units to the geology of the island; a metamorphic complex in the north-east, a sequence of sediments ranging in age from Permian to mid-Tertiary, a series of masses of ultrabasic and basic magmatic rocks, and the laterite cover that largely overlies the ultrabasics.

The metamorphic complex is composed of pelitic schists with some calcareous varieties and basic igneous rocks, of blue-schist facies along the coast, the remainder being of greenschist.

The main sedimentary sequence begins with a considerable thickness of immature clastic sediments (greywackes with some black shales) the base of which is probably in the Permian and the top in the Lower Jurassic. This is followed by a sequence of continental, esturine and nearitic-marine sediments, ranging in age from the Upper Jurassic to Middle Eocene. There are a few sediments that are post-Eocene, for the most part confined to the coastal plain.

The magmatic rocks are of two kinds. Firstly, there is a well-developed sequence of basaltic pillow-lavas associated with small amounts of pyroclastic rocks and manganiferous jaspers, which appear to lie on top of the Eocene sediments. Secondly, there are ultrabasic rocks that occur as a series of slab-like masses lying on top of all other formations. In some areas, these periodotites are relatively fresh, but in others, and in particular around the nickel-mining areas, they are serpentinized. Like the basalts the age of these rocks is uncertain because their present position may well have resulted from tectonic movements. Some seams and pods of chromite occur and are mined at the southern end of the island, and the Tiebagui mass in the north.

Laterite is developed over large areas of the island but particularly over the peridotite masses. The laterites occur on the inland plateau areas and on hill-tops, and usually have a capping of hard limonite crust or 'ferricrete' below which is a porous, iron-rich laterite. Underneath this is usually a softer more clay-rich layer that passes into a zone of weathered sepentinite with a spheroidal texture. In most places the upper part of the weathered serpentinite zone is impregnated with silica.

The land-surfaces on which the laterites have developed were formed by erosion during the early to mid-Miocene. On slopes below the hill-tops is found sedimentary material with the same composition as the laterites, and quite large areas of these 'transported laterites' occur in valleys and small basins draining the peridotite massifs. These sediments are magnesium-rich and sometimes contain concretions of magnesite.

The Geology of the Nickel Deposits

Most of the peridotites contain of the order of 0·25% nickel (with a little cobalt), but it occurs as a trace element in the silicate minerals of the rock.

In the laterite profiles the nickel is found in two forms. The first is in the ferruginous laterite and weathered rock zone. The second form is as distinct masses, veins, veinlets or pockets containing the mineral garnierite. These are found around residuals of unweathered serpentinite and in cracks and fissures running down into the underlying rock, often well below the base of the laterite profile. Generally the material that is mined is a mixture of the lower parts of the ferruginous laterite, and the weathered rock zone; the veinlets of garnierite enriching the lower grade laterite material. In some places relatively rich material lies directly on very low-grade bedrock, while in others, there is a gradual lowering of the grade at the base of the ore. Much of the high-grade ore mined consists of a rubble of blocks of serpentinite coated with nickel-rich material in a red clay matrix. Above the nickel-rich zone are found pockets of a black earthy material known as 'asbolane'; a fine-grained mixture of manganese oxides with some iron oxide, aluminous minerals and fair quantities of cobalt.

Size and Grade

The total amount of nickel-bearing material on the island has been estimated at 1·5 Gt containing on average a little over 1%. The original nickel ore was worked down to a cut-off grade of about 3%, and the amount of this higher-grade material is perhaps of the order of 30 Mt averaging 3·5%, most of which has been exploited. The remaining material is 'lateritic ore' originally treated as overburden waste, containing 1·5% nickel with some cobalt.

Geological Interpretations

The origin of these deposits is relatively well understood. Nickel is a normal, minor component of the magma that forms peridotite.

The formation of a laterite profile over a body of peridotite involves the breakdown of the silicates and the removal in solution of the magnesium and silica. The former is easily transported, and is often re-precipitated in the derived sediments as magnesium-rich clay minerals or as magnesite. Silica is removed more slowly and a proportion of it remains in the weathered bedrock zone as an impregnation of chalcedony or quartz. During the process nickel passes temporarily into solution, but is quickly re-precipitated in one of two ways. It may be absorbed on to the iron oxide minerals (that are the principal constituents of the laterite), or progressively leached from the laterite and re-precipitated in the weathered rock zone below the laterite as the mineral garnierite.

Manganese is one of the metals that remain insoluble in the laterite where it forms oxides like iron, but they have a tendency to absorb such metals as cobalt.

Like all laterites, these nickel-bearing ones take time to develop and on New Caledonia it is thought that their formation took place during a period of stability beginning in the Miocene. Later in the Tertiary the island was uplifted and new cycles of erosion took place which resulted in the transportation of some of the laterite on to lower ground.

The less well understood part of the island's geological history is the place and time of the formation of the peridotite masses. It now seems probable that they are slices of what was originally part of the Pacific Ocean crust that were thrust (or obducted) on to the metamorphic and sedimentary rocks during the Oligocene.

Mining

The garnierite deposits of New Caledonia are exploited by the Societé Anonymé Le Nickel. (Several other international mining corporations are involved with the development of the low-grade laterite deposits.) The ore is worked from open-pits or open-cuts depending on the topography. The raw ore is upgraded by a combination of hand sorting, washing and screening to remove the low-grade clayey material, and lumps of unweathered serpentinite. Some of the concentrated ore is exported, but the rest is treated on the island.

The traditional process for treatment of the concentrate is to smelt it with coke, limestone and gypsum to produce a sulphide 'matte', which may then be converted to the oxide by air, and to the metal by reduction with charcoal and producer gas. A great deal of the concentrate is now smelted in electric furnaces to ferro-nickel. The low-grade laterite ores are treated by a hydro-metallurgical process. New Caledonia currently produces about 7 Mt/a of ore yielding about 8000 t/a of nickel (in the form of matte) and over 50 000 t/a of ferro-nickel.

Further Reading

GLASSER, E. (1903), Les Principaux Gisements Connus en Nouvelle Caledonie. In: *Les Richesses Minérales de la Nouvelle Caledonie.* pp.397–536. Joul. des Mines 10 ser. **5**. [Although old, this is the most complete documentation of these deposits, full of useful factual information. It also sets out the basic principles of the accepted genetic interpretation.]

CHETELAT, E. de (1947), La Genèse et l'Evolution des Gisements de Nickel de la Nouvelle Caledonie. *Bull. Soc. Geol. Fr.* 5me ser. **17**, 105–60. [Pretty well the definitive genetic essay on the subject that develops the ideas of GLASSER.]

LACROIX, M. A. (1943), Les Peridotites de la Nouvelle Caledonie, leurs Serpentines et leurs Gites de Nickel et de Cobalt; les Gabbros qui les Accompagnent. *Acad. Sci. Paris, Mem.* **66**. [The major work on the ultrabasic rocks.]

ROUTHIER, P. (1953), Versant Occidental de la Nouvelle Caledonie entre le Col de Boghen et la Pointe d'Arama. *Mem. Soc. Geol. Fr. Nouv. Ser.* **67**, 271pp. et 25pl. [Probably the best-detailed description of the general geology of the island, describing the setting of one of the nickeliferous masses.]

BOLDT, J. R. (1967), *The Winning of Nickel.* Methuen, London. [This admirable book contains one of the few descriptions of the New Caledonian nickel deposits in English. Geology, mining and metallurgy are discussed; see particularly pp. 10–13, 61–4, 183–7, 389–97 and 403–7.] For a good brief description of the deposits, see Ch. 6 of ROTHIER, 1963, on p. 11 of this volume.

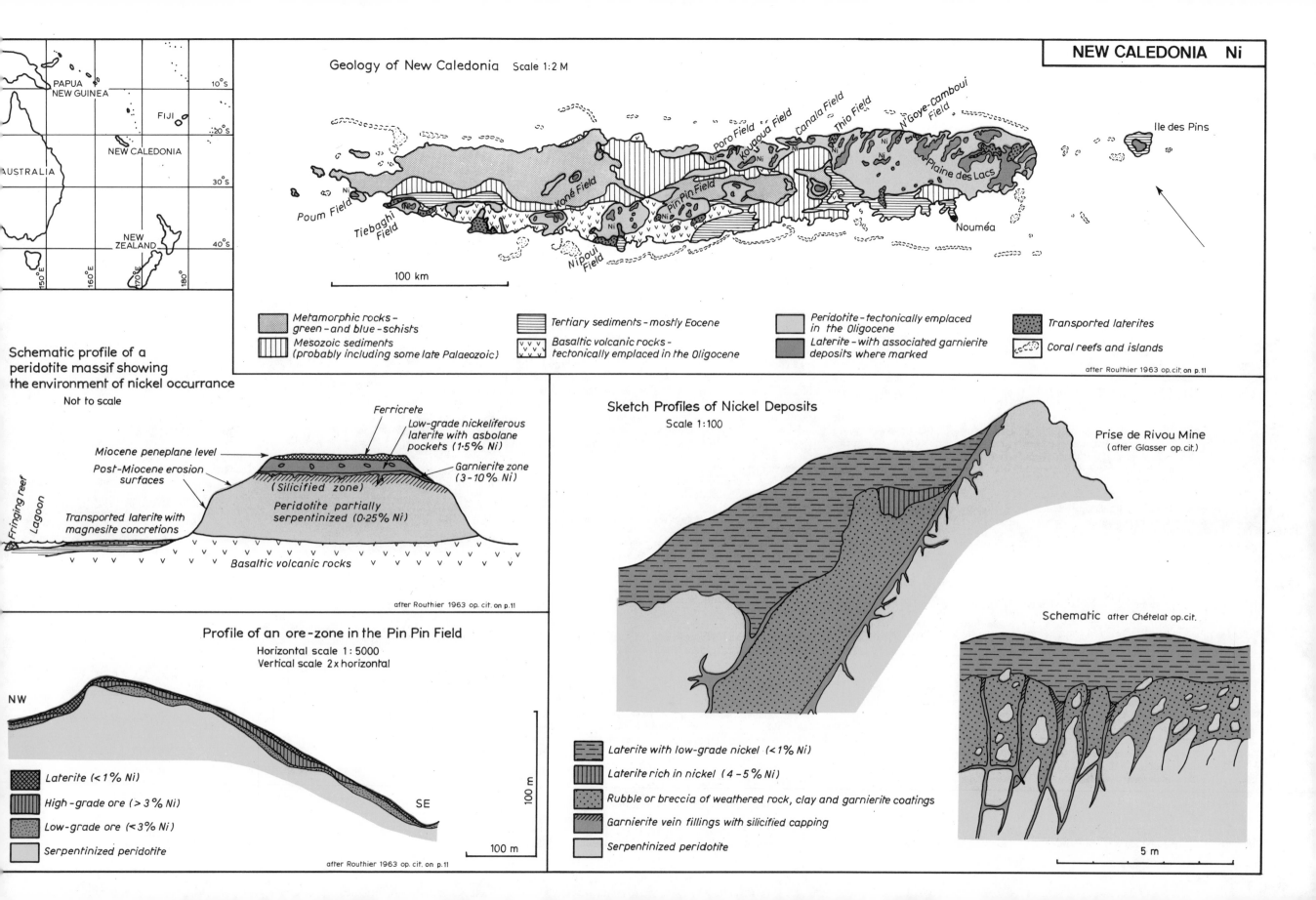

Geology of New Caledonia   Scale 1:2 M

PAPUA NEW GUINEA
FIJI
NEW CALEDONIA
AUSTRALIA
NEW ZEALAND

10°S
20°S
30°S
40°S
150°E 160°E 170°E 180°

Poum Field
Tiebaghi Field
Nipoui Field
Koné Field
Pin Pin Field
Poro Field
Kougoua Field
Canala Field
Thio Field
N'Goye-Camboui Field
Plaine des Lacs
Nouméa
Ile des Pins

100 km

Metamorphic rocks – green – and blue–schists

Mesozoic sediments (probably including some late Palaeozoic)

Tertiary sediments – mostly Eocene

Basaltic volcanic rocks – tectonically emplaced in the Oligocene

Peridotite – tectonically emplaced in the Oligocene

Laterite – with associated garnierite deposits where marked

Transported laterites

Coral reefs and islands

after Routhier 1963 op.cit. on p.11

Schematic profile of a peridotite massif showing the environment of nickel occurrance

Not to scale

Ferricrete
Low-grade nickeliferous laterite with asbolane pockets (1·5% Ni)
Miocene peneplane level
Post-Miocene erosion surfaces
Garnierite zone (3–10 % Ni)
(Silicified zone)
Peridotite partially serpentinized (0·25% Ni)
Fringing reef
Lagoon
Transported laterite with magnesite concretions
Basaltic volcanic rocks

after Routhier 1963 op.cit. on p.11

Sketch Profiles of Nickel Deposits

Scale 1:100

Prise de Rivou Mine (after Glasser op.cit.)

Schematic after Chételat op.cit.

Laterite with low-grade nickel (<1% Ni)

Laterite rich in nickel (4–5% Ni)

Rubble or breccia of weathered rock, clay and garnierite coatings

Garnierite vein fillings with silicified capping

Serpentinized peridotite

5 m

Profile of an ore-zone in the Pin Pin Field

Horizontal scale 1:5000
Vertical scale 2× horizontal

NW
SE

100 m
100 m

Laterite (<1% Ni)

High-grade ore (>3% Ni)

Low-grade ore (<3% Ni)

Serpentinized peridotite

after Routhier 1963 op.cit. on p.11

The Nsuta Manganese Deposit – Ghana

This relatively simple deposit illustrates the way manganese oxide minerals can accumulate in a laterite profile. The manganese originated as a metamorphosed manganiferous sediment; quite unworkable in itself, but which has been concentrated on an old erosion surface, and which also carries ferruginous and aluminous laterites.

Location

Nsuta lies at latitude 5° 16′ N, longitude 1° 59′ W and 100 m above sea level, conveniently on the Sekondi–Kumasi railway 63 km from the port of Takoradi, and is only 7 km away from the Tarkwa gold-mining district in south-western Ghana.

Geographical Setting

The area is one of relatively flat-topped or gently rounded hills that rise to about 200 m above sea level, separated by rather steep-sided, flat-floored valleys around 100 m in altitude, and these form a network of tributaries to the Bonsa river that drains the whole area southwards to the sea. Originally the area was a dense tropical forest with tall trees including many of fine hardwood. Much of the forest has been felled, both for the timber markets of Europe, and for the nearby gold mines. A little farming takes place around the local villages in forest clearings, but the main agricultural activity is the growing of cocoa. The climate is humid tropical with over 2 m of rain falling during two wet seasons, for the most part in the form of regular, late afternoon thunderstorms. Temperatures vary very little and remain in the high twenties throughout most of the year.

Nsuta is in the part of Africa once known as the 'white man's grave' and the early days of mining were affected to a large extent by malaria and yellow fever. There are numerous small villages in the area; Nsuta is no more than a mining camp, but a fair-sized town has grown up around the neighbouring gold mines of Tarkwa.

History

The commercial potential of Nsuta was first realized in 1914 by Sir Alfred Kitson during the preliminary work of the then Gold Coast Geological Survey, although the deposit was within a concession that had been granted to a prospecting company in 1910. By 1917 a company was formed to exploit the deposit, and production of direct-shipping ore began soon after, developing into one of the world's leading producers of the mid-1920s.

Geological Setting

Nsuta is at the eastern side of a large area of Pre-Cambrian crystalline rocks that extends from Ghana westwards through the neighbouring countries of West Africa. In this area there are two main groups of supercrustal rocks: the older Birrimian Group, and the younger Tarkwaian. The Birrimian Group is divided into lower and upper groups, the former consisting of schists derived from clastic sediments, and the latter containing a range of metamorphosed volcanic and pyroclastic rocks. In Ghana these rocks have undergone greenschist-facies metamorphism and are much invaded by granites, granodiorites and associated dyke rocks. Absolute ages show the group to have suffered its main metamorphism just over 2 Ga ago. Rocks of the Tarkwaian Group are found in a series of basins, and consist in the main of conglomerates, sandstones and shales, locally derived from the erosion of the Birrimian, which were probably formed not long after the main deformation of the Birrimian rocks.

There is little record of the late Pre-Cambrian and Phanerozoic history of the area. A long way to the east there are Palaeozoic sediments in the Volta River Basin, and Tertiary sediments are found in a number of small coastal basins to the south. It is clear that during the Tertiary the whole area was eroded down to a series of peneplains. In the region of Nsuta one such plain is represented by flat-topped hills of about 200 m altitude, and the flat-floored valleys with their meandering rivers are evidence of a lower plain that is still developing.

This part of Ghana is well known for vein-gold deposits that are found in the Upper Birrimian meta-volcanic rocks, and for gold-bearing conglomerates that occur in the Tarkwaian Group. In some parts of the country, bauxite and aluminous laterites are found on the flat-topped hills of the old erosion surfaces and diamonds are recovered from alluvium.

Geology of the Nsuta Deposit

Manganese ore is found as a series of elongated flat or irregular-shaped bodies, on the top of a series of hills. They are contained in the laterite profile overlying an extensive horizon of manganese-rich meta-sediments which seem to exist at a particular level in the stratigraphy of the Upper Birrimian Group. The ores are massive, patchy, or bedded accumulations of psilomelane, with numerous veinlets and cavity fillings of pyrolusite.

The ore changes with depth and to some extent laterally. At the top, and particularly over the areas where the manganese-rich sediments sub-outcrop, rich black-ore occurs just below the surface. This is cavernous with dense masses of psilomelane and large cavity fillings of pyrolustie often with stalactitic growths. Below this, and extending laterally, are zones, nodule-beds or irregular patches of similar material in a matrix of weathered rock. The grade of this material generally decreases downwards. The deeper parts of the ore become more and more transversed by quartz veins (but the manganese ores do not differ in this respect from the surrounding country rocks). On the sides of the main hills are found areas of rubble containing blocks and fragments of black-ore in a matrix of loosely consolidated rock fragments which may extend down the slopes well beyond the sub-outcrop areas of the manganese-rich sediments.

The bedrock beneath the deposit consists of a series of greenschist-facies metamorphic rocks. Most were originally volcanic, with a large proportion of tuffs. Some shaly rocks are present (now phyllites) and a few minor intrusions. Within this sequence occurs the manganese-rich horizon consisting mainly of gondite, a rather special metamorphic rock composed mostly of spessartite garnet with quartz, rhodonite, rhodocrosite and chlorite. The pure primary rock is rarely seen because it easily alters to a rock with garnets set in a black manganese oxide matrix.

The whole sequence is well-folded and has a strong cleavage, the crests of the folds in the manganese-rich bedrock horizon often being the most favourable sites for overlying manganese ore. The area is affected by some faults, with which are associated deeper weathering and therefore greater manganese concentration.

Size and Grade

No authoritative figures have ever been published for the size of Nsuta, but available plans and production statistics show that the whole deposit could have totalled 80–100 Mt, of which part was undoubtedly direct shipping ore (46–50% Mn) but a great deal was of lower grade and required washing and concentration before sale.

Geological Interpretations

No one has disputed that Nsuta was formed by the accumulation of manganese from the gondite bedrock during the development of the laterite profile on an old erosion surface. Under the tropical weathering that has persisted in this area (at least through most of the Tertiary and up to the present day) manganese is only locally redistributed, and remains as an insoluble residue along with iron and alumina. The manganese occurs in the bedrock as silicates which weather leaving the manganese as oxides, the silica being removed in solution. Thus, the ores at Nsuta represent a specialized laterite profile that has developed over a manganese-rich bedrock. Gondite is a metamorphic product, probably of an original manganese oxide or silicate-rich sediment. Unmetamorphosed sediments of such a composition are not unknown, and it is not so surprising to find them at the top of a sequence of basic or intermediate volcanic rocks; but the precise nature of the process of manganese deposition in the original sediment is not known.

Mining

Mining of Nsuta has been simple. The hills have been gradually benched away, originally by hand digging, later by mechanized methods. For most of its life the mine was operated by the African Manganese Company, later to become the National Manganese Mining Corporation under the control of the Ghanaian Government. In its hey-day it produced 800 000 t of ore or concentrates a year, but has since dwindled to 250 000 t with the exhaustion of the reserves.

Mined ore is crushed and washed so that the finer, lighter and non-manganiferous material is removed and it is then shipped by the railway to the coast. For some years in the late 1920s and during the Second World War, Nsuta was one of the key manganese mines of the world, because at the beginning of that period, the only other major high-quality deposit was that at Nikopol in southern Russia.

Further Reading

HIRST, T. (1938), The Geology of the Tarkwa Goldfield and Adjacent Country. *Gold Coast Geol. Surv. Bull.*, **10**, 24 pp.

JUNNER, N. R. *et al.* (1942), The Tarkwa Goldfield. *Gold Coast Geol. Surv. Mem.*, **6**.
[These two papers contain an account of the geology of the region.]

SERVICE, H. (1943), The Geology of the Nsuta Manganese Ore Deposits. *Gold Coast Geol. Surv. Mem.*, **5**, 32pp. [A full description of the deposit written at the peak of its importance.]

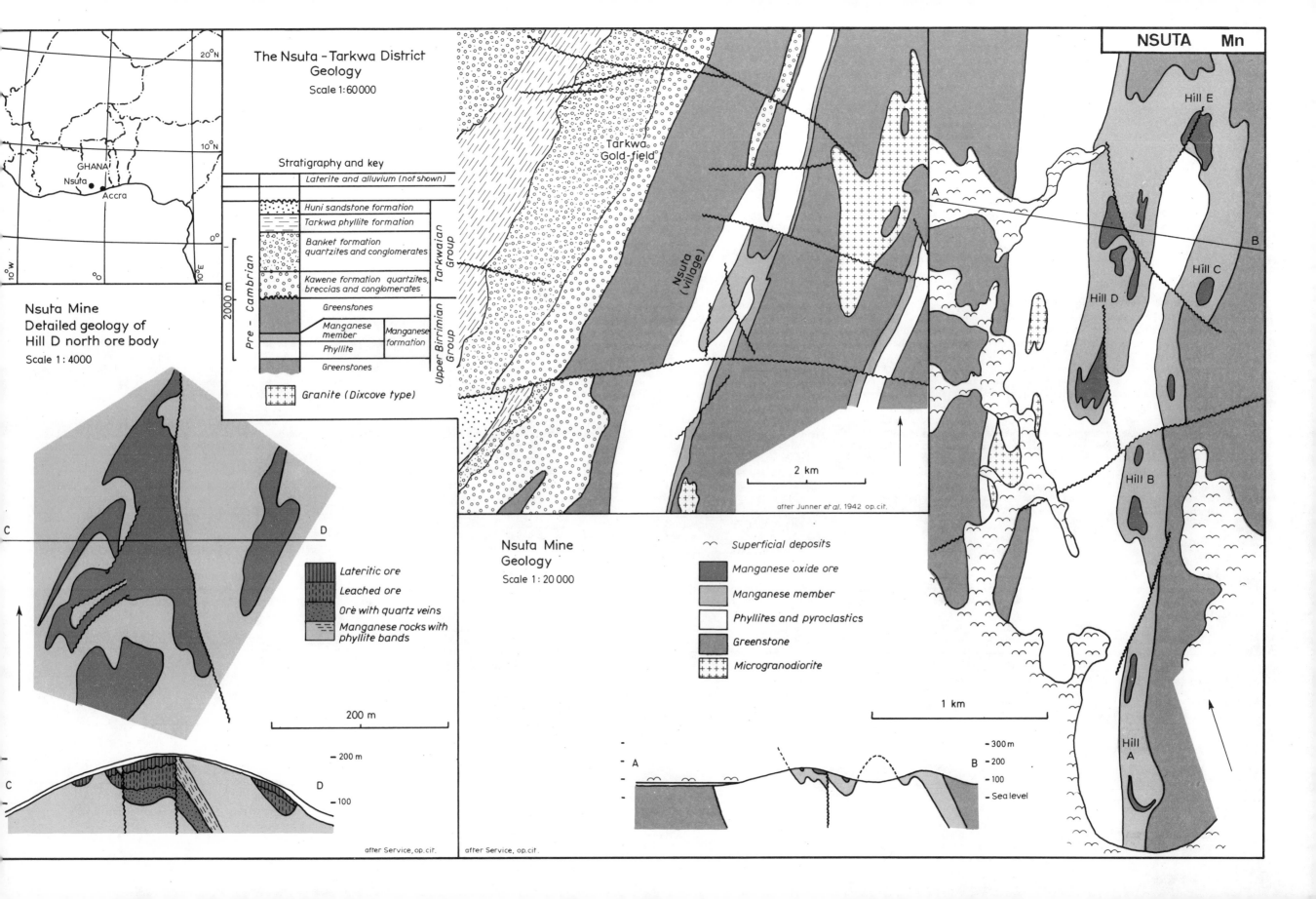

The Nsuta-Tarkwa District
Geology

Scale 1:60 000

Stratigraphy and key

Laterite and alluvium (not shown)	
Huni sandstone formation	Tarkwaian Group
Tarkwa phyllite formation	
Banket formation quartzites and conglomerates	
Kawene formation quartzites, breccias and conglomerates	
Greenstones	Upper Birrimian Group
Manganese member	Manganese formation
Phyllite	
Greenstones	

Pre - Cambrian

2000 m

+ + + Granite (Dixcove type)

Nsuta Mine
Detailed geology of
Hill D north ore body

Scale 1 : 4000

Lateritic ore
Leached ore
Ore with quartz veins
Manganese rocks with
phyllite bands

200 m

— 200 m
— 100

after Service, op.cit.

Tarkwa
Gold-field

Nsuta
(village)

2 km

after Junner et al. 1942 op.cit.

Nsuta Mine
Geology

Scale 1 : 20 000

Superficial deposits
Manganese oxide ore
Manganese member
Phyllites and pyroclastics
Greenstone
Microgranodiorite

1 km

— 300 m
— 200
— 100
— Sea level

Hill E
Hill C
Hill D
Hill B
Hill A

A
B

after Service, op.cit.

GHANA
Nsuta
Accra

20°N
10°N
0°
10°W
10°E

The Tin Deposits of the Kinta Valley – Malaysia

This remarkable valley cut into limestones between granite hills is perhaps the largest known alluvial deposit in the world, even if less famous than the Klondyke. Almost the whole valley floor is covered by thick alluvium, most of which contains cassiterite, probably representing a fifth of the world's known tin.

Location

The Kinta Valley tin-mining district is an area some 70 km by 15 km centred on the town of Ipoh at latitude 4° 36′ N, longitude 101° 05′ E and 39 m above sea level and is within Perak province on the Malayan Peninsula of Malaysia.

Geological Setting

The River Kinta flows southwards over an alluvial plain from an altitude of 75 m at the head, descending to 10 m where it joins the River Perak, and the valley opens out on to a broad coastal plain. On either side are rounded, forested hills rising up to 1100 m altitude to the west, and in the east to the 1700 m high peaks of the Cameron Highlands. On the valley floor are steep-sided hills of limestone with many caves, and much development of karst topography. The climate is wet tropical, averaging 3 m of annual rainfall, heaviest in the October to December monsoon. The temperature varies very little, averaging 28 °C, and the relative humidity remains high most of the year. The area was originally a jungle, but many trees have been cleared for tin smelting; little timber remains in the valley and only remains on the surrounding hills because of a system of state forest reserves. The area is well populated, a large proportion of the people being directly involved in, or dependent on, mining; but there is some agriculture; rubber grows well, tapioca, groundnuts and a little rice are grown. The valley contains several towns served by a railway and main roads that give access to the west coast ports of Telok Anson and Port Weld.

History

There is fragmentary evidence that tin from Malaya was exported to China for bronze making about 1000 B.C. A place called Temala ('land of tin') is shown on Ptolomy's map at about the right latitude for the tin fields known today, and it is quite clear from Arabic writings, that there was a tin industry there in the ninth century A.D. European influence began with tin trading by the Portuguese in the sixteenth century and the Dutch from the seventeenth. The area was occupied by the British in 1787, after which time European interests turned from trading with local miners to larger scale mining operations. It was reports by two French engineers, De la Croix and De Morgan in 1882 and 1886, that led to the formation of the Societé Anonyme des Etains de Kinta, the first of the modern mining operations in the valley. In 1890, 4600 ha were held under mining licences; sixty years later this had grown to 36 800 ha, well over half the total valley floor.

Geological Setting

Lower Palaeozoic rocks, found in various parts of the Malayan Peninsula, change from shallow water deltaic to deeper water limestones and shales by the end of the Silurian, at which time folding took place. The renewed sedimentation in the Upper Palaeozoic was quite different on eastern and western sides of the Peninsula. On the east side, thick sequences of clastic sediments, turbidites with ultrabasic intrusions are found; while on the west side, where the Kinta Valley is situated, a thick sequence of carbonate sediments with shales and sandstones was formed during the Devonian, Carboniferous and early Permian. Folding metamorphism and granite intrusion took place in the middle of the Triassic, after which there was a change to continental sedimentation for the remainder of the Mesozoic, with renewed granite intrusion in late Triassic, early and late Cretaceous. Tertiary rocks are of relatively minor importance, but during the Quaternary there was considerable erosion and continental sedimentation in response to sea-level changes, with the formation of the alluvium-filled valleys and areas of karst topography on the Upper Palaeozoic carbonate rocks. Mesozoic granites occur widely over the Peninsula, and mineralization is often found *in situ*, or evidenced by the accumulation of alluvial minerals in Quaternary cover. However, this mineralization varies across the Peninsula, tin being found on the eastern and western flanks, and gold in the central zone.

The most important primary mineralization is associated with granites of Upper Triassic age, but the most important deposits are found in derived Quaternary alluvium.

Geology of the Kinta Valley

The bedrocks of the valley floor are folded and metamorphosed carbonates, shales and sandstones of Upper Palaeozic age. Crystalline limestone is the dominating lithology, although significant thicknesses of schists and quartzites occur. The axis of folding is roughly along the north/south trend of the valley. The complexity of the folding and the depth of the weathering make the thickness of the rock sequences difficult to estimate. Granites form the hills on either side of the valley and some offshoots are exposed on the floor of the valley. They intrude the sediments, producing marked contact metamorphic aureoles.

Two kinds of mineralization are found in the bedrock: vein deposits of cassiterite and sulphide-bearing quartz in granite and associated schists, and irregular pipe-shaped deposits containing cassiterite, with some sulphides and calc-silicate minerals, in the limestones. However, although locally rich, none of these deposits is as important as the alluvial ones in the valley.

Four main units are recognizable in the Quaternary cover of the valley floor. At the sides are a series of so-called 'boulder-beds' and clay-rich deposits, which have acquired various local names; the 'Gopeng Beds', the 'Tekka Clays', etc. They are ill-sorted masses of rock fragments, usually of local provenance, that tend to occur round the edge of the valley at about 70 m altitude. Filling most of the valley bottom is the Older Alluvium, a sandy or muddy material with some pebble beds and lignite seams. The grain size and composition of the alluvium tends to reflect the rocks around, a sandy 'granite wash' being found at the edges of the valley, and a red mud or clay over the limestones. The alluvial cover begins near the head of the valley at 70 m altitude and becomes over 120 m thick at the mouth. The alluvium partially infills a pronounced karst topography in the limestone areas and is much disturbed by solution collapse. Broken stalactites are among the evidence of the formation of the valley by cave formation (and collapse) before the deposition of the main alluvium. The two other units are the better-sorted Younger Alluvium of the present river valley, and the organic-rich, or peat cover, of the land between. To some extent the bedrock, where exposed free of alluvium, is lateritized.

Cassiterite, with a host of minor heavy minerals, is found throughout the Quaternary deposits of the valley. The richer concentrations are in the boulder beds and marginal clays and in the Older Alluvium, where it infills the karst surface. However, the distribution of cassiterite is complicated and many local variations are encountered.

Size and Grade

It is notoriously difficult to give estimates of tonnage and grade of alluvial deposits of this kind; it is not unknown, for example, for an area of alluvium to be worked at a given grade, with high recovery, and subsequently to be worked at a similar grade. From 1876 to 1950 the valley produced concentrates containing 1·26 Mt of tin metal and there are still considerable reserves. Today, grades worked are between 150 and 350 g/m³ of tin, but much richer material was worked in the past. One estimate of the total amount of stanniferous alluvium in the valley is 11 000 000 000 m³.

Geological Interpretations

The folded carbonate rocks are thought to have formed the roof of an extensive and complex granite batholith intruded in the Upper Triassic, and which formed the hydrothermal and contact replacement cassiterite-bearing deposits. Faulting and the ease of weathering of the limestones caused the karst valley to form during the late Tertiary and early Quaternary, and, the marginal boulder beds are probably local landslides and mud-flows of rotted rock that formed along its side, or in cave systems. The valley was probably eroded, down to a base level now 120 m below sea level, during one of the main Quaternary glaciations. The valley began to fill with alluvium as the sea transgressed, so that for a time the sedimentation was esturine. It is probable that this unique concentration of cassiterite is due primarily to its occurrence in the bedrock, mostly carbonate rocks, large volumes of which have been dissolved away and secondly, due to the concentration by the valley's river systems by repeated concentration as the base level fluctuated with climatic change during the Quaternary.

Mining

There are two main mining methods now used in the Kinta Valley: dredging and gravel-pump mining. Bucket dredges operating in their own ponds are used in the flatter, lower-lying areas free from too many hard rock obstacles and limestone karst pinnacles. The largest dredges can dig to a depth of 40 m at a rate of 670 m³/h. They carry their own washing, screening and gravity concentration plants. The more irregular and higher-level deposits are worked by washing or digging down the material into a sump, most frequently with high-pressure monitors, from which the alluvium is pumped to a gravity concentration plant, traditionally the 'palong', a sloping trough with riffles along the bottom, supported on a bamboo frame. All operations produce cassiterite concentrates for shipment to smelters elsewhere. The valley produces 25 000–30 000 t of tin each year, about 10% of the world's total.

Further Reading

INGHAM, F. T. and BRADFORD, E. F. (1960), The Geology and Mineral Resources of the Kinta Valley, Perak. *Fed. Malaya Geol. Surv. Mem.* **9**, 347pp. [The authoritative work on the area.]

CHUNG, S. K. (1973), *Annual Report of the Geological Survey of Malaysia, 1973.* [Contains good review papers of the regional geology and mineralization of the Malayan Peninsula.]

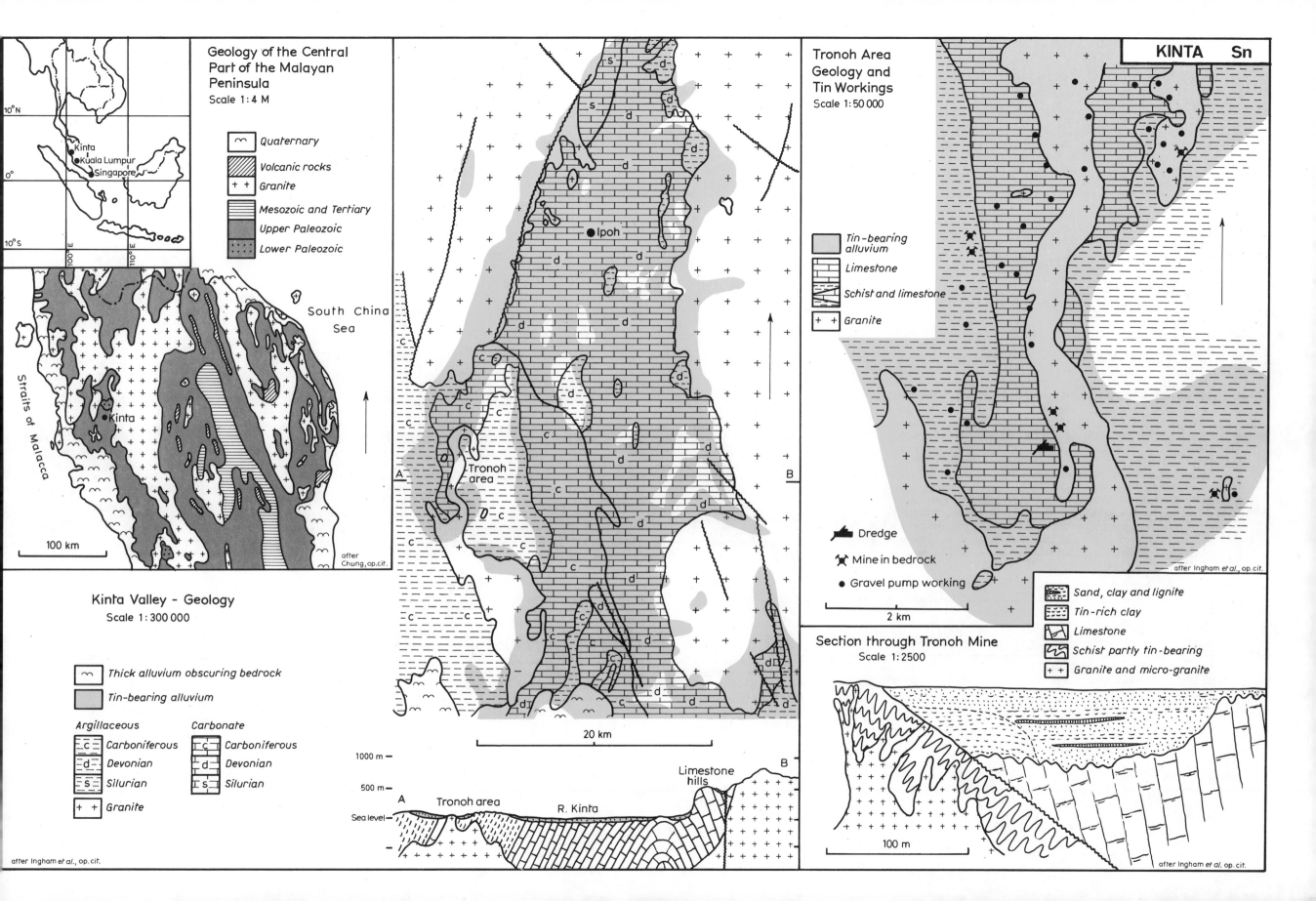

Geology of the Central Part of the Malayan Peninsula
Scale 1:4 M

10°N
0°
10°S

Kinta
Kuala Lumpur
Singapore

South China Sea

Quaternary
Volcanic rocks
Granite
Mesozoic and Tertiary
Upper Paleozoic
Lower Paleozoic

Straits of Malacca

Kinta

100 km

after Chung, op.cit.

Kinta Valley - Geology
Scale 1:300 000

Thick alluvium obscuring bedrock
Tin-bearing alluvium

Argillaceous
c Carboniferous
d Devonian
s Silurian
+ Granite

Carbonate
C Carboniferous
d Devonian
s Silurian

Ipoh

Tronoh area

A

B

20 km

1000 m —
500 m —
Sea level —

A Tronoh area   R. Kinta   Limestone hills   B

KINTA   Sn

Tronoh Area Geology and Tin Workings
Scale 1:50 000

Tin-bearing alluvium
Limestone
Schist and limestone
Granite

Dredge
Mine in bedrock
Gravel pump working

after Ingham et al., op.cit.

2 km

Sand, clay and lignite
Tin-rich clay
Limestone
Schist partly tin-bearing
Granite and micro-granite

Section through Tronoh Mine
Scale 1:2500

100 m

after Ingham et al. op.cit.

The Beach-Sand Deposits of North Stradbroke Island – Australia

Stormy seas seasonally breaking on to a long sandy shore are probably responsible for the accumulation of heavy minerals on the beaches and under the great sand-dunes of eastern Australia. This is a typical example of many deposits around the Australian coast and elsewhere in Africa, India and America. Such deposits, though low-grade, are easily worked and supply a substantial proportion of the world's ilmenite, and virtually all its rutile and zircon.

Location

The deposits are situated along the eastern side of North Stradbroke Island (latitude 27° 33′ S, longitude 153° 27′ E) just off the south-eastern coast of Queensland and 42 km from the city of Brisbane.

Geographical Setting

The south-east coast of Queensland is the site of some of the world's most spectacular sand-dunes. North Stradbroke is one of a number of islands just off this coast that are mostly built of sand and it (together with its neighbouring islands, South Stradbroke and Morton) forms a protective screen to the mouth of the Brisbane River. A long sandy beach, offering good surfing (even by Australian standards), faces the South Pacific Ocean and is separated along most of its length by a swamp from the main part of the island, which is a mass of scrub-covered sand-dunes rising to a height of 220 m. Along the west side, an escarpment cut into the sand overshadows a narrow sandy and swampy coastal plain were, on one of the rare outcrops of consolidated rock, stands Dunwich, the main settlement. The climate is warm and moist with summer (January) temperatures generally over 20 °C and winters rarely less than 13 °C. Over 1 m of rain falls in the year, primarily in summer storms coming with the prevailing, strong south-easterly winds. This wild scrub-covered island was practically uninhabited before working of beach-sands began, but today mining has to compete with recreational development for the growing population of Brisbane.

History

The black, heavy mineral sands of the Australian east coast have attracted attention since the last century, but beyond a few unsuccessful trials for gold, nothing was done until the development of the market for zircon and ilmenite in the 1930s. The Second World War stimulated the demand for these minerals (and rutile). Systematic exploration began in 1947 followed by production three years later.

Geological Setting

The hinterland of the region where the island is situated is formed of several groups of Palaeozoic rocks largely consisting of fine-grained sediments with a few volcanic rocks. All are folded and much intruded by granitic rocks of late Palaeozoic age. These rocks, which form the high ground, are interspersed by basins of Mesozoic sediments dominated by great thicknesses of soft sandstones, deeply dissected by the Clarence and Brisbane Rivers. Large areas are covered by basalt flows. Along the coast is a plain, built up of sand-dunes and swamps, that ranges in age from the early Pleistocene to those that are still forming at the present.

Geology of North Stradbroke Island

On the western shore of the island a few small outcrops of the consolidated Mesozoic, and perhaps older, rocks show that the sand that makes up most of the island rests on a platform of rock near the present sea level. Most of the island is made of the wind-blown sand in which several sequences may be recognized, separated by sand layers cemented by organic material. The sand probably ranges in age at least through the mid to upper Pleistocene and the Holocene. The high dunes culminate in long ridges crossing the island in a NW–SE direction. Along the eastern shore is a continuous beach over 20 m wide, terminated on the inland side by fore-dunes that rise to a height of 5–10 m. Usually there are several sand ridges parallel to this, breached in places by 'blow-outs' behind which sand-ridges are elongated to the north-west. Beyond these parallel dunes is a swamp, tidal in the south, and the land rises steeply from the west side of the swamp on to the high dunes.

Heavy Mineral Deposits

Mixed with well-rounded quartz grains is a variable proportion of other more dense minerals. Of the total heavy mineral fraction, the following are the most important:

Zircon	28–31%
Rutile	27–37%
Ilmenite	31–46%
Monazite	0.2–0.4

The size of the grains (median diameter) ranges from 0.11 mm to 0.13 mm. Traces of other minerals occur including tourmaline, garnet and lucoxene. Unfortunately, chromite occurs associated with the ilmenite, and some concentrates average 0.93% Cr, seriously affecting their marketing for pigment manufacture.

Commercially exploitable accumulations of heavy minerals occur in three environments; the modern beach, under the parallel dunes, and in the high dunes. On the modern beach, thin layers of rich black-sand occur under a thin covering of white sand on the upper part of the beach. Heavy minerals may make up 40% of the total, but quantities are limited. Layers of alternating black and white sand, over a thickness of 1–1.5 m, are found at the base of the fore-dunes and adjoining parallel dunes, at a level that rises gently inland. Seven per cent heavy mineral content is not uncommon in these deposits. In the upper parts of the high dunes are found less well-defined layers of heavy sand which, although lean, are present in immense quantities. The same minerals occur in all environments but the proportion of ilmenite is greater in the high dunes.

Size and Grade

Most of the deposits on the modern beach have been exhausted but were originally estimated at 16 000 m³ averaging 230 kg/m³ total heavy minerals. Under the parallel dunes were originally 1 million m³ averaging 160 kg/m³ and in the high dunes are over 200 million m³ with 34 kg/m³. The expected yield of these reserves is 1.7 Mt of zircon, 1.7 Mt rutile and 2.8 Mt of ilmenite, of which perhaps 10% could be refined to meet the chromium specification of the market.

Geological Interpretations

There is little or no disagreement about the formation of the heavy sand accumulations themselves. During storms, sand is thrown by the waves on to the berm, and the lighter quartz is returned preferentially to the sea in the undertow; at low tide, the wind blows sand from the beach to cover the black sand on the berm, and gradually the fore-dunes advance seaward over the layers built up by each storm. Exceptionally high winds carry sand from the near-shore dunes on to the central part of the island, carrying with it some of the heavy minerals. There is far less agreement about the source of the sand, and about the history of its accumulation. Most authors agree that the sand is carried to the sea by rivers where it drifts along shore, northwards from their mouths. It also seems probable that most of the sand is derived from the Mesozoic sandstones of the Clarence and Brisbane basins. However, some authors conclude that the ultimate source of the heavy minerals is the palaeozoic granites and the basalts, while others look towards the Pre-Cambrian shield of the interior of the continent, for the ultimate source. Some authors have found the enormous quantities of sand at the height of the dunes difficult to explain by present-day processes, and have suggested that eustatic changes in sea level during the Pleistocene glaciation account for the phenomenon; but this view has been challenged.

Mining

Two companies, Associated Minerals Consolidated, and Consolidated Rutile, operate on the Island, producing annually about 90 000 t of rutile concentrates and 70 000 t zircon. Production methods vary from hand or mechanical shovelling on the beach, to floating dredges in ponds cut in the dunes, or dry excavation on the high dunes. Normally a crude concentrate is made of the heavy minerals by gravity separation on site, which is shipped to Dunwich or on to the mainland for separation of the individual mineral constituents by a combination of gravity, magnetic and electrostatic separation techniques.

Further Reading

BEAZLEY, A. W. (1948), Heavy Mineral Beach Sands of Southern Queensland, Part 1. *Proc. Roy. Soc. Qd.*, **59** (2), 109.

BEAZLEY, A. W. (1950), *Idem*, Part II. *Ibid.*, **61**, 59. [A thorough study of the deposits and the provenance of the minerals, but based on work done before systematic commercial exploration.]

GARDNER, D. E. (1955), Beach Sand Heavy Mineral Deposits of Eastern Australia. *Min. Resources Aust. Bull.* **28**. [Contains results of a systematic exploration of the area, maps and a discussion of the origin of the deposits.]

CONNAH, T. H. (1961), *Beach Sand Heavy Mineral Deposits of Queensland*. Geological Survey of Queensland Publication No. 302. [A full documentation with many illustrations of the area and a summary of known reserves.]

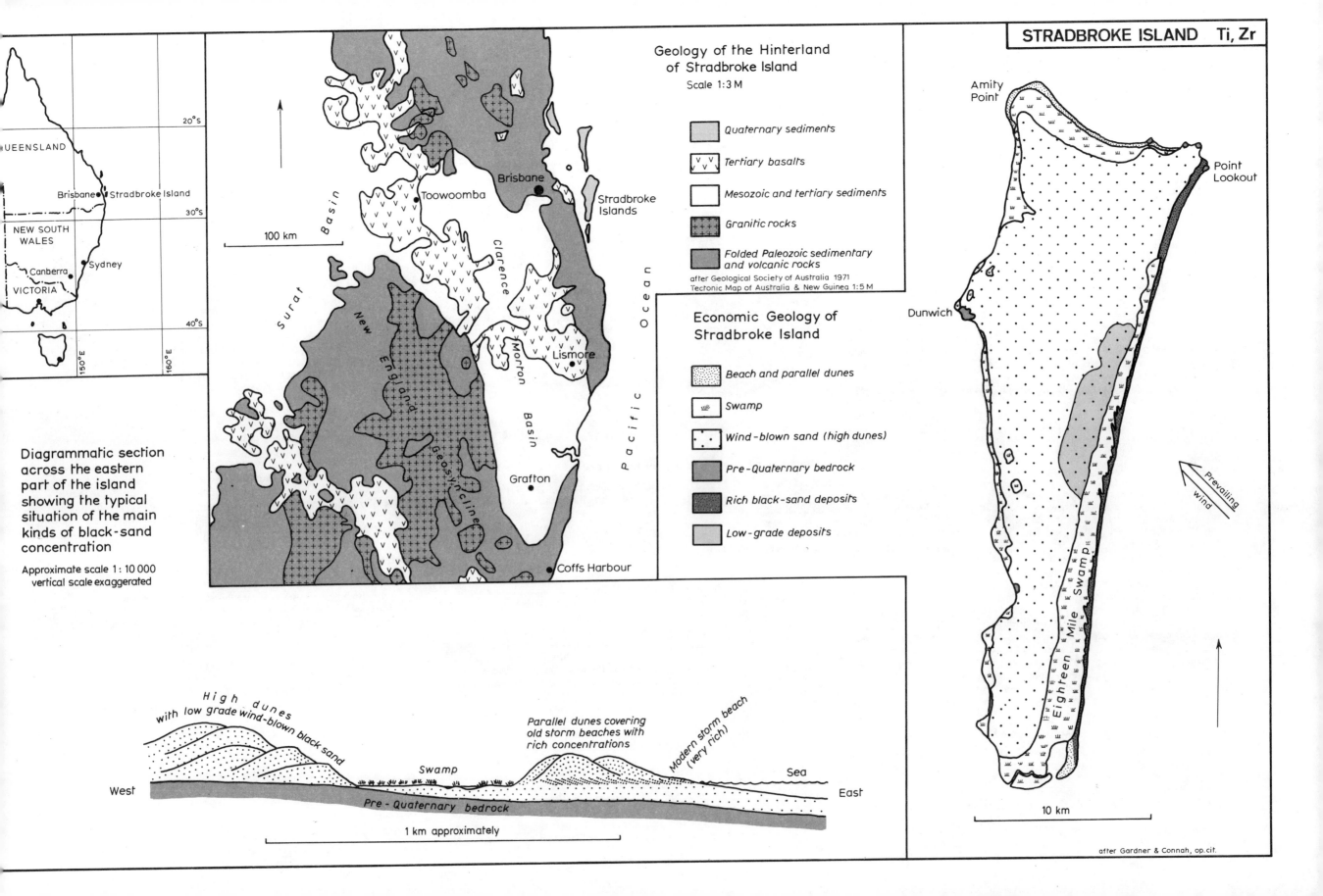

STRADBROKE ISLAND Ti, Zr

Geology of the Hinterland
of Stradbroke Island

Scale 1:3 M

Quaternary sediments

Tertiary basalts

Mesozoic and tertiary sediments

Granitic rocks

Folded Paleozoic sedimentary
and volcanic rocks

after Geological Society of Australia 1971
Tectonic Map of Australia & New Guinea 1:5 M

Economic Geology of
Stradbroke Island

Beach and parallel dunes

Swamp

Wind-blown sand (high dunes)

Pre-Quaternary bedrock

Rich black-sand deposits

Low-grade deposits

QUEENSLAND

Brisbane • Stradbroke Island

NEW SOUTH
WALES

Canberra
VICTORIA   • Sydney

20°S
30°S
40°S

150°E   160°E

100 km

Surat Basin

New England Geosyncline

Clarence Morton Basin

Pacific Ocean

Toowoomba
Brisbane
Stradbroke Islands
Lismore
Grafton
Coffs Harbour

Diagrammatic section
across the eastern
part of the island
showing the typical
situation of the main
kinds of black-sand
concentration

Approximate scale 1:10 000
vertical scale exaggerated

High dunes
with low grade wind-blown black sand

Parallel dunes covering
old storm beaches with
rich concentrations

Modern storm beach
(very rich)

Swamp

Sea

West   East

Pre-Quaternary bedrock

1 km approximately

Amity
Point

Point
Lookout

Dunwich

Eighteen Mile Swamp

Prevailing wind

10 km

after Gardner & Connah, op.cit.

The Witwatersrand Gold-Uranium Deposits – South Africa

These remarkable deposits take the form of a number of quartz-pebble conglomerate horizons that are found around the edge of what is probably the oldest major sedimentary basin. Although the subject of great controversy, they are probably some kind of sedimentary deposit.

Location

The Witwatersrand Basin is 260 km across at its maximum extent, centred approximately on latitude 27° S and 27° 30′ E, partly in the Transvaal, and partly in the Orange Free State of the Republic of South Africa.

Geographical Setting

The region of the Basin is a flattish plateau-like area known as the 'velt', elevated to between 1300 m and 1500 m above sea level. The area is traversed and drained by the Vaal River flowing westwards to join the Orange. The climate is typically subtropical modified by the high altitude, with a dry cool winter and a wet hot summer. The total rainfall is about 400 mm, enough to support extensive farming. The country's largest city, Johannesburg, lies on the northern edge of the Basin and there are several large towns on the area; Klerksdorp in the west, Odendaalrus in the south and Parys in the centre. The whole area is well populated and served by good roads and an extensive rail network leading to the Cape Province and to neighbouring Mozambique.

History

Although the Bantu, who were the first people with a metallurgical culture to inhabit the velt, were skilled workers in gold, it is doubtful if they ever worked these deposits, but their rich gold ornaments certainly stimulated the Voortrekkers to look for gold. In 1886, George Walker and George Harrison, exploring on a farm near the present city of Johannesburg, found and laid claim to gold-bearing conglomerate outcrops. Within two years similar conglomerate outcrops, some bearing gold, were found all over the region. For some years these deposits were thought to be small 'fossil placers' that would soon be exhausted, but in 1892 a borehole was sunk 1½ km south of the main outcrop in Johannesburg, which intersected rich gold 730 m below surface. The real significance of Walker and Harrison's discovery began to be realized. So developed the gold-mining region which began to extend east and west from Johannesburg. Since that time several concealed extensions of the Basin have been found. The discovery of these deposits is one of the greatest mineral exploration ventures of all time, involving over 2000 deep boreholes totalling over 2500 km of drilling, and one of the earliest successes of magnetic and gravity geophysical methods.

Geology of the Witwatersrand Basin

The Basin has the shape of an elliptical annulus 260 km by 130 km, although the southern half of it is obscured by later sediments. It is situated in between a series of massifs of granitic rocks of Archean age. The sediments in the Basin, which reach a maximum of 10 km in thickness, have been divided into several groups. At the base is the discontinuous Dominion Reef Group consisting of the erosion products of the granitic basement, conglomerate, greywacke and arkose, filling in the post-Archean topography, and passing up in places into andesitic and rhyolitic volcanic rocks.

Then come the clastic sediments of the Witwatersrand Super-Group. Many conglomerates occur in these rocks, often marking small unconformities. The first major unconformity in the Basin occurs at the base of the Ventersdorp Group, whose basal conglomerate is often gold-bearing and overlain by basic and intermediate lavas that extend over the suboutcrop of the Witwatersrand. Above, another major unconformity is the Transvaal Super Group, which overlaps the suboutcrop of the Witwatersrand. It is divided into the relatively thin Black Reef Group, which has a passably gold-bearing basal conglomerate overlaid by black shales, and gives way to the Transvaal Dolomite Group, mostly composed of cherty dolomites. Above this are the mixed sediments of the Pretoria Group that are better developed to the north and west, outside the Basin area.

The area is intruded by many dykes, some of which are feeders to the Ventersdorp Lavas; others are much younger. The structure is quite complex. In general the Witwatersrand rocks dip inwards around the Basin towards the centre where they dip up again round the Vredefort Dome. Around the edge, and particularly in the south, the Basin is faulted into a series of horsts and grabens roughly parallel to the Basin edge. On the whole the deformation is not extensive and the rocks are only feebly metamorphosed.

The southern half of the Basin is covered by a relatively thin layer of Karoo sediments. These generally begin with tillites followed by coal measures and sediments largely continental in character. One of South Africa's major coalfields occurs just to the north-east of the basin.

The age of the Basin is known in some detail. The granitic basement is generally 3·0 Ga or older, the lavas of the Dominion Reef Group are dated at between 2·6 and 2·8 Ga and those of the Ventersdorp Group at 2·3 Ga. The sedimentation in the Basin probably finished about 2·0 Ga ago, since the Pretoria Group is cut by the Bushvelt Complex that is dated at 1·96 Ga. There is then a very long gap before the Karoo.

Geology of the Gold–Uranium Deposits

Workable gold and/or uranium occurs at specific stratigraphic levels in the sequence and are locally termed 'reefs'. In general they are of two types; conglomerate or 'banket' reefs, and carbon-seam reefs. The first type consists of clean conglomerate of well-rounded pebbles usually of quartz, but sometimes including chert, quartzite and other hard materials. The matrix usually contains a lot of pyrite, itself sometimes in the form of small pebbles or pebble-like fragments termed 'buckshot pyrite', and with it, fine micaceous minerals and fine-grained quartz. Uranium is generally present as fine, rounded grains of uraninite, but the hydrocarbon material, thucholite, is common. The conglomerates are remarkable for the lack of heavy minerals such as are found in normal clastic sediments. Gold is by far the rarest trace mineral of all and occurs as small interstitial grains in the matrix or as a 'smear' on the surface of pebbles. The second type are semi-continuous bands of a carbonaceous material often only a few millimetres thick known as 'carbon leaders'. The 'carbon' is a fibrous, high-carbon, hydrocarbon complex which contains uranium and sub-microscopic gold. Electron microscopy has shown that this material has an organic-like structure as if it began life as a mould-like encrustation on the underlying rock surface.

There are many reef horizons, beginning in the Dominion Reef Group, and occurring in the Government Reef Group, but best developed on two levels; one in the Main–Bird Group and the other in the Kimberley–Elsburg Group. A further reef occurs on the unconformity with the Ventersdorp Group known as the Ventersdorp Contact Reef and yet another, less important one called the Black Reef, at the base of the Transvaal Super Group.

Within the reef horizons, gold and uranium tend to concentrate in ore shoots or 'pay-streaks', which are elongated and orientated towards the centre of the Basin and in the same direction as the orientation of various sedimentary structures and features such as pebble size distribution, pebble orientation, etc.

Size and Grade

It is almost impossible to give meaningful figures for the reserves of these deposits. The average grade of ore worked today is about 9 g/t but some operate at grades as high as 31 g/t. The Basin has already produced over 30 000 tonnes of gold bullion; that is, about three-quarters of the total world's production for all time, and it could probable produce as much, or more, again.

Geological Interpretations

Few mineral deposits have been the subject of more controversy than this. The main argument has been between those who hold that the deposits are essentially sedimentary, and those who regard this as impossible. The latter group of theorists proposed that the gold and uranium were introduced by solutions along the pebble beds.

At least a source of the gold and uranium might have existed in vein deposits in the Archean schist belts and granites that were eroded to supply the Basin with sediment. The balance of opinion seems to favour alluvial gold and uraninite grains being trapped between pebbles, or on some primitive organic growths, and then being progressively reconcentrated as the sedimentary sequence developed in some sort of neritic or littoral environment. It has been suggested that the relatively anoxigenic atmosphere of the time, allowed uraninite and pyrite to survive mechanical transport better than it does today. The root of the problem is that at the time of formation of this, the world's earliest major sedimentary basin, we simply do not know what were the environmental conditions, so that little help can be derived from the study of modern sediments.

Mining

There are about forty mines working on the Basin, owned and operated by a series of mining finance companies. Most of the mines operate through deep vertical shafts to one or more reef horizons that are generally worked by longwall stopping methods. Some of the mines are as deep as 3500 m. The ore has to be milled to a very fine pulp in order to liberate the gold, which is extracted by cyanidation. Uranium is extracted by acid leaching and ion-exchange recovery. Some pyrite is recovered by flotation and used to produce sulphuric acid, partly for uranium leaching.

Further Reading

HAUGHTON, S. H., ed. (1964), *The Geology of some Ore Deposits in Southern Africa*. Johannesburg. The Geological Society of South Africa. 625 pp. [The reader needs go no further than this volume for a comprehensive set of papers on almost all aspects of these deposits.]

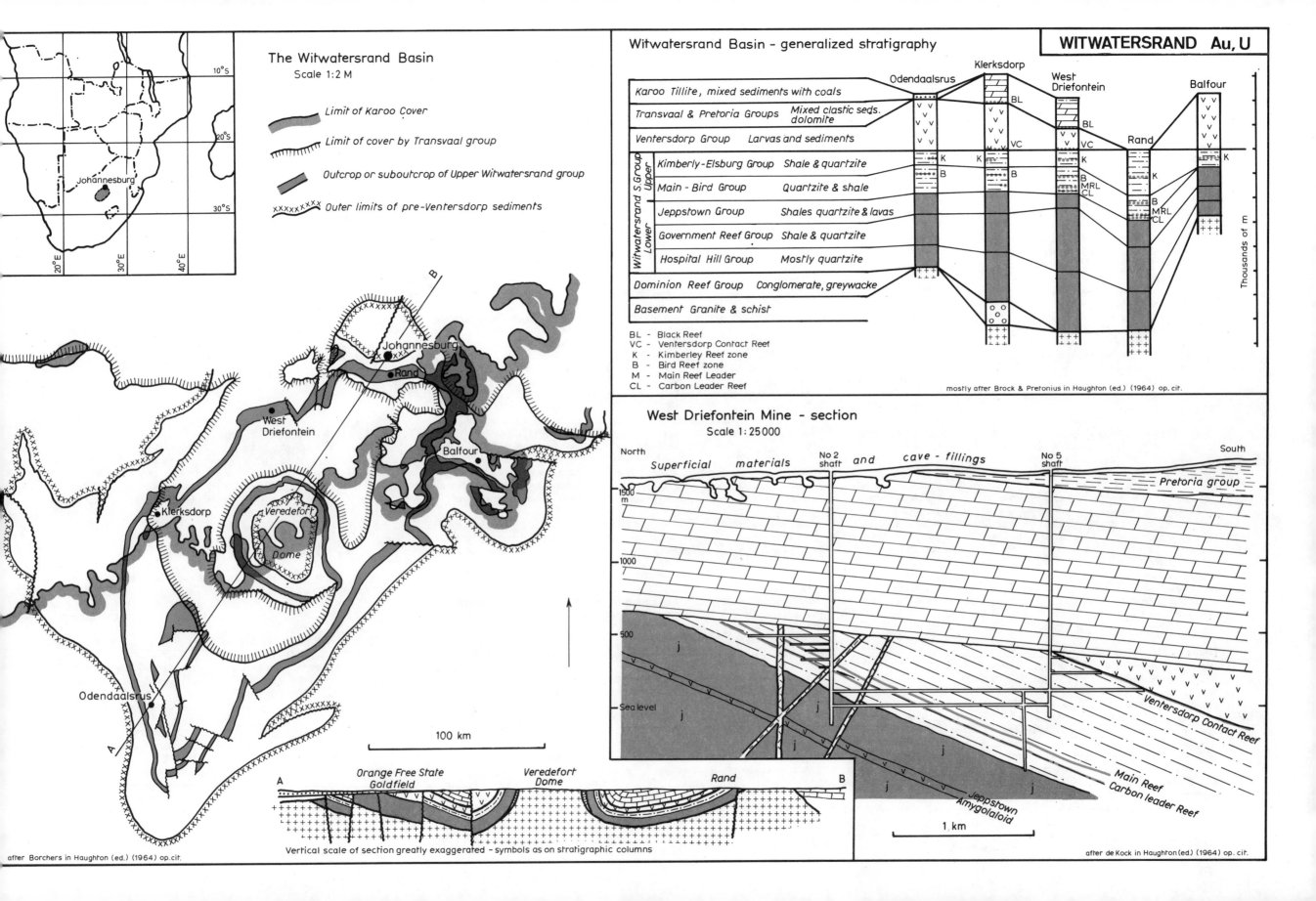

The Witwatersrand Basin
Scale 1:2 M

— Limit of Karoo Cover

⊤⊤⊤ Limit of cover by Transvaal group

▬ Outcrop or suboutcrop of Upper Witwatersrand group

×××××× Outer limits of pre-Ventersdorp sediments

Johannesburg

Rand

West Driefontein

Balfour

Veredefort Dome

Klerksdorp

Odendaalsrus

100 km

Orange Free State Goldfield · Veredefort Dome · Rand

Vertical scale of section greatly exaggerated – symbols as on stratigraphic columns

after Borchers in Haughton (ed.) (1964) op.cit.

Witwatersrand Basin - generalized stratigraphy

WITWATERSRAND Au, U

Odendaalsrus · Klerksdorp · West Driefontein · Balfour

Karoo Tillite, mixed sediments with coals		
Transvaal & Pretoria Groups	Mixed clastic seds. dolomite	
Ventersdorp Group	Larvas and sediments	

Witwatersrand S.Group Upper:
- Kimberly-Elsburg Group — Shale & quartzite
- Main - Bird Group — Quartzite & shale

Witwatersrand Lower:
- Jeppstown Group — Shales quartzite & lavas
- Government Reef Group — Shale & quartzite
- Hospital Hill Group — Mostly quartzite

Dominion Reef Group — Conglomerate, greywacke

Basement Granite & schist

Thousands of m

BL - Black Reef
VC - Ventersdorp Contact Reef
K - Kimberley Reef zone
B - Bird Reef zone
M - Main Reef Leader
CL - Carbon Leader Reef

mostly after Brock & Pretonius in Haughton (ed.) (1964) op. cit.

West Driefontein Mine - section
Scale 1:25 000

North · Superficial materials and cave - fillings · No 2 shaft · No 5 shaft · South

Pretoria group

1500 m

1000

500

Sea level

j

Ventersdorp Contact Reef

Main Reef

Carbon leader Reef

Jeppstown Amygolaloid

1 km

after de Kock in Haughton (ed.) (1964) op. cit.

The Uranium Deposits of the Blind River Area – Canada

This is one of the world's major sources of uranium. The ore-minerals occur in conglomerates at the base of the first sequence of epicontinental sediments to be deposited after the stabilization of the Archean Shield of Canada. They make an interesting comparison with the gold–uranium-bearing conglomerates of the Witwatersrand in South Africa.

Location

The area is named after a river and the town at its mouth, on the north shore of Lake Huron. It is also sometimes known by the name Elliot Lake after the main town at the centre of the area at latitude 46° 20′ N, longitude 82° 40′ W and an altitude of 400 m in the Algoma Province of Ontario.

Geographical Setting

The land rises steeply from the level of Lake Huron (177 m) to rough country around 400 m in altitude with steep rocky ridges, lakes and swamps connected by rivers draining south to Lake Huron. The area suffers a cool continental climate with sub-zero winters, summer temperatures averaging 23 °C and 800 mm of rain and snow. But there are 150 annual frost-free days on average. Originally, the area was mixed boreal forest of pine, spruce, maple and birch, but a great deal has been cleared and the thin soils so eroded that much of the area is almost a wasteland. However, some grain and cattle farming is carried on in the valleys and on old lake beds. There are several towns in the area, mostly along the Lake, connected to other parts of the country by the Canadian Pacific Railway and the Trans-Canada Highway. The principal town of Elliott Lake grew from a small village to a population of 20 000 in the first five years of the mining boom in the 1950s.

History

J. L. LeConte, a geologist, recorded a 'pitchblend-like mineral' in quartzites just north of Lake Huron in 1847, but it was only a curiosity until a century later when the Atomic Age stimulated the search for uranium. The area along the shore was prospected in 1948, but the many radioactive outcrops were thought to be due to thorium minerals. Franc Joubin, a chemical engineer turned prospector, was not convinced that there was no uranium; he secretly continued work and proved economic mineralization in 1953. The secret leaked and a rush started, 5000 claims being staked in that season. By 1956, Pronto mine began production and by 1958 was joined by ten others. However, the demand for uranium was quickly saturated and in the early sixties many mines were stopped, closed or kept on maintenance, to wait for the revival of demand in the seventies.

Geological Setting

Exposed to the north and forming a basement beneath the Blind River rocks, are the Archean rocks of the Superior Province of the Canadian Shield. Set in a 'sea' of granite-gneisses are the so-called greenstone belts, notably the richly mineralized Abitibi Belt. The Blind River area is terminated to the south by the Murray Fault Zone, beyond which the Pre-Cambrian rocks are covered by Palaeozoic sediments largely carbonates. The Blind River Area is part of a folded basin of clastic sediments of the Huronian Super-Group, laid down in the Aphebian Era, and represent the first sedimentary cover on the Archean Shield following its stabilization by the Kenoran tectonic events. The Huronian lies on a basement tectonized 2·45 Ga ago and is cut by intrusions 2·25 Ga old.

The Huronian is divided into four Groups named Elliot, Hough Lake, Quirk Lake and Cobalt, each of which begins with conglomeratic greywackes passing up through fine-grained sequences of argillites and terminates with an arenite sequence. It has been suggested that at least part of the conglomeratic greywacke sequences are tillites, and that each of the four groups represents a glacial and interglacial period, but the idea is much disputed. For an epicontinental sequence the Huronian is remarkably free of red rocks of any kind, being uniformly grey or buff. The uranium-bearing conglomerates are certainly either fluviatile or deltaic, and occur near the base of the conglomeratic greywacke sequences; but only those of the Elliot Group are economic.

The Huronian rocks infill a pronounced topography on the post-Archean surface, and the contact is often marked by evidence of palaeosol development. The Huronian gradually overlaps the shield to the east, and sedimentary structures show a provenance of the sediment from the north and north-west.

The area was gently folded and intruded by dolerite and gabbro and probably covered by Palaeozoic sediments, although little more of its history is known until the present topography was formed during the Quaternary Glaciation, which also left behind moderate amounts of till in the valleys.

Geology of the Uranium Deposits

The Elliot Group has been divided lithologically and the lowermost conglomeratic greywacke, Matinenda Formation, contains the main ore zones. The Matinenda is itself divided into three members: the Manfred, Stinson and Ryan which progressively transgress, from south to north, the hills and valleys of the pre-Huronian surface. These valleys tend to correspond to the contacts between schists and gnesses in the basement, and it is in these thicker, valley-fill sequences, that the ore zones are localized.

The ores are all much the same. They consist of moderately well-packed conglomerates with rounded quartz or chert pebbles from 60 mm to 25 mm size, in a matrix of sandy and shaly material; enhedral to subhedral pyrite of sometimes making up 20% of the rock. In the matrix are grains, usually somewhat rounded, of uraninite, uranothorianite, brannerite and thucholite. Uraninite grains are between 0·2 mm and 0·5 mm in diameter and frequently a component of small-scale sedimentary structures in the matrix. They have even been observed, pushed by pebbles into the underlying matrix. Titanium minerals, rutile and anatase, are found along with grains of monazite, zircon and a variety of other trace minerals (including galena formed from radiogenic lead). Gold is often present but normally only 0·2 g/t; at best 1·5 g/t.

Folding has given the ore-bearing strata an 'S'-shaped outcrop passing round the nose of the Quirk syncline in the north and the Chiblow anticline to the south, which is terminated by the Murray fault along the Lake shore. Around this structure there are three main ore zones. The largest is the Quirk Zone trending in the direction of sediment deposition and dipping about 40° south, down the north limb of the Quirk syncline. There are several conglomerates at, or just above, the base of the Manfred member of the Matinenda Formation, and the conglomerates are intimately interbedded with quartz-feldspatic arenite. Current bedding is general, pyrite lenticles often occupy the fore-sets and local concentrations of uraninite grains are clearly related to the sedimentary structures. Pebbles tend to become smaller towards the south-east.

On the other side of the syncline is the Nordic Zone with a similar orientation but shallower dip. There are three main conglomerates above the base of the Ryan member and the zone clearly occupies a valley along a schist-gneiss contact in the basement. In the south is the Pronto Zone, dipping from 15° to 20° south against the Murray Fault, which has a single 2·3 m thick conglomeratic zone (in the Ryan member), 15 m above a palaeosol on the Archean basement. The conglomerate here has larger pebbles and larger pyrite grains than is general.

Other (but sub-economic) conglomerate horizons are known in higher groups, and there are several (again sub-economic) hematite-bearing conglomerates containing thorium minerals in some of these formations. In one or two of the mines local, sub-economic concentrations of radioactive minerals have been found in other lithologies; in quartzites, and even in the gritty material of the Archean palaeosols.

Size and Grade

The published reserves of the eleven mines in the area that had been evaluated by 1958 total 206 Mt. Grades vary from 0·09% to 0·14% U_3O_8 averaging about 0·1%. The potential of the area may be double this at a similar grade and larger at lower grades.

Geological Interpretations

All the evidence points to the uranium-bearing conglomerates being fluvialite or deltaic sediments, and they may have been formed during a wet period preceding an ice-age. The colour of the rocks suggests deposition in anoxic conditions which, for a sequence of rocks that is apparently continental, needs some explanation. The uranium mineral grains look detrital, but some geologists prefer the idea that they are pseudomorphs of some earlier detrital mineral replaced by the action of uranium-bearing groundwater. Such a replacement process in these well-packed conglomerates over such a wide area, seems difficult to imagine. The pyrite may be diagenetic, although there is evidence that it is diagenetically modified from original detrital iron sulphide. A whole host of genetic models have been proposed involving circulating groundwaters through the conglomerates (seen as having been good aquifers before lithification) or involving hydrothermal solutions from some local or basement source. The problem is not solved, but taking into account the interpretations of the comparable conglomerates of the Witwatersrand, the detrital origin seems the best, and is fairly soundly based on detailed field observation. The provenance of the uranium is almost certainly the granitold rocks of the Archean, although the precise mechanism of its transport is not understood.

Mining

The largest mine in the area is the Denison, operated by a company of that name which has an underground mine with two shafts, an acid-leaching, ion-exchange recovery plant with a capacity of 5400 t/d and a plant which produces uranium metal and some yttrium as a by-product. Several other mines are producing, have produced, or are kept on a maintenance basis, of which the most important ones belong to the Rio Algom group who possess several plants of about 3000 t/d capacity.

Further Reading

DAVISON, C. F. (1957), On the Occurrence of Uranium in Ancient Conglomerates. *Econ. Geol.*, **52**, 668–93. [An admirable piece of geological polemic by one of that art's greatest exponents. A must!]

DERRY, D. R. (1960), Evidence of the Origin of the Blind River Uranium Deposits. *Econ. Geol.*, **55**, 906–27. [Among many, probably the most readable and intelligent genetic essay.]

HART, R. C., HARPER, H. G. et al. (1955), Quirk Lake Trough. *Can. Inst. Min. Met. Bull.*, **50**, 260–5.

HOLMES, S. W. (1957), Pronto Mine, In: *Structural Geology of Canadian Ore Deposits*, **2**, 305–16, Can. Inst. Min. Met. Montreal. [The only extant geological description of a mine in the area.]

JOUBIN, F. R. (1954), Uranium Deposits of the Algoma District. *Can. Inst. Min. Met. Tr.*, **57**, (Bull. No. 510), 431–7. [An historic paper, published by the discoverer only a year after his discovery.]

PIENAAR, P. J. (1963), Stratigraphy, Petrology and Genesis of the Elliot Group, Blind River, Ontario, Including the Uraniferous Conglomerates. *Geol. Surv. Can. Bull.*, **83**, 140pp.

ROSCOE, S. M. (1968), Huronian Rocks and Uraniferous conglomerates of the Canadian Shield. *Geol. Sur. Can. Papers* 68–40, 205pp. [Two papers which together make the best full description of the area.]

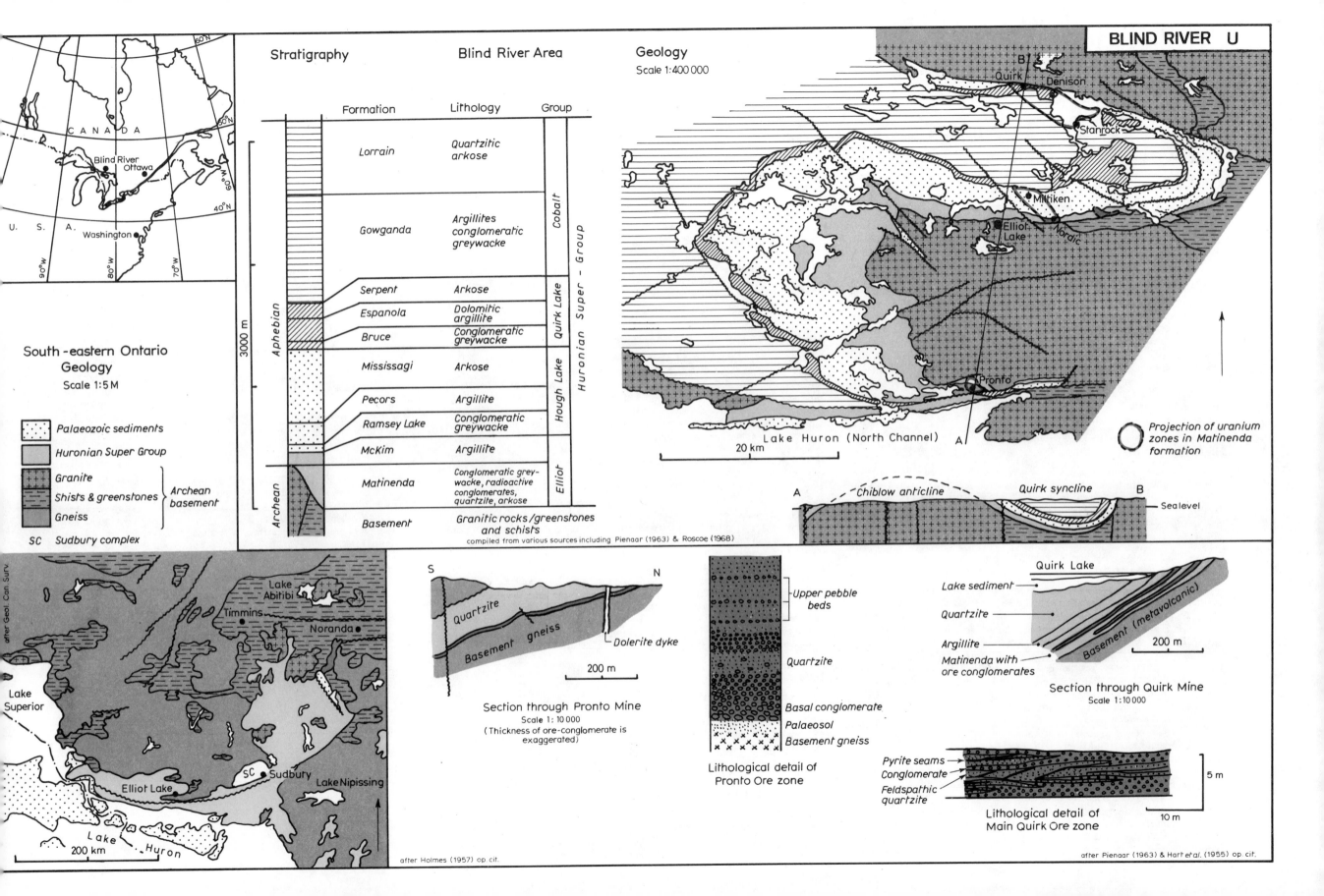

South-eastern Ontario Geology
Scale 1:5 M

Palaeozoic sediments

Huronian Super Group

Granite

Shists & greenstones } Archean basement

Gneiss

SC Sudbury complex

Stratigraphy   Blind River Area

Formation	Lithology	Group		
Lorrain	Quartzitic arkose	Cobalt		
Gowganda	Argillites conglomeratic greywacke	Cobalt		
Serpent	Arkose	Quirk Lake		
Espanola	Dolomitic argillite	Quirk Lake		
Bruce	Conglomeratic greywacke	Quirk Lake		
Mississagi	Arkose	Hough Lake		
Pecors	Argillite	Hough Lake		
Ramsey Lake	Conglomeratic greywacke	Hough Lake		
McKim	Argillite	Elliot		
Matinenda	Conglomeratic greywacke, radioactive conglomerates, quartzite, arkose	Elliot		
Basement	Granitic rocks/greenstones and schists			

Aphebian — Archean

3000 m

Huronian Super - Group

compiled from various sources including Pienaar (1963) & Roscoe (1968)

Geology
Scale 1:400 000

Quirk
Denison
Stanrock
Milliken
Elliot Lake
Nordic
Pronto
Lake Huron (North Channel)
20 km

⬭ Projection of uranium zones in Matinenda formation

A Chiblow anticline   Quirk syncline B
Sealevel

after Geol. Can. Surv.

Lake Abitibi
Timmins
Noranda
Lake Superior
SC Sudbury
Lake Nipissing
Elliot Lake
Lake Huron
200 km

S   N
Quartzite
Basement gneiss
Dolerite dyke
200 m

Section through Pronto Mine
Scale 1: 10 000
(Thickness of ore-conglomerate is exaggerated)

Upper pebble beds

Quartzite

Basal conglomerate
Palaeosol
Basement gneiss

Lithological detail of Pronto Ore zone

Quirk Lake
Lake sediment
Quartzite
Argillite
Matinenda with ore conglomerates
Basement (metavolcanic)
200 m

Section through Quirk Mine
Scale 1:10 000

Pyrite seams
Conglomerate
Feldspathic quartzite
5 m
10 m

Lithological detail of Main Quirk Ore zone

after Pienaar (1963) & Hart et al. (1955) op. cit.

after Holmes (1957) op. cit.

The Esterhazy Potash Deposits – Canada

In several parts of the world large basins containing thick accumulations of salt occur, some like the Elk Point in Canada, having beds of potassium salts near the top of the salt sequence. This example is large, and its commercial exploitation from below 1000 m of water-bearing sediment is a triumph of modern mining.

Location

Situated at latitude 50° 48′ N, longitude 101° 54′ W at 511 m altitude, the mine working the deposits is 19 km NE of the town of Esterhazy, and 180 km east of Regina, in the Canadian state of Saskatchewan.

Geographical Setting

Esterhazy lies in flat glaciated country between the Qu'Appelle River, and the Canadian National Railway running from Winnipeg to Saskatoon. The climate is cold continental with temperatures averaging 17 °C in summer and −20 °C in winter. A modest amount of snow falls from the strong west winds in winter, and the north-westerly ones bring rain in summer; total precipitation being between 400 and 500 mm each year. Originally, the area was boreal forest and, although large areas have been cleared, lumbering is still a local industry. The 120 frost-free days a year allow grain and grass to grow, and the area is at the extreme northern limit of the great grain and ranching zone of the North American Mid-West.

History

Potash was discovered at a depth of 2340 m in the Norcanol Radville No. 1 well drilled for petroleum in 1943. But it was not until three years later that it was discovered near Esterhazy, again in an oil well, at a much shallower depth and within reach of mining. The discovery of potash in the area was largely due to oil drilling, and to the introduction of gamma-ray logging which detects the natural radioactivity of potassium.

Geological Setting

Esterhazy is at the south-eastern edge of the Elk Point Basin, which occupies the land between the Canadian Shield and the foothills of the Rocky Mountains, and stretches from North Dakota in the United States to Northwest Territories. Sediments in this Basin were deposited at various ages, principally in the Devonian and Cretaceous.

A few relatively thin mixed formations of Lower Palaeozoic age were deposited over the Pre-Cambrian basement, succeeded by a thick sequence that began to be deposited late in the Lower Devonian and continued into the Mississippian. Dominating these are the rocks belonging to the Upper Elk Point Group (Mid Devonian) which are divided into the Winnepegosis Formation composed of arid-zone shelf carbonates and patch reefs, and the Prairie Formation which consists of anhydrite rock, halite rock with some shales, carbonate rocks and the potash rocks.

Three potash rock members are recognized, named, from the lowermost, the Esterhazy, Bell Plaine and Patience Lake. The three have a slightly different distribution, each extending further westward than the one beneath.

All three potash members wedge out to the north-west and south-east and were eroded along their north-east edge before the deposition of the higher members. Along the south-western edge, the members have irregular limits due to the effects of groundwater solution following deposition. The Bell Plaine Member underlays an area of 100 000 km², the other two members

being somewhat smaller, and all three average about 7 m in thickness. Potash rocks are of two types, one containing sylvite, the other carnallite; in both cases with halite, carbonates and clay minerals. The three members vary in the distribution of rocks containing the two main minerals but, in general, a central sylvite zone is surrounded by one dominated by carnallite. The clay content tends to increase in the west, and the upper beds tend to be coarser-grained.

The Prairie Foundation is overlain by a very persistent shale and carbonate horizon called the Dawson Bay Formation, the shale content of which has to some extent preserved the salts from solution by groundwater. A red shale member at its base is a valuable marker in drilling.

A series of carbonate-dominated formations lie on top of the Dawson Bay ending with the Lodgepole Limestone of Mississippian age. A major disconformity exists here, and there follows a mixed shale, silt and sand formation (the Blairmore) that marks the base of a thick and shale-dominated Cretaceous sequence. (In parts of the Basin, a few thin Jurassic beds are found.) The whole area is covered by a thick mantle of glacial till and outcrop is very scarce.

The Elk Point Basin has large reservoirs of oil, mostly in the Mesozoic formations, but rocks equivalent to the Devonian Elk Point Group carry oil in the north-west, which rocks are also host to the lead–zinc deposits of the Pine Point District (q.v.).

Geology of the Esterhazy Deposits

The Esterhazy deposits comprise a relatively small area of about 1500 km² in the east of the potash region where the Esterhazy Member is rich, thick and shallow enough to be mined. (In this area the Patience Lake Member is not present, probably eroded away, and the Belle Plaine is thin and of poor grade.) The productive beds are beneath 945 m of sediments beginning with 90 m of glacial till which contains several water-bearing fluvioglacial sand zones and a thick sequence of shales. At the base of the shales is the mixed Blairmore Formation whose sandy members contain water under pressures of 3·1 M Pa. Most of the carbonate sediments below the Blairmore are water-bearing down to the Dawson Bay with its basal red shale marker (the so-called Second Red Bed). The productive beds are 25 m below.

The Esterhazy Member averages 9 m in thickness and has been divided into 50 separate beds of differing composition. Individual beds thin, thicken, merge into one another and pinch out altogether. The top 29 are poor in potash content. Mining is confined to a high-grade bed, No. 40 and parts of those above and below. The potash beds dip gently southwestwards at an inclination of 1 : 130.

The potash rock, generally called sylvinite, contains about 40% sylvite, 49% halite, 9% carnallite and 2% of insoluble clay and carbonate. The potash beds are interrupted by zones of low-grade material in the forms of channels and upwellings, and over the field there are sharp transitions from sylvinite to carnallite rock, usually accompanied by a colour change. Like most examples in the world, the sylvinite is pink or red in colour.

Size and Grade

The potash content of the Elk Point Basin is enormous, but one normally counts as resources only areas where the bed is at least 2·5 m thick, contains over 40% sylvite and is overlain by at least 15 m of solid sound halite rock. Even with these restrictions and a 40% extraction rate and a 90% mill recovery, the resources are well in excess of 6 Gt. Esterhazy has reserves to support its present production for several centuries.

Geological Interpretations

From a combination of oil exploration, potash mining and lead–zinc exploration and mining around Pine Point, there is a fairly clear picture of the palaeogeographic history of the Elk Point Basin. During the Mid Devonian the Basin was similar in shape and size to the present day, but palaeomag-

netic measurements show it to have been within a few degrees of the equator. At that time it was a sea connected to the ocean in the present north-west, which became filled with carbonate sediments while being surrounded by a denuded continental area with a dry climate. In the north-west, a barrier of reefal carbonates developed (the Presqu'ile Barrier) that restricted circulation of water into the Basin, and the salinity gradually rose causing the deposition of anhydrite, halite and the accumulation of potassium in the highly concentrated brines. The history of evaporite accumulation was long and complicated as variations in the influx of ocean water through the barrier continued.

Eventually the Basin dried out, and was covered by red continental sediments, and it was probably at this time that the potassium-rich brines formed the potash beds. There is some disagreement about the precise nature of the formation of sylvite and carnallite since they are clearly not primary evaporite minerals, but formed by a series of diagenetic processes.

Mining

The Esterhazy deposits are mined by International Minerals & Chemical Corporation who began planning the operation in 1955. Their major achievement (though they were not the first in Canada) was the sinking of a 1000 m deep shaft which took five years, including two and a half years to sink through the Blairmore Formation by freezing the ground with cold brine to seal out the water. Almost the whole shaft had to be either frozen or grouted with cement, and lined with steel 'tubbing'. Mining continues at a rate of a million tonnes a year using continuous-mining machines that extract the potash rock leaving 60% behind as roof support. The hoisted sylvinite is crushed, and the sylvite recovered by flotation in hot saturated brine, producing a concentrate that is transported via the nearby railway.

The product goes all over the world, including large quantities to Europe where farmers often spread as much as 30 kg/ha on arable land each year.

Further Reading

HOLTER, M. E. (1969), *The Middle Devonian Prarie Evaporite of Saskatchewan*. Sask. Dept. Mineral Resources Rept. No. 23.

HOLTER, M. E. (1971), Geological Criteria for the Location of Economic Potash Deposits. *C.I.M. Trans.*, **74**, 146–51. [Two good papers which together describe the whole field.]

KEYES, D. A. and WRIGHT, J. Y. (1966), *Geology of the IMC Potash Deposit, Esterhazy, Saskatchewan*. Second Symposium on Salt. Northern Ohio Geological Soc., Cleveland, Ohio. [The only extant description of the deposit.]

McCROSSAN, R. G. *et al.*, ed. (1964), *Geological History of Western Canada*. Alberta Soc. Petroleum Geologists, Calgary, 232pp. [A vast and comprehensive documentation containing a full description of the sedimentation in the Elk Point Basin.]

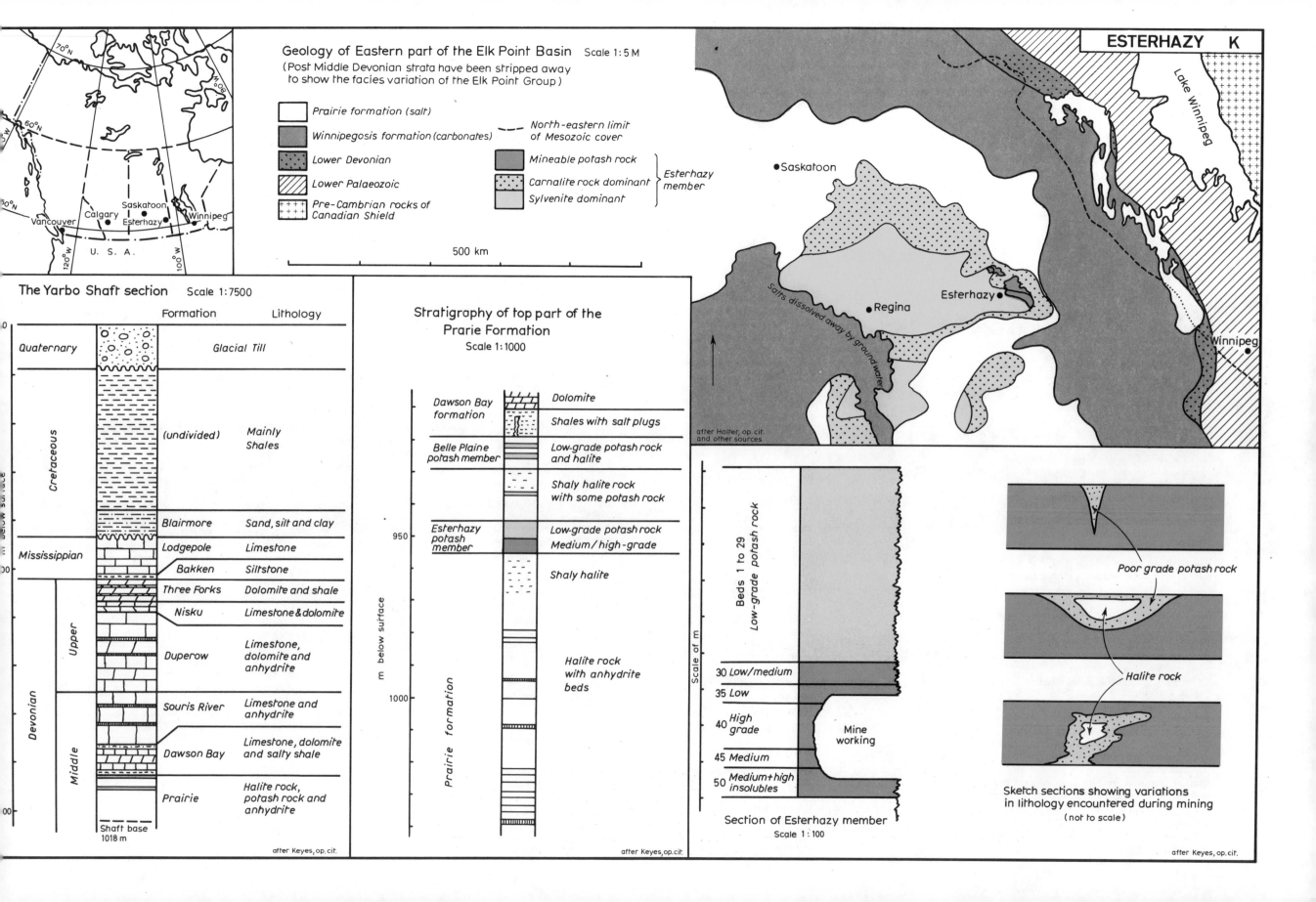

ESTERHAZY K

Geology of Eastern part of the Elk Point Basin   Scale 1:5 M
(Post Middle Devonian strata have been stripped away to show the facies variation of the Elk Point Group)

Legend:
- Prairie formation (salt)
- Winnipegosis formation (carbonates)
- Lower Devonian
- Lower Palaeozoic
- Pre-Cambrian rocks of Canadian Shield
- North-eastern limit of Mesozoic cover
- Mineable potash rock ⎫
- Carnalite rock dominant ⎬ Esterhazy member
- Sylvenite dominant ⎭

500 km

Salts dissolved away by ground water

after Holter, op. cit. and other sources

The Yarbo Shaft section   Scale 1:7500

Formation	Lithology
Quaternary	Glacial Till
Cretaceous (undivided)	Mainly Shales
Blairmore	Sand, silt and clay
Mississippian Lodgepole	Limestone
Bakken	Siltstone
Three Forks	Dolomite and shale
Nisku	Limestone & dolomite
Devonian Upper Duperow	Limestone, dolomite and anhydrite
Souris River	Limestone and anhydrite
Middle Dawson Bay	Limestone, dolomite and salty shale
Prairie	Halite rock, potash rock and anhydrite

Shaft base 1018 m

after Keyes, op. cit.

Stratigraphy of top part of the Prarie Formation   Scale 1:1000

Dawson Bay formation	Dolomite
	Shales with salt plugs
Belle Plaine potash member	Low-grade potash rock and halite
	Shaly halite rock with some potash rock
Esterhazy potash member	Low-grade potash rock
	Medium/high-grade
	Shaly halite
Prairie formation	Halite rock with anhydrite beds

m below surface: 950, 1000

after Keyes, op. cit.

Section of Esterhazy member   Scale 1:100

Beds 1 to 29 Low-grade potash rock

Scale of m: 30 Low/medium, 35 Low, 40 High grade, 45 Medium, 50 Medium+high insolubles

Mine working

after Keyes, op. cit.

Sketch sections showing variations in lithology encountered during mining
(not to scale)

Poor grade potash rock

Halite rock

The Sulphur Salt Dome – U.S.A.

In 1903, the first commercial sulphur to be produced by the Frasch process came from the cap-rock of this salt dome. A decade later the Gulf Coast had become the most important sulphur-producing area in the world. Salt domes are diapiric or 'intrusive' structures that are found in several parts of the world, wherever thick wedges of sediment have been deposited on top of salt sequences. They are important as loci of petroleum reservoirs, sometimes worked for the salt itself, but the main interest here is the occurrence of sulphur in the anhydrite cap-rocks of some of them. Sulphur and sulphur chemical are obtained from various sources: some is produced from pyrite, mined on its own or as a by-product of other metal sulphides; it is also produced from smelter gases and recovered from coal gas, petroleum and natural gas. But half the world's supply is produced from salt-dome cap rocks.

Location

Sulphur Dome is situated at latitude 30° 17′ N, longitude 93° 18′ W beneath land lying at 6 km above sea level, in the parish of Calcasieu, Louisiana, 20 km west of the town of Lake Charles.

Geographical Setting

The strip of land, from 10–100 km wide, inland from the coast of the Gulf of Mexico is largely swamp. Sulphur Dome is situated in this swampy area beside the Calceaieu River. The climate is sub-tropical and wet. Over 1 m of rain falls annually, more in summer than in winter. Average summer temperatures are in the high twenties (°C) and the low twenties in winter. The land above swamp level is (or was) thickly forested with sub-tropical pines, magnolias, etc. and the swamps are crowded with lush bamboos. When drained this land can be cultivated very successfully. There is a small township called Sulphur near the deposit which is on the main railway and road from New Orleans to Houston.

History

The Red Indian inhabitants of the Gulf Coast knew about the strange muddy mounds of the Five Island region and used the brine from the many springs to produce salt. Early European settlers did the same, and throughout the nineteenth century these 'domes' were the object of much interest and speculation. The presence of sulphur and sulphurous gas and water was reported by travellers as early as 1812. In 1862, during the drilling of a brine well at Five Islands, rock salt was encountered which revealed for the first time the underlying nature of the strange mounds. About the same time a local doctor, Kirkman by name, became interested in an oil seepage on the island now known to overlie the Sulphur Dome deposit. Kirkman drilled a well in the adjoining marsh but found nothing. At the end of the Civil War a local company drilled a deeper well to investigate this same oil seepage and while finding no oil, they hit pure sulphur at a depth of 137 m. During the years that followed an attempt was made to sink a shaft through the mud, clay and sand. The attempt came to an end with the loss of several lives (and a good deal of money) when the shaft was flooded by quicksand exuding poisonous sulphur gases. A second attempt was made to sink a shaft by another company but, on encountering the same problem, they called in a well-known chemical engineer Herman Frasch for advice. Frasch conceived the idea of drilling into the sulphur, melting it with hot water and pumping it to the surface. The process took ten years to perfect and in 1903 commercial production began. Within a decade the sulphur industry of the world was revolutionized.

Geology of the Gulf Coast Region and the Salt Domes

The exploration for oil and gas between the Mississippi Delta and the Mexican border has yielded a great deal of geological information, and only a very brief summary can be attempted here. The geological history of the area begins with downwarping of the basement of the American Mid-West, and the formation of a semi-landlocked sea during Permian times. Evaporites formed at this time which are represented by the Louann Salt, deep in the Gulf Basin, and the Castile Anhydrite of the Delaware Basin further west. From the Cretaceous, and ever since, the Gulf Basin has been filled with sediment, and immense thicknesses of deltaic and nearitic marine material accumulated (over 10 km thick in places). During sedimentation a series of faults and minor folds developed, making the whole area somewhat complicated and a happy hunting-ground for the oil prospector. Along a series of zones of instability, cylindrical masses of salt, presumed to be derived from the Louann, penetrate the overlying strata sometimes almost to the surface. Most of these 'salt plugs' have a capping of a rather particular rock consisting of anhydrite, calcite and gypsum known as 'cap-rock', and it is in this that sulphur occurs.

On the surface the salt plugs are seen as circular areas of disrupted strata associated with salt springs, oil and gas seepages, sulphurous springs, mud volcanoes and a variety of other phenomena. When the dome of the salt plug is deep, the indications are not so obvious, or totally lacking, but they can be detected down to great depths, firstly because salt has a low density and this gives rise to a negative gravity anomaly, and secondly because the cap-rock acts as a good seismic reflector.

Oil occurs in small quantities in cap-rocks. More important, however, is the formation of oil and gas traps in the uptilted strata alongside the salt plugs. Salt is mined from some of the shallower domes, and exposures in these mines have provided vital information; they have shown that, internally, the salt is banded and the bands are highly distorted as if the salt had flowed upwards. Of more than 200 salt domes known along the Gulf Coast, 25 contain commercially exploitable sulphur in their cap-rocks.

Geology of the Sulphur Dome Deposit

The Sulphur Dome salt plug is a relatively small one, being only 750 m in diameter, but its cap-rock is exceptionally thick, reaching 300 m at its maximum. The dome lies beneath a small island surrounded by marsh. At the east side of the island is a small oil seepage and, on the opposite side, a spring of sulphurous brine charged with some natural gas. Between the surface and the top of the cap-rock is 100 m or more of sand and clay; the Beaumont Clays. The upper part of the cap-rock consists of coarsely crystalline calcite with some organic matter and cavities, some quite large, in places filled with sulphur. Towards the base of this zone gypsum begins to appear, and it passes eventually into a zone of anhydrite (usually with some gypsum) with veinlets and cavity fillings of calcite and sulphur. Eventually these are lost, and pure anhydrite overlies the main mass of halite. The halite rock is strongly banded, coarsely crystalline, the bands consisting of narrow layers of anhydrite.

A few small pools of thick oil were found in the cap-rock, and some oil and gas occurs along the side of the salt plug, but the amounts have never been commercially important.

Size and Grade

One cannot speak about the 'grade' of sulphur in a cap-rock deposit because the proportion of sulphur in a tonne of rock is far less important than the amount which can be recovered; and this depends largely on the details of its distribution and the morphology of the cap-rock. In any case, it is rarely possible to extract more than about one-third of the total sulphur. By the time Sulphur Dome became exhausted it had produced 9 Mt of sulphur. One estimate of the sulphur resources of the Gulf Coast region is 200 Mt.

Geological Interpretation

Despite many years of (at times bitter) controversy, there is now general agreement about the origin of salt domes. In the Gulf region it is believed that as sediment accumulated on top of the Louann salt the pressure of the sediment began to deform the underlying salt, exploiting original irregularities on its surface. The salt having a density of 2·2 t/m³ is lighter than consolidated rock, with its density of 2·8 t/m³, so the salt began to rise and flow as plugs, lubricated by formation water in the overlying rocks. These salt plugs penetrated the overlying formations, some of which, being permeable and water-saturated, caused the salt to dissolve, leaving behind a residue of the less soluble anhydrite at the top of the plug. The plug would go on rising at least until it encountered unconsolidated sediment with a bulk density equal to that of the salt. This process was a gradual one taking place throughout the Cretaceous and the Cainozoic. On their way upwards these plugs bend up the surrounding sediments forming suitable traps for hydrocarbons. Once formed, some of the accumulated hydrocarbons leaked through the cap-rock from the traps below and, at the rock temperatures encountered in the Basin, reacted with anhydrite to form calcite and reduced sulphur compounds (such as hydrogen sulphide or sulphur). There is some evidence that the reduction of anhydrite was accelerated by bacteria. This interpretation is supported by studies of isotopes. The ratio of those of sulphur are characteristic of the sulphate in sea water, the ratio of those of carbon in the carbon of petroleum, and the isotopes of strontium in the calcite, leave little doubt that the calcite did not originate as a normal limestone. The amount of sulphur in the cap-rock seems to depend on its thickness, on the length of time that the dome was stable, and on the amount of hydrocarbon that leaked into the cap.

Mining

Sulphur Dome was developed by the Union Sulphur Company, and was the site of the development and first commercial use of the Frasch process. In this process, the cap-rock is divided into blocks into each of which a well is drilled and a series of pipes cemented into the rock. Superheated water (> 160 °C) is pumped down the well which melts the sulphur (melting-point 110 °C) and the liquid sulphur is lifted to the surface by compressed air. At the same time the excess water is 'bled off' from the ground via another well. The liquid sulphur is purified at the surface and was originally allowed to solidify in tanks, from which it could be broken up and removed. Today, it is more common to ship it in liquid form in special insulated tankers or barges. One problem with the process is the large amount of water needed; it must be fresh and the quantity used varies between 6 m³ and 26 m³ per tonne of sulphur recovered (which poses a particular problem for the offshore deposits). Fortunately, this is to some extent compensated for by the ready availability of cheap natural gas.

Further Reading

KELLEY, P. K. (1926), The Sulphur Salt Dome, Louisiana, pp. 452–69. In: *Geology of Salt Dome Oilfields* (Moore, ed.), Am. Assn. Petro. Geol. [The main descriptive work on the deposit in a volume full of interesting papers.]

HALBOUTY, M. T. (1967), *Salt Domes of the Gulf Region of the United States and Mexico.* Gulf Duf. Co. 415pp. [A more up-to-date work than Kelley, pretty well the definitive work on the subject.]

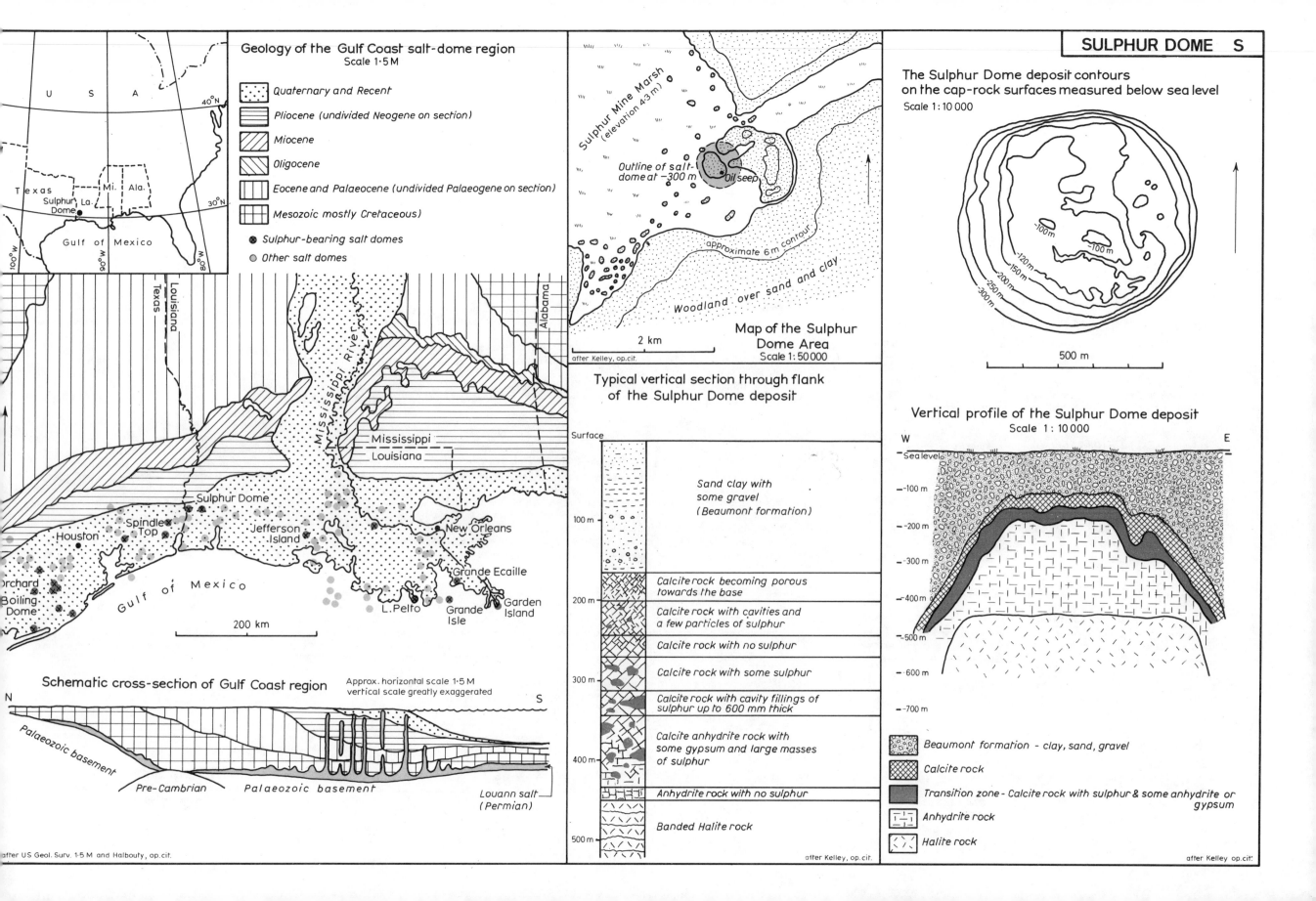

SULPHUR DOME S

Geology of the Gulf Coast salt-dome region
Scale 1·5 M

- Quaternary and Recent
- Pliocene (undivided Neogene on section)
- Miocene
- Oligocene
- Eocene and Palaeocene (undivided Palaeogene on section)
- Mesozoic mostly Cretaceous)
- ⊗ Sulphur-bearing salt domes
- ○ Other salt domes

Schematic cross-section of Gulf Coast region
Approx. horizontal scale 1·5 M
vertical scale greatly exaggerated

Palaeozoic basement
Pre-Cambrian
Palaeozoic basement
Louann salt (Permian)

after US Geol. Surv. 1·5 M and Halbouty, op. cit.

Map of the Sulphur Dome Area
Scale 1:50000

Sulphur Mine Marsh (elevation 4·3 m)
Outline of salt-dome at −300 m
Oil seep
approximate 6 m contour
Woodland over sand and clay
2 km
after Kelley, op.cit.

Typical vertical section through flank of the Sulphur Dome deposit

Surface

Sand clay with some gravel (Beaumont formation)

Calcite rock becoming porous towards the base

Calcite rock with cavities and a few particles of sulphur

Calcite rock with no sulphur

Calcite rock with some sulphur

Calcite rock with cavity fillings of sulphur up to 600 mm thick

Calcite anhydrite rock with some gypsum and large masses of sulphur

Anhydrite rock with no sulphur

Banded Halite rock

after Kelley, op.cit.

The Sulphur Dome deposit contours on the cap-rock surfaces measured below sea level
Scale 1:10 000

−100 m
−120 m
−150 m
−200 m
−250 m
−300 m
500 m

Vertical profile of the Sulphur Dome deposit
Scale 1:10 000

W
E
Sea level
−100 m
−200 m
−300 m
−400 m
−500 m
−600 m
−700 m

- Beaumont formation – clay, sand, gravel
- Calcite rock
- Transition zone – Calcite rock with sulphur & some anhydrite or gypsum
- Anhydrite rock
- Halite rock

after Kelley op.cit.

The Iron Deposits of the Northampton District – U.K.

This is one of the best examples of an oolitic, sedimentary iron deposit (sometimes called 'minette'), a type which occurs in marine sediments from late Pre-Cambrian to mid-Tertiary. This example is one of a group that occur round the edge of a large basin complex in north-west Europe, filled with sediments of Permian to Recent age, the deposits being largely Mid or Lower Jurassic. In other parts of the world, notably southern Russia, a very similar type of deposit occurs, but contains manganese rather than iron.

Location

The district is 100 km long and 25 km wide, centred on the town of Corby (latitude 52° 29′ N, longitude 0° 41′ W) in the county of Northamptonshire, England, although the district stretches into the neighbouring counties of Lincolnshire and Leicestershire.

Geographical Setting

The Northampton district lies in rolling green country, interrupted by several steep escarpments, one of the best and richest farmlands in western Europe. The county is well-wooded and drained into the North Sea by three rivers, the Trent, Welland and Nene. The climate is temperate and moist, temperature averaging 5°C in winter and 17°C in summer. The rainfall is 750 mm per annum, mostly carried on the westerly winds from the Atlantic, but cold, polar weather with snow can affect the area during the worst winters. Like most of England, the area is densely populated, and there is a series of major towns along the length of the district. The main railway from London to the north-east passes through the district, and gives access to the industrial centres of the Midlands and North.

History

Smelter sites of Roman age are found in the district and iron forges are mentioned in the Domesday Book (twelfth century A.D.) but the discovery of a method of using coal to smelt iron, led to its abandonment in the early years of the Industrial Revolution.

However, by the middle of the nineteenth century the railways had developed and the demand for iron greatly increased. As a result of the building of canals and later the cutting of the railways through the hilly country of the East Midlands from London to the industrial heart of the country, many sections of iron-rich rocks were discovered, including a major one at Desborough, north-west of Kettering in 1857. From that time on the district was steadily opened up. The building of large ocean-going ore ships in the 1960s began to make imported high-grade iron ore more attractive, and the district began to decline in importance, although it remains in production.

Geological Setting

The Northampton district is on the edge of a large basin complex of Mesozoic and Cainozoic sediments that covers a substantial part of north-west Europe and underlies the North Sea. The basement of this basin consists of rocks that were tectonized either in the late Pre-Cambrian, or the end of the lower and upper Palaeozoic. Round the edge of the basin are a series of upland areas, composed of more or less deformed sediments with intrusives that were tectonized during the Hercynian event; while below the sediments, in England, is a Pre-Cambrian Shield.

The rapid erosion of the Hercynian mountainlands under desert conditions produced the extensive tracts of red beds of Permo-Triassic age, associated with which are thick evaporites (now largely under the North Sea and northern Germany). From then onwards a series of marine transgressions and regressions affected the area and the basin became filled with shales, sands and limestones. Several tectonic events affected the area, which led at times to the emergence of various parts, and a large peninsula of land stretching from south-east England to the Rheinisches Schiffergebirge almost separated the basin into two for much of the time.

The Jurassic was the main period of iron deposition. Two main groups of iron districts are known; one in eastern England, and the other in eastern France and Luxembourg. In both areas the sedimentation begins with dark marine shales, containing some calcareous siderite-rich rocks that continue up to (or almost to) the top of the Toarcian. The greatest concentration of iron is of uppermost Toarcian, Aalenian or lower Bajocian age. Following these there are much more variable lithologies that include esturine, deltaic and shelf limestone facies, that preceded a return to marine shale deposition in the Upper Jurassic. Calcareous iron-rich rocks occur throughout the sequence up to the lower Cretaceous.

The main ironstone horizon is very persistent round the basin and is known, from oil wells, to occur in the North Sea. It is only, however, in a few places that the thickness and grade necessary for economic exploitation is found, in local sub-basins, of which the Northampton district is one of the largest.

Sedimentation in the North European Basin was interrupted for a while during the Quaternary Ice Age, and the Jurassic outcrops in England, down to about latitude 51° 30′ N, are extensively covered by glacial till (mostly boulder clay) that can vary in thickness from less than a metre up to over 30 m.

The Geology of the Iron Deposits

Stratigraphically, the iron deposits are situated at the base of the Northington Sand, the lowermost formation of the Inferior Oolite Group. Below it is the Lias clay, a dark marine shale containing much organic matter, pyrite and siderite nodules. Above the iron horizon there is a discontinuous sequence of sandy rocks and then an unconformity over which are silts, sands, pebble beds and a thin lignite seams of the Lower Esturine Formation, followed by the cream-coloured oolitic Lincolnshire Limestone.

The Northampton Sand is a persistent formation that occurs over a wide area, and consists for the most part of ferruginous sandstone. The iron ore represents a local change of facies to rocks rich in siderite and chamosite, with a largely oolitic texture.

There are three iron ore minerals in these rocks; chamosite, siderite and goethite. Associated with them are a great variety of other minerals, clay (principally kaolinites), calcite, quartz, pyrite, colphane, micas, feldspar and organic materials. There are two principal types of ore; oolitic and the so-called 'mudstones'. The most common type consists of ooliths of chamosite cemented by siderite and containing on average about 32% Fe. Some varieties, found particularly at the base of the formation, contain nodules of pyrite and collophane and may contain over 2% of phosphorus and sulphide (and not as a rule workable as iron ore). Other types contain calcite in the matrix, commonly as beds of shell fragments. Near the top of the formation, another type containing kalinite is found, while on the margins of the field increasing quantity of detrital grains of quartz and feldspar tend to dilute the iron content.

The mudstone ores are non-oolitic, fine-grained rocks more properly called 'siderite microspatites' or sidero-lutites. They are composed largely of siderite and, on average, they contain 37% Fe. Much of the siderite, both in the microsparites and the oolitic rocks, shows diagenetic textures. Many of the ores are very broken, and brecciated. Goethite may occur as a primary mineral, but for the most part it is a result of modern weathering.

The iron ore bed averages 6 m in thickness, and dips from its outcrop towards the east at gradients of about 0.4%. The continuity of the ore formation is interrupted by several features. It is eroded away in the deeper valleys and by valleys cut into it before the deposition of the Lower Esturine Formation. It is also interrupted by a set of north-west-tending faults and, to some extent, by solution pipes.

Size and Grade

The total amount of ore in the district has been estimated to be of the order of 2 Gt, but only 1.44 Gt is classifiable as measured or indicated ore. Of this about 440 Mt has been exploited. Of the remainder, by no means all can be mined under prevailing conditions of economics and environmental constraints. The grades worked range from 25% Fe to 35% and are probably the lowest grade iron ores worked anywhere in the world.

Geological Interpretations

There is no doubt that these ironstones are sedimentary. Palaeographic evidence shows that this part of England was situated in an island strewn straight between the Tethys Ocean and the Boreal Sea. If the origin of the ooliths is interpreted correctly, the deposits must have been formed in or near the tidal zone. All the primary iron minerals are ones containing reduced ferrous iron, which only remains stable where a sediment is protected from oxidation, and this normally exists where decomposing organic matter is in sufficient abundance to neutralize the effect of the atmosphere. Bearing in mind that the ore horizon occurs in a transition from an organic-rich marine shale environment to that of a swampy estuary, it is not too difficult to imagine that these deposits were laid down in off-shore pools or depressions rich in organic activity.

The problem that has never been satisfactorily resolved is the source of the iron. The most likely source seems to be detrital fragments of iron-rich material or iron-coatings of minerals, brought into the sea from the surrounding land. This would be likely to occur if the surrounding land was well-vegetated, sub-tropical and undergoing laterite formation. There is not much direct evidence that this was the case (there are some laterized basalt flows known from boreholes in the Central North Sea), but it is perhaps the most plausible explanation.

Mining

Iron ores of the district are mined by various sections of the State-owned British Steel Corporation. Underground mining was formerly practised but is not favoured because the ore is so shallow that high rates of extraction cause too much subsidence and disruption of agriculture. The predominating method is strip-mining. Stripping ratios of 10:1 or 12:1 have been successfully employed by the use of very large power shovels and draglines. The ore is usually blended with high-grade concentrates from other sources, and sintered before smelting, and many of these ores contain sufficient calcite to produce a slag without further addition of limestone and are known as 'self-fluxing'.

Further Reading

HALLAM, A. (1975), *Jurassic Environments*. Cambridge University Press, Cambridge, 269pp. [A very good, up-to-date essay on the subject with a lot to say on the problems of ironstones – sets the subject in its local, regional and global perspective.]

HOLLINGWORTH, S. E. and TAYLOR, J. H. (1951), *The Northampton Sand Ironstone – Stratigraphy, Structure and Reserves*. Mem. Geol. Surv. Great Britain, London, H.M.S.O. 211pp. [The principal work on the district and a comprehensive one.]

TAYLOR, J. H. (1949), *Petrology of the Northampton Sand Formation*. Mem. Geol. Surv. Great Britain, London, H.M.S.O. 111pp. [Very detailed account of the ores themselves.]

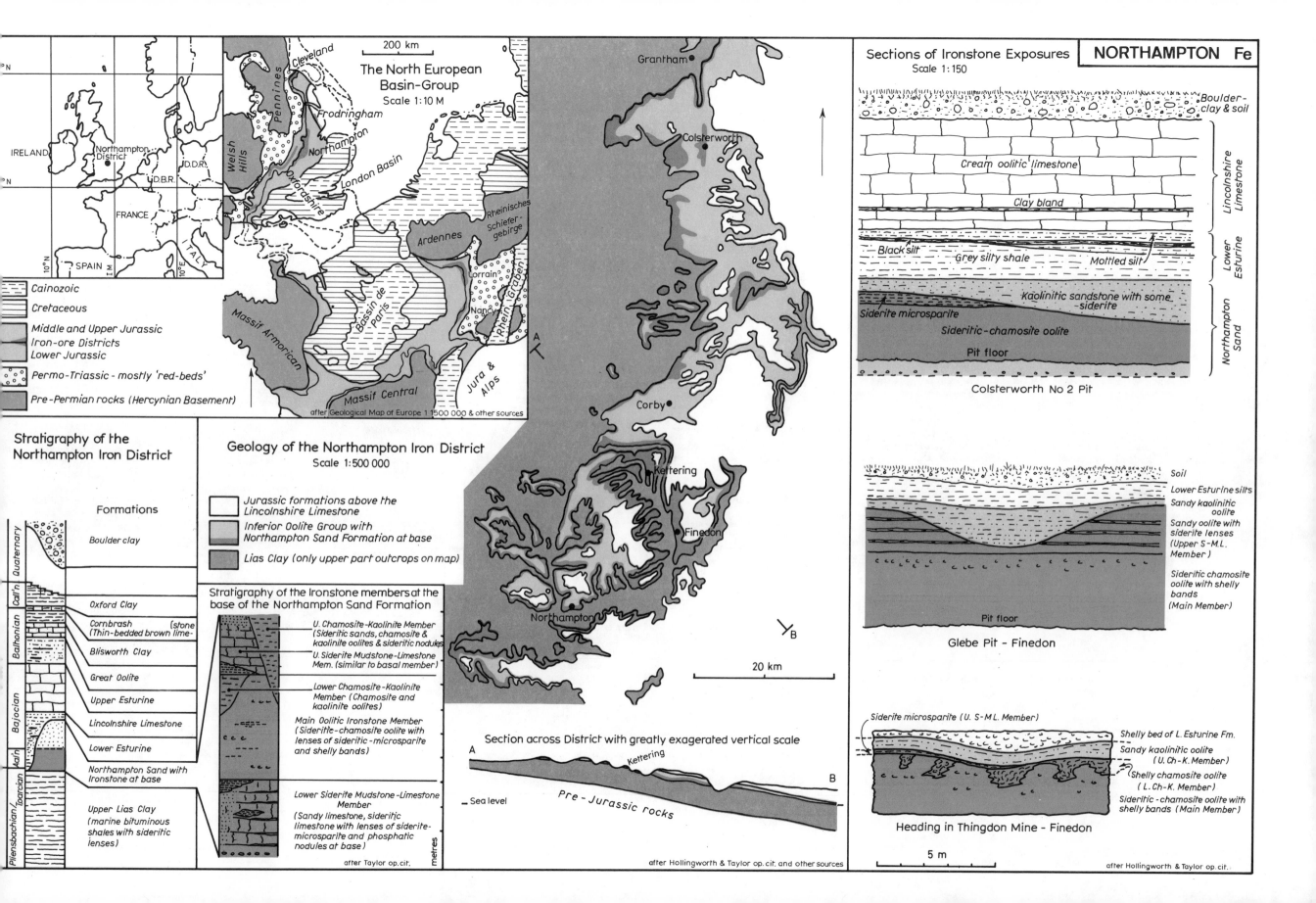

The Mesabi Iron Range – U.S.A.

This is the largest of the iron ore ranges of the Lake Superior area. The ores are, or are derived from, taconite; a cherty iron-rich chemical sediment, possibly of organic origin. Some of the ores are rich enough to be shipped directly, but others are magnetite-bearing tactonite from which a concentrate can be made. The Lake Superior is one of a number of very large Proterozoic iron-rich sedimentary basins.

Location

The Mesabi Range trends approximately north-eastwards for 170 km, centred on the town of Hibbing (latitude 47° 25′ N, longitude 92° 56′ W and 450 m altitude) passing through Itasca and St. Louis counties of Minnesota.

Geographical Setting

The iron deposits of Mesabi occur on the flank of the Giants Range, that is about 580 m above sea level, and a fairly pronounced feature at the north-eastern end, gradually becoming less pronounced and of lower altitude to the south-west. The range divides the Mississippi drainage basin from two sets of streams that flow into the Hudson Bay and St. Lawrence. The area has a cool continental climate with a dry winter averaging -20 °C and a summer averaging 24 °C with a wet autumn. The total precipitation is about 600 mm, largely as summer rainstorms, plus autumn rain and snow. The area was originally forested, mixed but predominantly coniferous, but much timber has been felled during the course of mining. Several towns have grown up along the range which are served by an extensive network of railways built to carry away the ore, most of which converge on Duluth, a port on Lake Superior, 100 km to the south-east.

History

A government surveyor, William Burt, is credited with first reporting the iron ores of the Marquette Range close to Lake Superior in 1844, and Mesabi was certainly known shortly afterwards. In 1890, the brothers of the Merritt family began to develop the area, and built a railway to Duluth shipping the first ore in 1892. Realizing that the reserves of direct shipping ore were small compared with the vast quantities of lower grade taconite, the Mesabi Syndicate was set up in 1915 to develop some process of concentration. The research was unsuccessful for many years and it was not until 1945 that a potentially economic benification process was designed and tested. In 1956, the first concentrator began operation, producing magnetite concentrates.

Geological Setting

The iron ranges of the Lake Superior area, although interrupted by faults and intrusions, are probably the remnants of a basin of sedimentation 500 km long and 300 km wide. The basement rocks are the greenstone belts and granites of the Archean Superior Province of the Canadian Shield. The supercrustal rocks of the greenstone belts can be divided into two; a lower largely volcanic group that contains the important Soudan Iron Formation that is mined on the Vermillion Range, and an upper, largely sedimentary group. This latter is represented at Mesabi by the Knife Lake Group; silts, quartzites, greywacke and conglomerates. The Archean era ended with tectonism and the intrusion of granites, about 2·5 Ga ago, which at Mesabi are represented by the Giants Range Granite (2·67 Ga).

The eroded remnants of the Archean rocks are overlain by sediments of the Huronian Super-Group (also known as the Animikie by some authors). In general, the sequence is thicker and more easily subdivided in the east and south and tends to thin towards the north.

Where complete, the Huronian begins with a group of dolomites and quartzites, and then passes up into the taconite-bearing group. The taconite is usually underlain by quartzites and overlain by argillaceous rocks in which further taconite may occur. In the Marquette Range the whole sequence is repeated. The taconite-bearing formations contain both massive and strongly banded taconite, and can be either of cherty silicate or carbonate-rich facies. In addition there are limestone and dolomites, usually revealing algal structures, and a minor development of argillaceous sediments, sometimes tuffaceous.

The Huronian was folded and intruded by granite at about 1·7 Ga following which there was an important episode of vulcanism, forming the lavas and pyroclastics of the Keweenawan Super-Group in which occur the famous native copper deposits of the Keweenawan peninsula and, in the overlying sediments, the White Pine copper deposit.

The whole Lake Superior area was refolded, faulted and intruded by granite and gabbro at about 1·1 Ga, forming the present structure. Among the intrusions is the Duluth Gabbro Complex that cuts the Mesabi Range at its eastern end, separating it from what is probably its continuation, the Gunflint Range.

Platform sediments were deposited over the area in the Palaeozoic, but it is probable that much of the area was exposed throughout a large part of the Phanerozoic. Small deposits of conglomerate of Cretaceous age are found in depressions on the flanks of the iron ranges, the only record of the long period of erosion and weathering which may have been responsible for forming some of the iron ores.

In the Quaternary the whole region was heavily glaciated, forming the present topography and leaving behind an extensive and, in places, thick mantle of till.

Geology of the Mesabi Iron Ore Deposits

The Huronian of the Mesabi Range lies over the Knife Lake Group and the Giants Range Granite and consists of three formations. Below the ore-bearing formation, known here as the Biwabik, is the Pokegama Quartzite, generally about 50 m thick but not always present. Above the Biwabik is the Virginia Formation, over 600 m thick and composed of dark thin-bedded argillaceous sediments. The Biwabik Iron Formation is between 100 m and 230 m thick and divided into four members. A few thin beds of limestone, shaly sediment and algal chert are present, but by far the majority of the formation consists of taconite. The taconite may be divided on the basis of texture, composition and mineralogy. There are massive thick-bedded varieties that consist of granules 0·25–1·0 mm in diameter composed of iron silicates, oxides or carbonate set in a matrix of chert. They contain between 28 and 33% iron and frequently exhibit sedimentary textures such as cross bedding. Both the members contain extensive zones where the dominant iron mineral is magnetite. The other rock type is the laminated taconite, consisting of thin, evenly bedded layers of iron minerals and chert, with a little fine clastic material. They contain 20–30% iron and, again, a facies change can be seen from carbonate to silicate with local development of oxides. In the east, the taconites are partially recrystallized, the chert being replaced by quartz and the iron minerals by magnetite whereas, at the south-western end, the formation thins and becomes shaly.

Several kinds of ore have been worked along the range. The first type is a mixed hematite–limonite ore that is found in irregular lenses near the surface at the sub-outcrops of the granular taconite members, down to a depth of 120 m or more. These ores are of high grade and termed natural or direct shipping ore. The ore bodies grade into a zone of decomposed taconite which is sometimes workable but, particularly where laminated taconite occurs, becomes of poor quality because of increased alumina content. Most of the bodies of natural ore occur where the taconite is extensively faulted and folded, in the Hibbing area and around 'The Horn'.

The major type of ore worked today is magnetite–taconite. This consists of the richer zones of the granular taconite, in which much of the iron is present as magnetite.

A little ore occurs in Cretaceous conglomerates on the lower slopes of the range, sometimes lying on the Virginia Formation. This ore generally consists of cobbles and pebbles of iron ore in a limonitic matrix, and occurs largely in the western part of the range.

Size and Grade

The natural ores have been almost exhausted and the Range produced about 2·5 Gt containing between 51 and 57% iron. Under present working conditions there is production and planned production that would indicate the presence of well over 2 Gt of magnetite-taconite grading at about 33% iron and, in time, the total may be much larger. The natural ores were generally shipped with 7–8% SiO_2, 1% Al_2O_3, 0·5% Mn, 0·04% P_2O_5 and 0·01% S. The pellets from the taconite concentration contain 61% Fe and rather less phosphorus and aluminium than the natural ores.

Geological Interpretations

It is generally agreed that the natural ores were formed by the extensive leaching of the taconite during the Phanerozoic, removing the silica and other impurities and leaving a residue of iron oxides. Thus these are essentially residual deposits, albeit derived from a parent rock originally rich in iron. Fracturing of the original taconite seems to have been important in promoting the weathering process. The magnetite-taconite ores were probably formed by the metamorphism. Both the periods of folding may have been accompanied by metamorphism, and the intrusion of the Duluth Gabbro Complex is generally regarded as having had an important effect at the eastern end of the range.

The origin of taconite itself is more of a problem and has been dealt with in the introductory section and in Section V.

Mining

Two companies predominate in the mining of the range; U.S. Steel Corporation and Pickards Mather. Both have extensive open-pit operations along the range. In the early days there was some underground mining of the richer natural ores, but these have all been replaced by large open pits. The taconite concentrators grind the ore to liberate the magnetite (about 85% passing 325 mesh) and recover the mineral by magnetic separation. The resulting concentrate is formed into pellets which are then fired before shipping. Most of the ore from Mesabi is sent by rail to Duluth where it is shipped to the iron and steel works around Lake Michigan. The Range currently produces about 65 Mt/a of pellets and concentrates.

Further Reading

LEITH, C. K. (1903), *The Mesabi Iron-bearing District of Minnesota*. U.S. Geol. Surv. Mono. **43**, 316pp. [This beautifully produced, if old, volume is still worth a look.]

MARSDEN, R. W. *et al.* (1968), The Mesabi Iron Range, Minnesota, pp. 519–27. In: *Ore Deposits of the United States, 1933–1967*. (Ridge, J. D., ed.), Am. Inst. of. Met. and Pet. Eng., New York, 1968. [Good modern summary with bibliography of all the other relevant literature to that date; the same volume also contains a general paper on the Lake Superior iron ores by R. W. Marsden, as well as papers on the other ranges.]

MOREY, G. B. (1973), The Mesabi, Gunflint and Cuyuna Ranges, Minnesota, pp. 193–208. In: UNESCO, *Genesis of Pre-Cambrian Iron and Manganese Deposits*. Proceedings of the Kiev Symposium, August 1970. UNESCO Earth Science Series No. 9, UNESCO, Paris 1973.

MOREY, G. B. (1973), The Mesabi Range. In: *Geology of Minnesota, A Centenary Volume*. Geol. Surv. of Minnesota, pp. 204–17. [These two papers give a modern review of the deposits and their geological setting. Note also that *Econ. Geol.* **68** (7) is devoted to Pre-Cambrian iron deposits and contains several good review papers.]

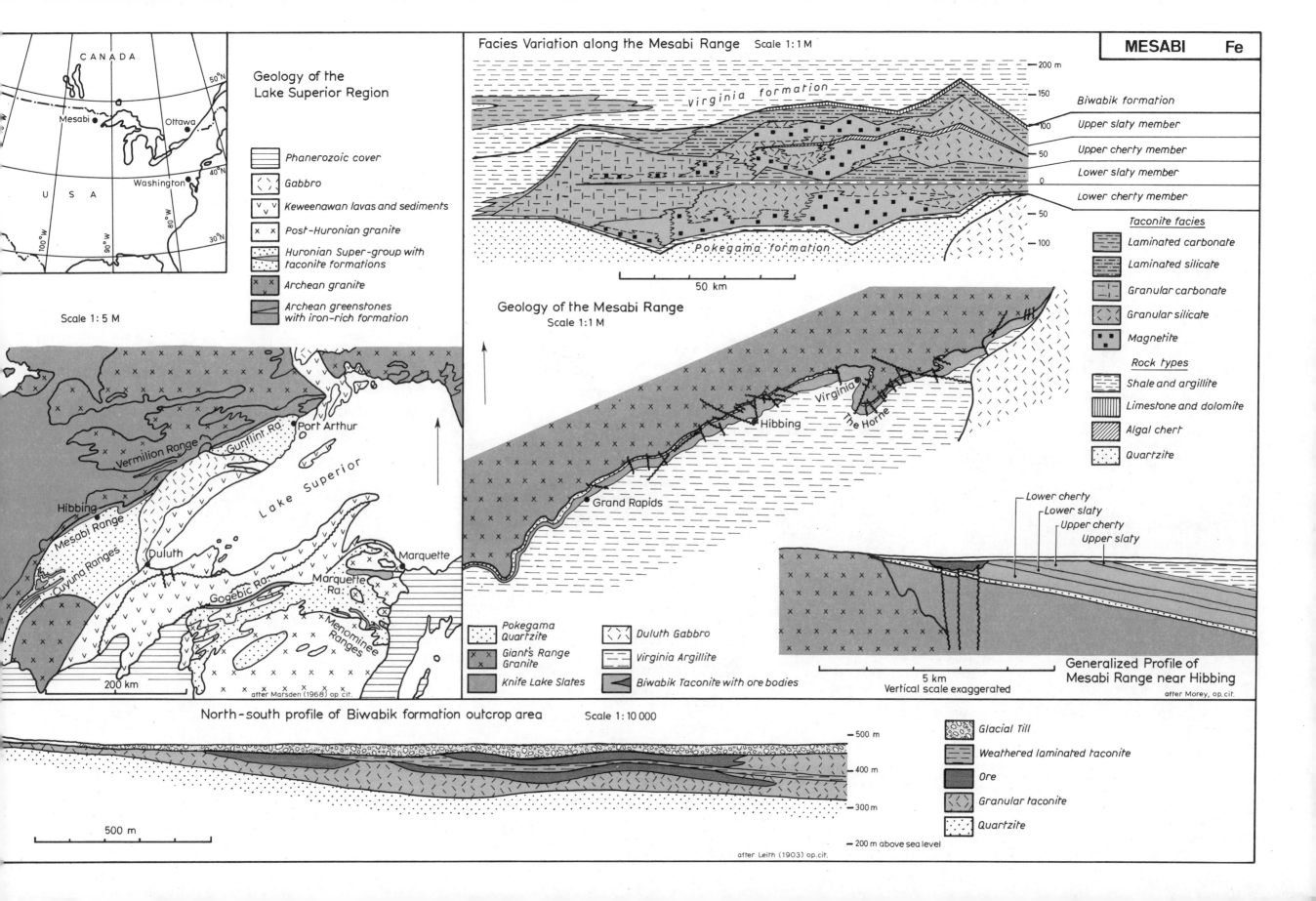

Geology of the Lake Superior Region

Scale 1:5 M

Phanerozoic cover

Gabbro

Keweenawan lavas and sediments

Post-Huronian granite

Huronian Super-group with taconite formations

Archean granite

Archean greenstones with iron-rich formation

CANADA

Mesabi

Ottawa

Washington

USA

50°N

40°N

30°N

100°W

90°W

80°W

Port Arthur

Vermilion Range

Gunflint Ra.

Hibbing

Mesabi Range

Cuyuna Ranges

Duluth

Lake Superior

Marquette

Marquette Ra.

Gogebic Ra.

Menominee Ranges

200 km

after Marsden (1968) op.cit.

Facies Variation along the Mesabi Range   Scale 1:1 M

Virginia formation

Pokegama formation

Biwabik formation

Upper slaty member

Upper cherty member

Lower slaty member

Lower cherty member

200 m
150
100
50
0
-50
-100

50 km

Taconite facies

Laminated carbonate

Laminated silicate

Granular carbonate

Granular silicate

Magnetite

Rock types

Shale and argillite

Limestone and dolomite

Algal chert

Quartzite

Geology of the Mesabi Range

Scale 1:1 M

Virginia

The Horne

Hibbing

Grand Rapids

Pokegama Quartzite

Giant's Range Granite

Knife Lake Slates

Duluth Gabbro

Virginia Argillite

Biwabik Taconite with ore bodies

Generalized Profile of Mesabi Range near Hibbing

Lower cherty

Lower slaty

Upper cherty

Upper slaty

5 km

Vertical scale exaggerated

after Morey, op.cit.

North-south profile of Biwabik formation outcrop area   Scale 1:10 000

500 m
400 m
300 m
200 m above sea level

Glacial Till

Weathered laminated taconite

Ore

Granular taconite

Quartzite

500 m

after Leith (1903) op.cit.

The Iron Deposits of the Itabira District – Brazil

In addition to being one of the world's leading centres of iron-ore production, this district displays many of the characteristic features of Pre-Cambrian stratiform iron deposits which have suffered metamorphism, deformation and modification by tropical weathering, and includes type localities of two important rock types: 'itabirite' and 'canga'.

Location

The Itabira Mining District is centred on the town of Itabira (19° 37′ S, 43° 14′ W and 800 m above sea level) in the state of Minas Gerais and is in the north-east corner of the so-called Quadtilàtero Ferrífero.

Geographical Setting

The country is mountainous, resulting from the dissection of a (Cretaceous?) peneplain, rising to peaks on the iron-ore outcrops (Pico de Cauê, 1330 m and Pico de Conceição, 1348 m) and descending to 750 m in the adjoining Rio do Peixe. The climate is humid, sub-tropical, with maximum temperatures between 20 °C and 30 °C and a rainfall of 1500 mm, falling mostly in two seasons. Originally thickly forested, the area is now covered by dense scrub and woodland growing on a deep residual soil.

Itabira is connected by road to Belo Horizonte, the main town of the Quadrilàtero Ferrífero and by rail to the port of Tubaro.

History

From the early days of European colonization gold was worked in the area, but the first recorded iron mining began in 1814. Interest in large-scale mining began early in the twentieth century, but was delayed by title disputes and lack of capital until the Second World War when modern production began.

Geological Setting

The Quadrilàtero Ferrífero forms part of the Brazilian Pre-Cambrian Shield and is composed of a series of long belts of folded and metamorphosed super-crustal rocks, largely sediments, in an area of granite and granite-gneiss. Wherever seen, these granitic rocks have intrusive contacts with the super-crustal rocks, and thus cannot necessarily be regarded as a basement complex. Among the super-crustal rocks three super-groups are recognized. The Rio das Velhas Super-Group is a varied sequence of sediments and volcanic rocks, extensively metamorphosed and deformed, intruded not only by granitic rocks, but also by basic and ultrabasic bodies.

The Minas Super-Group, separated from the Rio das Velhas by a major unconformity, is composed at the base of platform sediments including an important group of chemical sediments followed by a thicker flysch sequence. The chemical sediments include carbonates and iron-rich sediments with which the ore deposits are associated. The flysch sequence is terminated by an unconformity representing a period of deformation accompanied by metamorphism; that was followed by the deposition of the largely clastic Itacolomi Super-Group.

Little is known of the Phanerozic history of the area, but for some evidence of a late Mesozoic peneplain; the formation, probably in the Cainozoic, of laterites and small areas of thin continental sediments.

Geology of the Itabira District

The district consists of a down-folded and faulted block of meta-sediments in granite-gneiss. The Rio das Velhas is represented by mica schists, quart-zites and dolomitic schists containing some bodies of serpentinite, soapstone and amphibolite. The Minas Super-Group is represented by a relatively thin, impersistent series of phyllites belonging to the Caraca Group, itabirites of the Itabira Group, and by quartzites of the Picacicaba Group.

A few highly altered dykes cut the whole sequence. Although direct local evidence is lacking, its higher grade of metamorphism shows the Rio das Velhas rocks were probably deformed before the deposition of the Minas, but the main deformation into three en-echelon synclines took place at the end of the Minas deposition.

Geology of the Iron Deposits

The ores of the district comprise the iron-rich sediments of the Cauê Formation, and a variety of ore types associated with it. The Formation outcrops over 22 km and averages 150 m in thickness, doubled in places by folding. Two main deposits occur, centred on the two peaks of Cauê and Conceiação. Both are composed of various ore types as follows.

Itabirite. Though not always regarded as ore, the primary rock of the Cauê Formation is worked when rich enough or easily concentrated. Itabirite is a locally derived name, now widely used, for a rock composed of bands of specular hematite alternating with bands of granular quartz. The bands vary between 1 m and 10 mm in thickness and are very persistent. Locally, the rock may contain euhedral magnetite grains and minor amounts of mica. Itabirite may be soft and friable which aids concentration, and is generally regarded as ore if it has the following compositional range:

50–62%	Fe
2–28%	SiO_2
3– 8%	Al_2O_3
0·02%	P
0·01%	S

Hematite Ore. Lenticular and irregular bodies of hematite ore occur in the itabirite. These generally show a massive texture where they occur on the crests of folds, while on the flanks they tend to be schistose; both may occur hard or soft, even powdery in places. The ore bodies vary in length from a few kilometres to a few metres and in thickness up to 300 m. They are largely composed of hematite, sometimes quite coarse-grained with crystals of between 1 mm and 10 mm, but in some of the soft varieties the grain size may be as low as 75 μm. Magnetite and martite occur sometimes as pockets of very large crystals. The hematite ores are the richest in the district, averaging:

67%	Fe
0.35%	SiO_2
0·67%	Al_2O_3
0·02%	P
0·01%	S
0·06%	Mn

Canga and Rubble Ore. Canga is a local term, now widely used, for an ore, composed partly of limonite, that occurs as a surface mantle to the iron-ore outcrops. It may be simple cavenous limonite, or have a proportion of hematite ore or itabirite fragments. 'Canga Rica' is a breccia of iron-ore fragments cemented by about 5% limonite. At the edges of the deposits canga grades into the laterite that covers the area. Canga can be a valuable iron ore grading from 56% to 59% Fe and, although containing relatively high SiO_2 and Al_2O_3, much of this may be removed by simple washing.

At the surface, extensive areas are covered by rubble ore, consisting of loose blocks of hematite ore and itabirite, laying directly on outcrops, or landslid over the canga on the lower slopes.

Size and Grade

Present published reserves (including past production) for the Cauê deposit are 450 Mt of hematite ore at 67% Fe and 650 Mt of itabirite at 50% Fe, and a further 600 Mt of hematite and 800 Mt of itabirite at the Conceiação deposit further south.

Geological Interpretations

Itabirite is widely regarded as the metamorphosed derivative of a cherty iron-rich sediment or banded taconite, the presence of magnetite being due to local variations in the oxidation state during metamorphism. The hard hematite ore is thought by most authors as being derived by leaching of silica and recementation of itabirite or its parent taconite, but opinion varies as to the precise mechanism; some favour weathering followed by metamorphism, others a purely metamorphic and metasomatic process. The soft varieties of hematite ore and itabirite are regarded as having been derived from the hard varieties by weathering, while canga originates by more intensive weathering involving solution and redeposition of iron between the wet and dry seasons.

Mining

The deposits of the district are currently exploited by the Companhia Vale do Rio Doce SA which operates open-pits on the two main deposits. The ore is crushed and screened and depending on its type and grade, is then shipped direct, or concentrated by gravity or magnetic methods. Seventeen different specifications include lump ore, fines, concentrates and pellets. The mines have a productive capacity of 60 Mt/a of ore, producing 42 Mt/a of products.

Further Reading

ARGALL, G. O. (1974), How Vale do Rio Doce Expands to Largest Iron Ore Shipper. *World Mining,* **21** (1), 34–7. [Description of mining operations at time of publication.]

DORR, J. van N. and BARBOSA, A. L. de M. (1963), *Geology and Ore Deposits of the Itabira District, Minas Gerais, Brazil.* U.S.G.S. Prof. Paper 341-C, 110pp. [The principal geological description of the district.]

DORR, J. van N. (1964), Supergene Iron Ores of Minas Gerais, Brazil. *Econ. Geol.,* **59** (7), 1203–40.

DORR, J. van N. (1965), Nature and Origin of the High Grade Hematite Ores of Minas Gerais, Brazil. *Econ. Geol.,* **60** (1), 1–46.

DORR, J. van N. (1969), *Physiographic, Stratigraphic and Structural Development of the Quadrilatero Ferrifero, Minas Gerais, Brazil.* U.S.G.S. Prof. Paper 641-A, 110 pp. [Description of the regional geology.]

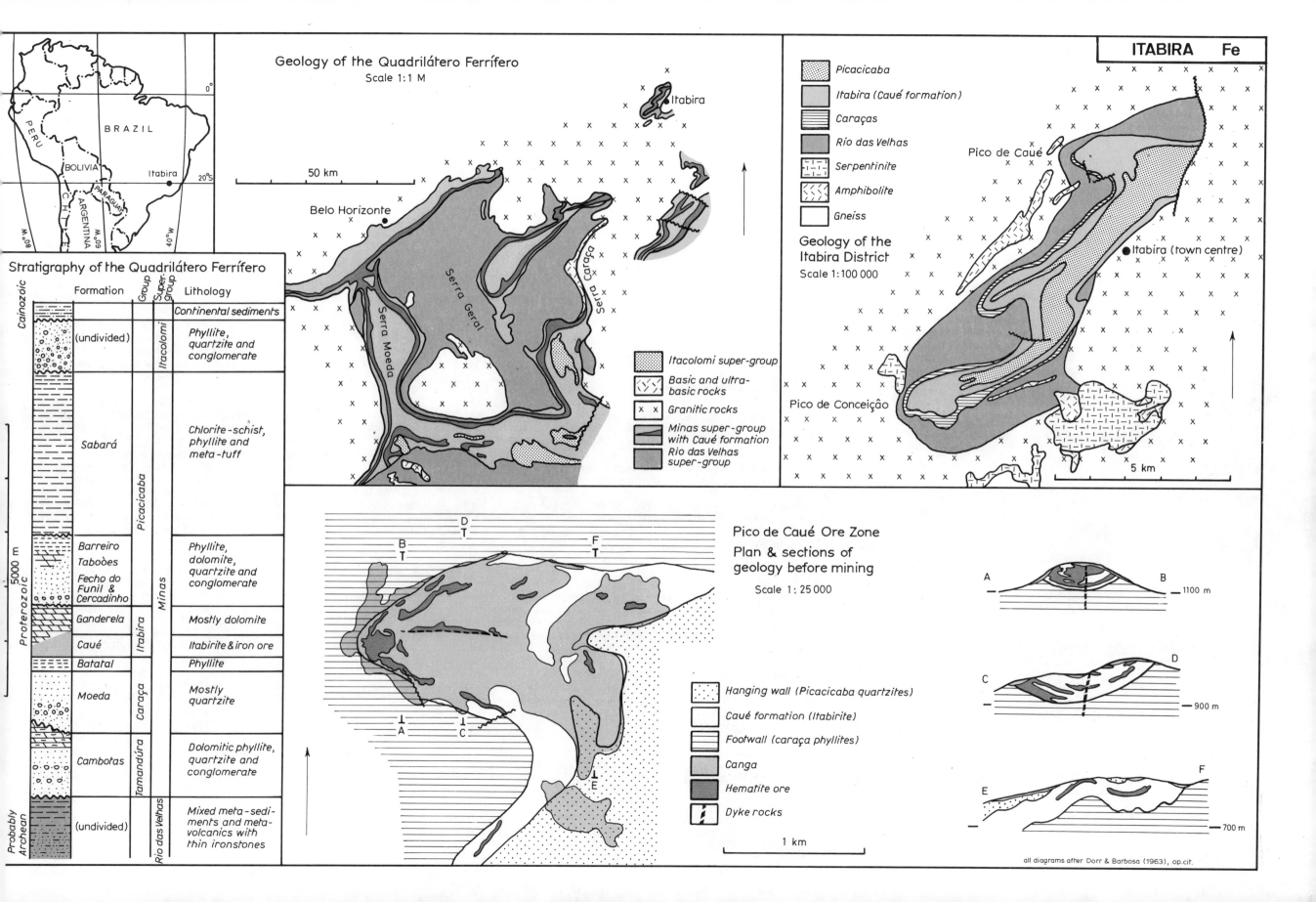

Section Two: Mineral Deposits in Sedimentary Rocks

Introduction

In this section are included a variety of deposits that are contained in sediments, but the mineralization is exotic to the environment of deposition of the sediments. There are two morphological terms commonly used to describe such deposits; 'stratiform' which literally means having the form of a stratum, that is to say the exotic mineral constituents form beds interlayered with the host sediment; the other term, 'strata-bound' refers to deposits which do not have the form of strata, but are wholly contained within a bed or member of a host sedimentary formation. Two further terms, genetic in meaning, are also commonly used in discussing these as well as other types of deposit. A 'syngenetic' deposit is one in which the exotic constituents were deposited along with the normal ones; the converse is an 'epigenetic' deposit in which the mineralization was introduced into the sediment after its formation. These two terms have proved extraordinarily difficult to use because they can become subject to minute differences of interpretation, particularly where the time-span between deposition of the sediment and the introduction of the mineralization was short; or where the rocks were subsequently subjected to alteration, deformation or metamorphism (processes which alter the texture of the mineralized rock destroying or rendering ambiguous the original relationships between the normal rock constituents and the exotic minerals). Yet another source of ambiguity in the use of these terms is that everything depends on what is regarded as the host rock of the mineralization.

A body of mineralization can appear epigenetic with respect to its immediate surroundings. However, it may be there as a result of a series of processes that took place during diagenesis, from material that was deposited in a sequence of sediments of which the immediate host is a part. It is quite possible, as will be seen in later discussion, for mineralization to take place over such a length of time that part may be syngenetic and part epigenetic. Similarly, a syngenetic deposit may be modified by some later process, redistributing some of the mineralization in such a way as to make it appear epigenetic.

One can perhaps say, as a general truism, that the origin of a body of mineralization in a sediment is best sought among its relationships to the environment of deposition of the host sediment, be it of the individual sedimentary unit that forms the host, or the whole basin, or a region of which it forms part. Only after this approach has been thoroughly tried need external sources be invoked.

Before describing the types of deposit that fall within this group, it is as well to review the environments of sedimentary deposition and the lithologies which result. One can begin by dividing the earth's surface into three physiographic domains; continents, coasts and the sea. Continental environments are often ones of net erosion, but sediments of considerable volume and complexity are deposited on land. Firstly, there is the environment of scree, talus, outwash-fans, landslides and mudflows, generally yielding ill-sorted and immature clastic sediment; secondly, there are glacial and peri-glacial materials also ill-sorted and heterogeneous. There are also wind-blown sediments, loess in the cold perglacial zones and dune sands and pediments in hot desert regions. Sediments, usually much more mature and well-sorted, are deposited in rivers, valleys and lakes. Among these there are differences depending on the sediment – water ratio and on the permanency or seasonal nature of the water flow. In warm-wet and cold-wet· climates, both rivers and lakes can be swampy and much organic matter may be incorporated into the sediments. Rivers can be braided or meandering, and lakes can be saline in hot-dry climates.

There are two major types of coastal environment; linear, and interrupted or lobate (the latter usually resulting from the intersection with the mouths of rivers). Both types can be dominated by terrigenous materials or by products from the sea itself, carbonates and evaporites. Both kinds of coast can be swampy.

Marine environments can be classified according to water depth, distance from the coast, physiography (that is to say, open ocean or gulf) and by sediment type. Broadly, one can distinguish a shelf environment which can be a continental shelf between land and the open sea, or a relatively shallow gulf. These may be dominated by terrigenous sediment, by carbonates or sometimes evaporites and, in the warm regions, there will often be reefs, built up by organisms. Beyond the shelves there is a zone of steeper slopes where unconsolidated sediment can become unstable and the characteristic product formed will be turbidites. Beyond this again is the region of pelagic sediments, which include fine-grained shales (usually with a high content of organic matter) or almost pure organic sediments such as radiolarites and calcareous oozes.

Among the mineral deposits to be described in this section are examples found in a variety of these environments, but the most important ones are in terrigenous and carbonate shelf-deposits and the fluvatile and lacustrine environments. However, in making some preliminary generalizations, it is more convenient to discuss them in groups according to the nature of the exotic constituents which they contain.

Copper minerals occur in several types of sediment. Copper is an element that is readily soluble in the presence of chloride and sulphate ions, but is precipitated in alkaline conditions and by the presence of sulphide (although it can remain in solution as a polysulphide complex at elevated temperatures). It does form carbonates under certain conditions, but does not normally co-precipitate with the common carbonates that form from sea water. It can be absorbed on to certain phylosilicates such as clay minerals, particularly the vermiculites. It plays a minor role in biochemical processes, being found in certain proteins, for example, in the crustaceans; but even in quite low concentrations it is toxic to many organisms.

Concentrations of copper are found in black pelagic shales along with organic material, iron sulphides and other metals such as lead, zinc, cobalt, silver and uranium. Deposits of this type are found today in the depths of the Black Sea. In older rocks it occurs in very extensive horizons such as the Kupferscheifer at the base of the Permo-Triassic sequence that formed in the semi-landlocked Zechstein Sea of northern Europe. There are horizons of what are called 'alum shales' in several places, an example being at the base of the Palaeozoic sequence along the edge of the Caledonides of Scandinavia. Alum shales get their name from the alum which develops on weathering because of their high aluminium and iron sulphide content, and they contain a large number of metals in small but anomalous quantities.

Copper is also found in fluvatile, lacustrine and esturine sediments. A common type are the so-called red-bed deposits, found in sequences of continental, highly-oxidized sediments, made up of fanglomerates, eolian sands and 'wadi-sediments'. Within such sequences there are sometimes intercalations of less oxidized fluvatile, lacustrine or esturine, muds, silts and sands (often containing organic matter and even seams of lignite) that display structures like rain pits, mud cracks, ripple marks and mud pellets. Copper minerals may be found over wide areas in such rocks (although under present economic conditions they rarely make ore-grade unless they have been enriched by oxidation).

The most important examples of copper deposits of this group are those of the Central African Copper Belt, and although other examples of the type occur in the world, none are so well developed or so well known. In the Copper Belt a few very extensive horizons containing copper sulphide minerals are found in a sequence of late Pre-Cambrian marine sediments. The copper minerals are associated with iron sulphide, a little cobalt and uranium, but very little else, and the copper-rich beds pass between different facies of sediment. Some of the beds are proximal, ferruginous, clastic sediments (including arkoses and sandstones) associated with aeolian sand and displaying similar structures to the 'red-bed' type; but exactly similar copper minerals are found in dolomitic shales and in shaly intercalations in thick dolomitic sequences. There is often evidence of an arid environment (mud cracks, dune sands and anhydrite for example) and, although deformation and metamorphism makes the evidence difficult to interpret, it seems that the copper content and mineral zoning in these beds has a relationship to palaeo-islands or 'basement heighs' in the sediment immediately beneath them.

The features common to all these copper deposits in sediments is that they have not been found in rocks formed earlier than the late Proterozoic; that they are of great lateral extent; and that the source of copper remains a mystery. It seems most probable that the copper was precipitated by contact with sulphide (reduced by organic matter from sea-water sulphate) either at the same time as the sediment was forming, or during early diagenesis. Indeed, in some examples, it looks as if copper may have been precipitated as sulphide in the upper few centimetres of mud, and then locally concentrated by reworking. Copper sulphides are found adhering to mud pellets, or along the forsets of current bedding, for example. The age limitation perhaps suggests that these deposits only began to form after some critical stage in organic evolution had passed.

It has been suggested that the origin of the copper is terrigenous, which could be true in some cases, but usually suitable provenance areas are lacking. It has also been proposed that the copper is derived from volcanic debris, or from hot springs of magmatic affiliation, but the evidence for these is sporadic (even though it is plausible in the case of the White Pine deposit in the United States). Another theory is that the copper is primarily scavenged from sea water by planktonic organisms, and becomes incorporated into the sediment by the wholesale death of the organisms. Such phenomena are not unknown in places where nutrient-rich currents meet arid shorelines, but it has yet to be shown that this process can concentrate copper from the micro-gram per litre quantities found in sea water to the several per cent found in the ore beds (and in sufficient quantity). The lateral extent, variable host lithology, relationship to sedimentary textures and palaeographic features, argue strongly against the older idea that these deposits were emplaced by solutions after the lithification of the sediments.

There are mineral deposits where the evidence for the introduction of the ore-minerals into the sediment after deposition is more convincing. In particular the so-called uranium-in-sandstone or Colorado Plateau type. Uranium is a metal with a very different chemistry from copper. It is mobile in sulphate and chloride solutions in its oxidized, hexavalent form, the uranyl ion $(UO_2)^{2+}$. In this form it is readily precipitated by reaction with a number of pentavalent anions, particularly those of phosphorus, arsenic and vanadium. It is also easily reduced to its quadravalent state, in which it forms a number of minerals such as uraninite and coffinite. It can be incorporated into organic-rich sediments by direct reduction, rather than by reaction with sulphur as is the case with the base metals. Uranium is found in small quantities in organic-rich marine shales, in marine phosphate beds and in coal seams. The much richer Colorado Plateau type deposits are, however, found in fluviatile and lacustrine sediments containing organic material in the form of fossil wood fragments and oil-like coatings on sand grains. The morphology of these ore bodies strongly suggests control of deposition by the presence of this organic material, and by variations in permeability of the host rocks. Some show evidence of repeated solution and redeposition in the sediment over a long period, and on a large scale. The distribution of ore often resembles that of an oil reservoir. On exhumation these deposits are often oxidized, and the uranium reprecipitated in its hexavalent form by reaction with phosphatic nodules in the sediment, or with vanadium which is often present in these types of sediment (as montroseite, a mineral that forms from the diagenesis of clays with adsorbed vanadium).

The source of the uranium in these deposits is the subject of much discussion. Although the concentration of the metal in river water is low and very dependent on the occurrence of uranium-bearing rocks in the drainage area, it has been suggested that the scavenging power of organic matter is efficient enough to provide a primary concentration which becomes re-concentrated by moving groundwater in the young sediment. Other theories have preferred the feeding of uranium into groundwater by volcanic springs or similar magmatic sources; in particular those associated with alkaline-acid rocks.

Another case of deposits that occur in fluviatile sands and show strong evidence of permeability control, are the lead-zinc ores of the Laisvall type.

These are of widespread occurrence along the edge of the Scandinavian Caledonides; stratigraphically just above a tillite, and below an important horizon of alum shales. The host sandstones in this case were probably highly permeable because they were formed during the transition from cold-continental conditions to a warmer, marine pelagic environment and were well washed by the torrential rains of the time. However, the precipitating mechanism of the lead and zinc sulphides is not clear. The favoured interpretation is the cooling or dilution with fresh water, of a solution carrying the sulphides as a complex, even though the source of such a solution is not easily found. Its most probable origin is in magmatic activity in the centre of the adjoining fold belt, but selective leaching of the alum shales has been suggested.

There are a few lead-zinc sulphide deposits, usually containing some copper, iron sulphides and a range of other elements in small quantity, that occur in marine terrigenous sediments of quite a different kind. Among them are the great deposits such as Sullivan. These are characteristically stratiform (at least in part) consisting of layers of near-solid sulphide interbedded with shale, mudstone, siltstone or sandstone. Whilst their form strongly suggests syngenetic deposition, their compositional complexity suggests a magmatic parentage, and some show evidence such as alteration or volcanoclastic material in the sediment, that would suggest a magmatic derivation. In some respects, they resemble the so-called 'Kuroko' deposits, but these are almost entirely associated with volcanic rocks, though in a marine environment. Several large stratiform lead-zinc-copper-iron sulphide deposits similar to the Sullivan type are found in carbonate sequences, such as the large Mt. Isa and McArthur River deposits in Australia. There is much discussion about these deposits and it is difficult to offer any sensible generalizations beyond noting that they occur in sedimentary environments that are rather peculiar and specialized. Sullivan, for instance, occurs in the unusually thick shallow water clastic sequence of the Belt Super-Group.

There are many examples of lead-zinc deposits that occur in warm sea shelf carbonate rocks about which more can be said. Lead and zinc are metals that are readily soluble as simple cations in rich chloride brines. Both are easily precipitated by reaction with sulphide complexes. Lead forms a carbonate cerussite, similar in structure to aragonite, a property that it shares with barium and strontium, and these three heavy metals may be co-precipitated with aragonite, a common constituent of marine shells. Zinc also forms a carbonate smithsonite, but one with a structure similar to calcite, and the metal may be co-precipitated with that mineral along with magnesium, iron, manganese, cadmium and cobalt.

Lead-zinc sulphide deposits that are found in shelf carbonate sequences vary in their form. They tend to occur as groups of ore bodies in districts or fields, often grouped in a way that has a relationship to the palaeogeography of the time of deposition, or to the post-depositional structure of the sedimentary sequence. Several important districts are situated in Palaeozoic carbonate sequences of the American Mid-West, and this type of deposit is sometimes referred to as the Mississippi Valley Type. In some cases the lead and zinc minerals seem to have been deposited along with the sediment in bedded form, in sediments that suggest a lagoonal or back reef environment. Others are strata-bound, and seem to have accumulated in spaces in the rock that were either created as part of the mineralizing process, or by some other influence. The host lithology is variable and includes reefs, bioherms, carbonate mud-bank deposits, calcarenites, calcilutites, oolites and mixed lithologies with sand or pebbles. The distribution of mineralization is strongly influenced by impermeable materials such as shales, marls, chert beds and the like. The processes that have been suggested as creating the space are: dolomitization, both early diagenetic and late; solution collapse, both subsurface and associated with karst development; slumping of reef or mudbank deposits, fossil decay or borings. Various forms of fracturing have been observed controlling the mineral distribution, induced by differential compaction or solution collapse of the underlying rocks, hydraulic fracturing by overpressured water, as well as by folding and faulting. Thus, the deposits may take many forms, depending on the relative chronology of the porosity creation process and of the arrival of the mineralization.

Such deposits are usually associated with iron sulphide, and not infrequently with small quantities of carbonaceous material, usually pyrobitumen. Most examples of the type contain minerals that have fluid inclusions that indicate a low temperature of deposition, normally no higher than one would expect from the normal temperature at the presumed depth of burial. They commonly have sulphur isotope ratios consistent with an origin of that element from sea water, and not a few have lead isotope ratios which are difficult to interpret and which yield ages inconsistent with the stratigraphic age of the host rocks.

It has been suggested that these deposits were formed by the reaction between metal-bearing brines and reduced sulphur of quite separate origins. One possible source for the sulphide is the reduction of sulphate in ground water (or as anhydrite) by hydrocarbons (which is also thought to be the origin of the hydrogen sulphide found in some natural gas fields). Another possible source could be iron sulphide precipitated by organic activity in marine muds. It is interesting to note that some of these deposits are associated with iron carbonate or iron oxide deposits, that could result from the replacement of iron sulphide by those of lead and zinc. This can be quite a useful guide in prospecting for such deposits. Three main sources of metal-bearing brines have been postulated: they could be of magmatic origin and fed into the groundwater via faults from depth or, fed into the sea as volcanic springs. Whereas this is plausible in some cases, many such deposits do not contain, or only contain very small amounts of, the other metals like copper that are usually associated with such solutions. Another possible source is from metals adsorbed on to clays that are released into the pore-waters during diagenesis and expelled on compaction. This also seems a plausible mechanism in some cases where shelf carbonates pass latterly into marine shales, but there is no evidence that shales selectively adsorb lead-zinc, or that they are selectively released on diagenesis, so the relative lack of copper, nickel, arsenic, antimony, etc. in many examples, poses a problem. The third hypothesis has the advantage that it does explain the relatively simple composition of many of these deposits. Lead, zinc, cadmium, barium, may all be co-precipitated with aragonite or calcite, and both of these minerals tend to undergo a series of changes during diagenesis and, since this takes place in a relatively concentrated brine, at least in warm arid climates, a pore-fluid of the right composition can be formed. In this case some process must induce the circulation of such fluids.

Barite and fluorite are often found in association with these deposits, and there are many deposits of these two minerals which are found in similar environments and contain small quantities of lead and zinc sulphides. Barium is an element that could originate from sea water by a similar mechanism to that suggested for lead. Fluorine is certainly concentrated from sea water in evaporites, and could be transported into carbonate rocks by diagenetic brines. However, both fluorine and barium can originate from magmatic sources.

The examples that have been chosen in the pages that follow have been selected to display the range of mineral deposits occurring in sediments, where the mineralization is essentially exotic, and where evidence is lacking for sources of the mineralization outside the sedimentary sequences in which they are found.

Further Reading

SELLEY, R. C. (1976), *An Introduction to Sedimentology*. Academic Press, London.

The Luanshya Copper Deposit – Zambia

Still sometimes known by its old name of Roan Antelope, Luanshya was the first deposit to be opened up in modern times, on the famous Central African Copperbelt. Luanshya is an example of the deposits found in this area, remarkable for their great lateral extent and restricted stratigraphic distribution. These deposits contain about 27% of the world's copper.

Location

Luanshya Mine is situated at latitude 13° 08′ S, longitude 28° 23′ E and 1250 m altitude, 35 km south-west of the town of Ndola in the Western Province. It is the southernmost deposit of the Copperbelt which stretches for 500 km to the north-west into neighbouring Zaire.

Geographical Setting

This part of Central Africa is rather flat, although elevated. Most of the Copperbelt is rolling grassland with a few low rounded hills. The seasonal River Luyanshya flows across the mine area on its way to join the Zambesi.

The climate is tropical tempered by the altitude. There are two distinct seasons, one wet with rain almost every day, from November to March during which time 1·2 m of rain falls. The remainder of the year is dry and hot with daytime temperatures around 32°C. Agriculture is not very productive, but cattle are raised and some crops grown around villages. Since the mines were opened up, the Copperbelt has become moderately well populated, and there are towns with over 100 000 inhabitants at Kitwe and Ndola. There is a good network of roads, but mineral production depends on the railways that connect the mines to the world markets. The Zambian railway system is connected to South Africa via Rhodesia, to Benguela in Angola, and to Mikindami in Tanzania.

History

There are no written records of the early history of mining in the Copperbelt, but there were substantial mining operations, already abandoned, before the first Europeans penetrated the area. Luanshya was no more than a small digging of malachite for use as a pigment when W. C. Collier was taken there in 1902. (He is said to have named the deposit Roan Antelope because he first saw copper-stained outcrops beneath the body of an antelope he had just shot.) It was not until 1922, after the British South African Company decided to give long-term exclusive rights to major companies, that the area became more interesting to investors. Selection Trust made a detailed investigation of Luanshya in 1925 and, two years later, began the development of an underground mine, which began production in 1931. In 1970, after Zambia became an independent state, the mine was partially nationalized and is now operated as a division of the Zambian Industrial and Mining Corporation.

Geological Setting

There are three elements to the geology of the Central African Copperbelt; an Upper Proterozoic crystalline basement, a folded sequence of very late Pre-Cambrian sediments known as the Katangan (containing the ore horizons) and a cover of much younger sediments.

The basement is composed of granites, and two groups of metamorphic rocks, an earlier group of schists, the Lufuba, and a younger series, largely of quartzites, known as the Muva Group. The Katangan rocks overlie an irregular topography on the basement, and its lowermost members fill in the hills and valleys of the pre-Katangan land surface.

Copper-rich horizons occur at various levels in the Katangan, but in Zambia they are confined to one situated a few metres above the level at which the pre-Katangan basement topography became filled in. The ore formation is in some places an arkose or greywacke, in others a shale or dolomitic shale. The Footwall Formation largely consists of conglomerate and sands which show evidence of both aquatic and aeolian sedimentation. Clastic sediments, with some dolomite, make up the Hangingwall Formation and these with the Ore Formation, make up the Lower Roan Group.

The Katangan rocks outcrop in two parallel belts on either side of a large basement culmination sometimes referred to as the 'Kafue Anticline'. These rocks were extensively folded and metamorphosed and the Katangan rocks are found today as a series of synclical remnants, folded into the basement. The age of the folding is thought to correspond to the beginning of the Cambrian, although stratigraphically useful fossils are lacking.

Almost the whole of the Ore Formation contains an anomalous amount of copper, but the economic copper ores are confined to certain areas. Most of these are associated with pre-Katangan basement hills and with facies changes in the Lower Roan indicating proximity to palaeo-shorelines. The mineralization consists of a dissemination of grains of sulphides, and is much the same in the argillaceous facies as the arenaceous. Cobalt is notable as a trace element in some of the ores. One striking feature of these deposits is that the sulphide minerals are zoned in a way that seems to relate directly to facies changes in the host sediments.

Some of the deposits are extensively oxidized and have a different mineralogy; malachite, cuprite, chrysocolla and a locally important copper-bearing vermiculite clay mineral. Small amounts of vein copper mineralization occur in some of the basement rocks.

The Geology of the Luanshya Deposit

The deposit is situated at the eastern end of an isolated synclinorium of Katangan rocks. Both biotite schists and granite occur in the basement, and the overlying Footwall Formation consists of conglomerate and quartzite with some argillite of very variable thickness (from 0–250 m) depending on the basement topography. The Ore Formation is 15–50 m thick, composed of dolomitic argillite, with dolomite concentrated near the base. The Hangingwall Formation and the remainder of the Roan are much as found elsewhere on the Copperbelt.

The workable ore-horizon makes up about half the thickness of the Ore Formation. There is usually a variable thickness of barren dolomite at the base followed by the disseminated ore. The deposit is in the form of a syncline that is almost isoclinal, plunging in an undulating fashion towards the west. The degree of flattening of the fold varies along strike being particularly intense (with a profusion of parasitic folds) in the central part where a granite is present in the basement.

The mineral assemblage of the ore is normal for the Copperbelt. There is a stratigraphic sequence to the sulphide mineral content of the ore-horizon, beginning with chalcocite and bornite at the bottom through chalcopyrite to pyrite. There is also a lateral facies variation in the mineralogy with chalcocite concentrated at the nose of the syncline in the east, bornite in the centre, and chalcopyrite and pyrite becoming more dominant in the west. Sedimentological evidence suggests that the original Lower Roan shoreline was a little to the east of the nose of the syncline. Just west of the centre of the ore-zone there is an almost barren gap in the mineralization at a point where dolomite-rich beds rise up through the ore-horizon. This is just to the west of a granite basement hill that rises through the Footwall Formation. This feature has been interpreted as a reef, or carbonate mudbank.

The ore is partially oxidized to carbonates in particular, and in places the oxidization penetrates to considerable depth.

Size and Grade

The total size of the ore zone in the Luanshya syncline is of the order of 200 Mt containing about 2·5% sulphide copper and a small additional amount of 'oxide' copper. More than half this has been mined. The deposits of the Zambian section of the Copperbelt have a total of over 30 Mt of copper metal.

Geological Interpretations

Originally, the Copperbelt ores were interpreted as hydrothermal replacement deposits, which is not surprising because many of the early finds were noticeably associated with granites. Over the years, however, increased mine development and exploration revealed first, that the Roan Group was not intruded by the granites and, that the controls on the ore distribution were fundamentally sedimentological and palaeogeographic. Interpretation is much hampered by the intensity of folding. The problem of origin is fundamentally one of the time of the introduction of the sulphide minerals in relation to the host rocks. The observed relationships of the ore minerals to the host rocks, and the wide lateral extent of the ore bodies argue against introduction later than the folding, or even lithification, of the rocks. There are some workers who interpret the ore as having been precipitated in the sediments from formation waters during diagenesis. The more popular interpretation is that the sulphides were formed at the time of sedimentation in local euxinic lagoons near the shoreline. Adherents of the sedimentary theory point to other copper-shale deposits, such as the Kupferscheiffer in Central Europe, as parallels.

The source of the copper has always posed a problem. Detrital or dissolved copper from the basement is a possibility, but there is relatively little of it to be seen in present outcrops. Sea water, or thermal waters, have also been suggested; the former has the difficulty that its content of copper is very low, and it is not easy to see how the concentration process could have taken place, although it has been suggested that planktonic organisms effected a preliminary concentration. As regards thermal waters, there is virtually no evidence for their existence, and no evidence of vulcanicity with which such waters are normally associated.

Mining

Luanshya has been developed by a series of shafts (the deepest of which is 1250 m) in the footwall of the southern limb of the syncline. The ore is mined from large stopes that have a clearly defined footwall, and a variable hanging-wall which is fixed by drilling and sampling. As a rule some sub-grade mineralization is left. Ore is hauled into the footwall and up the shafts to a flotation concentrator, and the mixed pyrite/copper sulphide concentrate is smelted on site to blister-copper before shipping to a refinery at Ndola. The mine is capable of producing 98 000 t/a of copper metal.

Further Reading

MENDELSOHN, F., ed. (1961), *The Geology of the Northern Rhodesian Copperbelt*. MacDonaly; London, 523pp. [The major work on the whole area with a good section on Luanshya (called Roan Antelope in the book) by the editor. It contains a full bibliography up to that date and a good summary of the genetic arguments.]

BRUMMER (1954), Geology of the Roan Antelope Copper Mines Ltd., N. Rhodesia. *Inst. Min. Met. Trans.*, **64**, 257–318, disc. 458–71. [The most complete work on this particular deposit.]

LOMBARD, J. and NICOLINI, P. (1962), Stratiform Copper Deposits in Africa. *Symp. Ass. African Geol. Surv.* Pt. 1 Lithology, Sedimentology, 212pp. Pt. 2 Tectonics, 265pp. Paris [Contains a great deal of factual information and discussion on the whole belt including the part in Zaire.]

SALES, R. H. (1962), Hydrothermal Versus Syngenetic Theories of Ore Deposition. *Econ. Geol.*, **57**, 721–34, disc., 1963 **58**, 145, 444–6, 447–56, 609–14, 614–18. [Perhaps the most eloquent attack on the sedimentary theory; should be read in conjunction with Ch. 7 of Mendelsohn.]

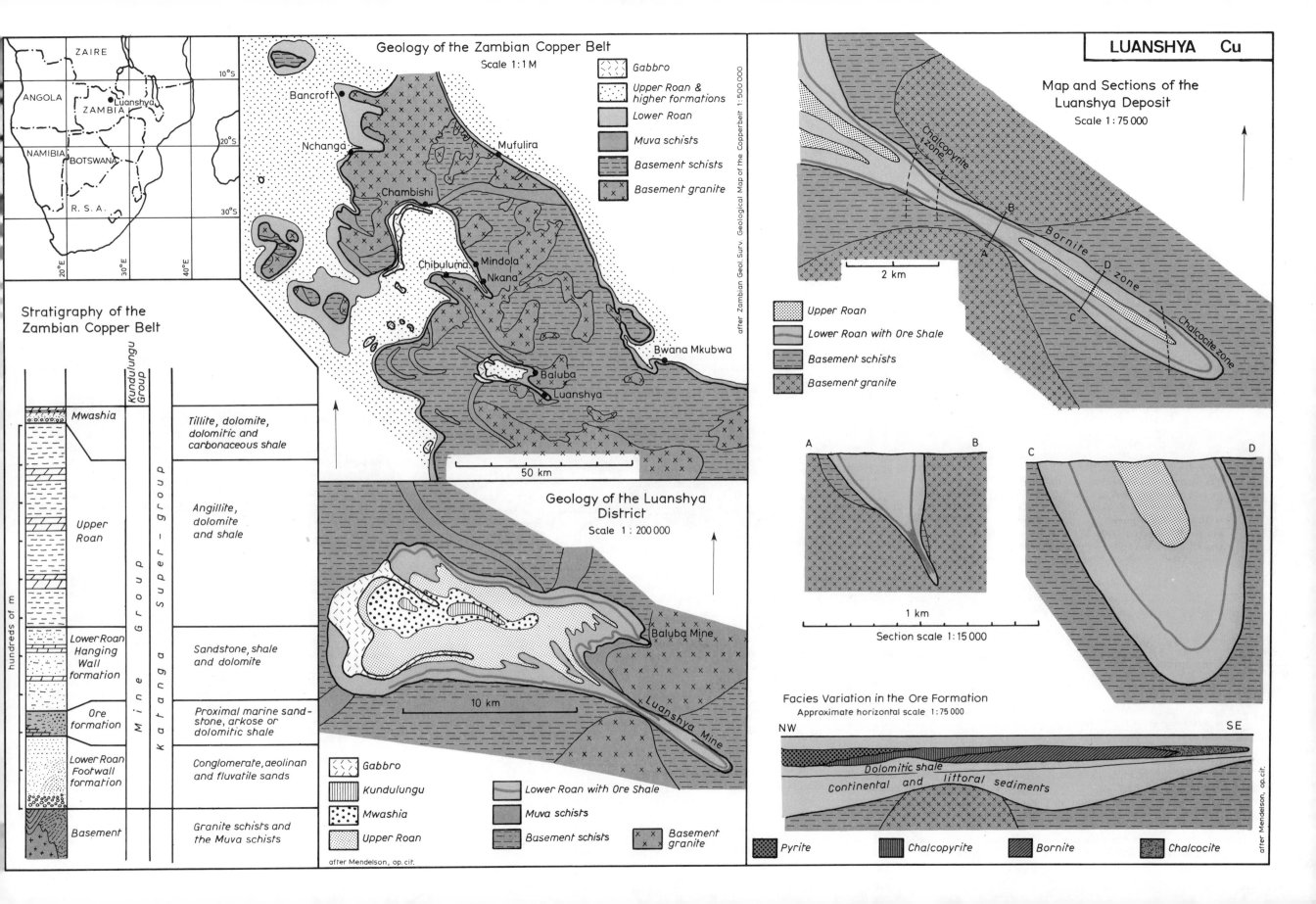

Geology of the Zambian Copper Belt
Scale 1:1 M

ZAIRE
ANGOLA
ZAMBIA
Luanshya
NAMIBIA
BOTSWANA
R.S.A.

10°S
20°S
30°S
20°E 30°E 40°E

Bancroft
Nchanga
Chambishi
Chibuluma   Mindola
Nkana
Bwana Mkubwa
Baluba
Luanshya

after Zambian Geol. Surv. Geological Map of the Copperbelt 1:500 000

Gabbro	
Upper Roan & higher formations	
Lower Roan	
Muva schists	
Basement schists	
Basement granite	

50 km

Stratigraphy of the Zambian Copper Belt

hundreds of m

Mwashia	Kundulungu Group		Tillite, dolomite, dolomitic and carbonaceous shale
Upper Roan			Argillite, dolomite and shale
Lower Roan Hanging Wall formation	Mine Group	Katanga Super-group	Sandstone, shale and dolomite
Ore formation			Proximal marine sand-stone, arkose or dolomitic shale
Lower Roan Footwall formation			Conglomerate, aeolinan and fluvatile sands
Basement			Granite schists and the Muva schists

Geology of the Luanshya District
Scale 1 : 200 000

Baluba Mine
Luanshya Mine

10 km

Gabbro	
Kundulungu	
Mwashia	
Upper Roan	
Lower Roan with Ore Shale	
Muva schists	
Basement schists	
Basement granite	

after Mendelson, op.cit.

LUANSHYA   Cu

Map and Sections of the Luanshya Deposit
Scale 1 : 75 000

Chalcopyrite zone
Bornite zone
Chalcocite zone
A B C D

2 km

Upper Roan	
Lower Roan with Ore Shale	
Basement schists	
Basement granite	

A   B   C   D

1 km
Section scale 1:15 000

Facies Variation in the Ore Formation
Approximate horizontal scale 1:75 000

NW   SE

Dolomitic shale
Continental and littoral sediments

Pyrite	Chalcopyrite	Bornite	Chalcocite

after Mendelson, op.cit.

The Ambrosia Lake Uranium Field – U.S.A.

This is the largest of the uranium ore fields on the Colorado Plateau. It is typical of so-called 'sandstone type' deposits that occur widely in the Plateau at several stratigraphic levels in Mesozoic sediments, as well as in several adjoining sedimentary basins.

Location

The deposits lie on the flanks of the Ambrosia Lake Dome situated at latitude 35° 26′ N, longitude 107° 52′ W and at an altitude of 2400 m in McKinley County, New Mexico.

Geographical Setting

Ambrosia Lake is in the more mountainous south-eastern corner of the Colorado Plateau. The landscape is dominated by Mt. Taylor, a 3471 m high extinct volcano that overlooks an area averaging 2200 m in altitude. It is at the head of a valley system that drains down to the Rio Grande, but is only a few kilometres from the continental watershed. The area is rather dry, receiving most of its 300 mm of rain in the hot summer that alternates with a long, dry, freezing winter. The land is sparse scrubland, with pine-wooded uplands, and is only useful for grazing or ranching. The main town of the area is Grants, 32 km south of the deposit on US Highway 66, and a main railway where they pass through the mountains on their way from Albuquerque to Los Angeles.

History

Bright yellow vanadium–uranium ore was discovered in western Colorado in 1898. Up to the Second World War mines in Colorado and Utah produced this material as a vanadium ore, although some plants were built to separate radium from the tailings. In 1948, prospecting was stimulated by the demands of the U.S. Atomic Energy Commission and, two years later, a Navajo Indian, known by the delightfully un-Indian name of Paddy Martinez, found outcrops in limestone north of Grants. In the prospecting and mining rush of the early fifties, ore was produced from sixty small open pits and shallow mines. The area was intensively mapped and prospected by radiometric methods, but curiously it was a wildcat drill hole put down because of the attractiveness of the Ambrosia Lake Dome as an oil target, that led to the discovery in 1955 of the Dysart Mine. A year later, seventy drilling rigs were on the site and large orebodies discovered, although it took some years of consolidation and development before serious mining could begin.

Geological Setting

Deep in the Grand Canyon and in a few other deeply eroded or upfaulted parts of the Colorado Plateau, one can see a basement of Pre-Cambrian schists and granitic rocks. Above this is a great thickness of sediment representing all of the Phanerozoic systems that, though faulted and gently folded, never suffered the intense deformation and magmatic activity of those in the ranges that surround the Plateau. In the Lower Palaeozoic the sedimentation was marine and even in thickness over the basement, but from Pennsylvanian times onwards, the sedimentation became thicker in certain parts of the Plateau forming a series of basins separated by belts of uplift. Mesozoic sedimentation was largely continental or proximal marine, and was dominated by clastic sediments derived from the rising mountain belts to the south and west. It is in sandstones, and locally in a few other rock types of this Mesozoic sequence, that most of the uranium and vanadium mineralization is found.

The whole of the Plateau was submerged during the late Cretaceous with the widespread deposition of the marine Mancos Shale, after which the Plateau was uplifted as the Laramide tectonic events were taking place in the surrounding areas.

This period of uplift was also the time of a great deal of the faulting, and the emplacement of a number of intrusives, particularly a series of lacoliths in the centre of the Plateau. In the south are several volcanic centres that erupted during the Tertiary and almost to the present day. The uplift of the Plateau was accompanied by erosion the formation of the Grand Canyon belonging to this period.

Uranium mineralization is found all over the Plateau. There are some vein deposits and deposits in pipe-shaped collapse structures, but most of it is in the form of irregular lenses in sandstones. These occur as several different stratigraphic levels, but most of the reserves are either in the Triassic, Chinle Formation, or the Morrison Formation at the top of the Jurassic. Most of the ore bodies contain both uranium and vanadium minerals with some iron sulphide, but in south-eastern Utah a little copper is found.

Geology of the Ambrosia Lake Deposits

The Ambrosia Lake area is underlain by a series of sediments dipping generally to the north-east. The original discoveries of uranium–vanadium ore were oxidized deposits in the Todilto Limestone, but the largest deposits are situated around the Ambrosia Lake Dome; an anticlinal structure that is buried beneath the Mancos Shale and Dakota Sandstone. The ore occurs in the Morrison Formation, which is here divided into three members. The basal Recapture Member is almost barren, and is composed of variegated mudstones with occasional thin sandy channel fillings. The main ore bodies occur in the overlying Westwater Canyon Member, which consists of large lenses or tongues of sandstone with thin shale or mud partings. The sandstone is reddish brown, yellow, or dark grey where the organic material content is high. The sand is not well-sorted and contains a certain amount of feldspar and other materials, probably of volcanic derivation. The Westwater Canyon sandstone lenses into the overlying Brushey Basin' Member which, though similar to the Recapture, contains several large tongues of sandstone originally thought to form a distinct member of the sequence, and locally called the Poison Canyon Sandstone. The ore bodies occur for the most part in the Westwater and the Poison Canyon sand tongues.

There are three types of ore body. The first are elongated lenticular bodies parallel to the bedding known locally as 'runs' and 'rolls'. These are not associated with and are only displaced by faults, and are thus known as 'pre-fault' ore bodies. The post-fault ore bodies are more irregular, transgressive and associated with faults. Some bodies are known as 'stacks' which are clearly transgressive and show evidence of having been remobilized from original 'runs' or 'rolls'. The third type are secondary ores, in which the primary ore has been changed to a new mineralogy.

The primary ore consists of zones of grey sandstone in which the sand grains are coated with hydrocarbon material, and coffinite. Fossil wood fragments occur replaced by coffinite. Coatings of the 'pitchblend' variety of uraninite also occur. Vanadium occurs as roscoelite and montroseite. Pyrite is generally present together with small amounts of other sulphide and selenium minerals. The secondary ores are largely sandstone impregnated with the bright yellow minerals carnotite, tyuyamunite and metatyuyamunite.

Size and Grade

Grades and tonnages of uranium ores are very sensitive to the price that buyers are prepared to pay for the metal and, since these are largely governments, it is notoriously difficult to make clear statements about reserves. At the workable grade of the mid-sixties, about 0·25% U_3O_8, the Ambrosia Lake Field contained about 40 Mt, but it is certain that at higher prices and lower grades the reserves would be much greater.

Geological Interpretations

Most people who have studied these deposits have concluded that they were formed by the precipitations of uranium from formation-waters flowing through the permeable sandstones. It is thought that the right chemical environment for precipitations was created by the organic matter. The resemblance of the Ambrosia Lake Dome to an oilfield is striking. There is some evidence that the primary 'runs' and 'rolls' were present early in the history of the sediment, at least before the burial of the Morrison Formation below the thick Cretaceous shale.

The Laramide events at the end of the Mesozoic and beginning of the Tertiary included vulcanism, faulting and gentle folding; and this would seem to have been a time of widespread formation-water circulation and was probably the period of redistribution and concentration of the primary ores. It may also be possible that during the uplift and erosion of the Plateau, downward-moving formation-waters caused redistribution. Once the rocks were eroded to the extent that the primary deposits came into the region of oxygenated groundwater, they were changed to the yellow secondary ores.

There is less agreement about the source and origin of the uranium. Some authors think that it was traces of uranium in the sediments that were leached over long periods of time, others preferring volcanic or igneous sources, point to the volcanic component of some of the host rocks, and the proximity of many of the deposits to volcanic centres or sub-volcanic intrusions.

Mining

Several major companies operate in the Ambrosia Lake Field, and there are a large number of mining operations sending ore to five main processing plants. Most of the ore is extracted by room and pillar methods, but cut and fill stopes are used where the ore is very thick. Both local alluvial sand, and mill-tailings, are used for backfill. As is the case with most uranium ores, the metal is recovered by acid leaching, and recovery by ion-exchange methods. The product produced by the mines is a very pure uranium oxide known generally as 'yellow cake'.

Further Reading

GRANGER, H. C. et al. (1961), Sandstone-type Uranium Deposits at Ambrosia Lake, New Mexico. Econ. Geol., 56, 1179–1210. [The major work on the area written at a time when the field was under active development.]

PAGE, L. R., STOCKING, H. E. and SMITH, H. B., eds (1955), Contributions to the Geology of Uranium and Thorium by the United States Geological Survey and Atomic Energy Commission for the U.N. International Conference on the Peaceful Uses of Atomic Energy, Geneva, Switzerland 1955 U.S. Geol. Surv. Prof. Paper 300, 739pp. [A mammoth volume which includes a number of important papers on the Colorado Plateau region.]

RIDGE, J. D., ed. (1968), Ore Deposits of the United States, 1933–1967. American Inst. of Min. Met. and Petrol. Eng. New York. [Chapter 6 of this volume is devoted to the Colorado deposits, shorter and more up-to-date than PAGE et al. (1955) and contains a good description of the Ambrosia Lake area in a paper by Kelley et al., pp.752–69.]

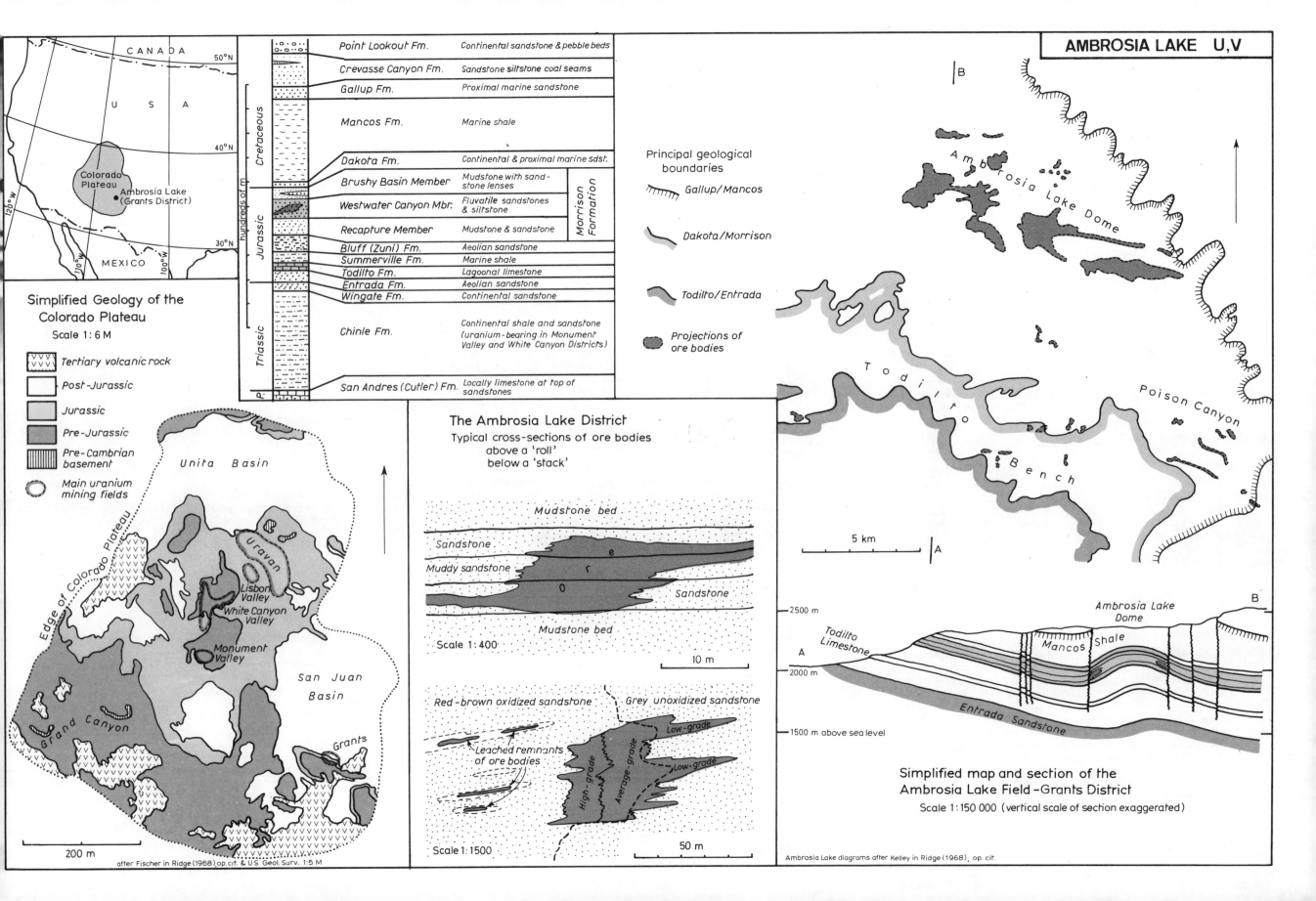

Simplified Geology of the Colorado Plateau

Scale 1:6 M

(v)	Tertiary volcanic rock
(blank)	Post-Jurassic
(grey)	Jurassic
(dark)	Pre-Jurassic
(lined)	Pre-Cambrian basement
○	Main uranium mining fields

Point Lookout Fm. — Continental sandstone & pebble beds
Crevasse Canyon Fm. — Sandstone siltstone coal seams
Gallup Fm. — Proximal marine sandstone
Mancos Fm. — Marine shale
Dakota Fm. — Continental & proximal marine sdst.
Brushy Basin Member — Mudstone with sand-stone lenses
Westwater Canyon Mbr. — Fluvatile sandstones & siltstone
Recapture Member — Mudstone & sandstone
Bluff (Zuni) Fm. — Aeolian sandstone
Summerville Fm. — Marine shale
Todilto Fm. — Lagoonal limestone
Entrada Fm. — Aeolian sandstone
Wingate Fm. — Continental sandstone
Chinle Fm. — Continental shale and sandstone (uranium-bearing in Monument Valley and White Canyon Districts)
San Andres (Cutler) Fm. — Locally limestone at top of sandstones

Morrison Formation

Cretaceous / Jurassic / Triassic / P.

hundreds of m

CANADA
U S A
MEXICO
Colorado Plateau
Ambrosia Lake (Grants District)
50°N / 40°N / 30°N
120°W / 110°W / 100°W

Principal geological boundaries

Gallup/Mancos
Dakota/Morrison
Todilto/Entrada
Projections of ore bodies

Unita Basin
Uravan
Lisbon Valley
White Canyon Valley
Monument Valley
San Juan Basin
Grand Canyon
Grants
Edge of Colorado Plateau

200 m

after Fischer in Ridge (1968), op.cit. & U.S. Geol. Surv. 1:5 M

The Ambrosia Lake District

Typical cross-sections of ore bodies
above a 'roll'
below a 'stack'

Mudstone bed
Sandstone
Muddy sandstone
Sandstone
Mudstone bed

Scale 1:400

10 m

Red-brown oxidized sandstone
Grey unoxidized sandstone
Leached remnants of ore bodies
Low-grade
High-grade
Average-grade
Low-grade

Scale 1:1500

50 m

Ambrosia Lake Dome
Todilto Bench
Poison Canyon

5 km

Simplified map and section of the Ambrosia Lake Field–Grants District

Scale 1:150 000 (vertical scale of section exaggerated)

2500 m
2000 m
1500 m above sea level

Todilto Limestone
Mancos Shale
Entrada Sandstone
Ambrosia Lake Dome
A
B

Ambrosia Lake diagrams after Kelley in Ridge (1968), op.cit.

The Laisvall Lead–Zinc Deposit – Sweden

This is one of the world's major lead deposits and certainly Europe's largest, and it is interesting in that it has the paragenesis of the well-known Mississippi Valley Type deposits and yet strong environmental similarities with the Colorado Plateau and Central African deposits that respectively contain uranium and copper. It is one of a series of deposits that occur in sediments lying on the stabilized shield rocks of Sweden, along the edge of the great Caledonian orogenic belt.

Location

Laisvall is situated at latitude 66° 13′ N, longitude 170° 25′ E and 430 m altitude beneath Lake Laisan in Norrbotten County, Swedish Lapland.

Geographical Setting

Laisvall is in a remote and rather inaccessible part of north-west Sweden at the foot of mountains that rise to the watershed and the Norwegian frontier. It is a land of glacial valleys filled by lakes and a combination of mixed pine/birch forest and mountain grassland. Being just in the rain-shadow of the mountains, it suffers a sub-arctic continental climate, a long, cold, but relatively dry winter, and a cool wet summer, albeit tempered by long days. Fishing is good; a little game roams the hills and some summer pastures are used; but generally there is little reason for humans to go there except for the mine. Laisvall is connected by road to the more populated and larger mining towns of the Skellefteå district, and to the Baltic coast 250 km to the south-east.

History

In 1898, a boulder of sandstone spotted with galena was found at Dalecarlia some distance south of Laisvall, and people speculated for many years as to where it might have come from. Prospecting in Lapland was greatly stimulated in 1929 by the discovery of the Boliden deposit, a gold-rich massive sulphide body 40 km inland from the Baltic port of Skellefteå. In the thirties, systematic tracing of mineralized boulders was undertaken by the Boliden company, and the area near the northern end of Lake Laisan was indicated as the possible source of the strange spotted lead-rich sandstones. Drilling on the lake shore in 1939 revealed the ore body, which was rapidly developed. Mining began in 1943.

Geological Setting

Scandinavia is traversed by a major geological break between the great tract of old crystalline rocks known as the Baltic Shield, and the highly deformed Palaeozoic cover rocks (which because of their continuation through Scotland are known as the Caledonides). In most parts of northern Sweden the basement shield rocks consist of granite and gneiss, interspersed with the eroded remnants of volcanic and sedimentary rocks forming long belts oriented in a north-westerly direction. This basement was stabilized during the Proterozoic era between 1·5 Ga and 1·7 Ga ago. Very late in Pre-Cambrian times a thick wedge of sediments and volcanics began to form along the present site of the Caledonides and, almost everywhere from the extreme north of Scandinavia to the Atlantic coast of Ireland, this begins with conglomeratic rocks that are widely believed to be tillites. In Sweden these seem to have a radiometric age of 640 Ma. Above the tillites are clastic sediments which include the well-washed white sandstones in which the Laisvall ore occurs. Above these, and persisting over great parts of Scandinavia, is the alum shale formation. This is a dark, carbonaceous,

pyritic shale containing anomalous quantities of uranium, thorium and other metals, which (given a higher price than the world is at present prepared to pay) could be one of the largest resources of uranium. Towards the west, the sediments which range in age up to the Silurian become much thicker and include very considerable thicknesses of volcanic rocks. During the tectonic development of the Caledonides large masses of these Palaeozoic rocks mixed with parts of the basement, were thrust towards the east (as a series of nappes) over the basement. In places erosion has cut 'windows' and along the eroded edge of the nappes is exposed a thin layer consisting of the basal part of the sedimentary sequence, that has remained relatively undeformed and autochthonous with respect to the basement.

Lead–zinc mineralization is found in the autochthonous sandstones at Laisvall, and along the outcrop belt to the north and south over a distance of 2000 km. Only at a few localities has it proved economically interesting.

Geology of the Laisvall Deposit

Granite forms the basement at Laisvall, exposed at surface 6 km west of the mine and in the deepest of the mine workings. It is a pink foliated granite that becomes altered at the top and overlain by the conglomeratic arkoses and pebbly shales that represent the tillite formation in this area. Above this is the Laisvall Formation, 40 m of sandstones divided into a lower member which is white and clean but with a few shaly partings, a middle brown shaly sandstone member, and an upper member similar to the lower but having a thin zone of phosphatic nodules and calcareous lenses at the top. A siltstone and shale formation follows, containing limestone lenses that yield a trilobite and brachiopod fauna of uppermost Lower Cambrian age. The alum shale follows and, away from the mine, this is succeeded by greywacke but, in the area of the mine, the alum shale is cut by the first of two thrust planes that separate the autochthonous sequence from the Kaskejure Nappe. For the most part this consists of broken and fractured rocks derived locally from the basement and its immediate cover. Higher up is a second thrust that separates this from the Yvaf Nappe, composed of chlorite and mica schists.

The Laisvall ore bodies are contained in the upper and lower members of the Laisvall Formation, but 7 km north-east of the main body is a smaller one, the Maive, that is in the basal arkoses. The main body in the lower member is elongated in north-easterly direction and is mostly lead mineralization, while a second body in the upper member spreads towards the north-west and tends to be more rich in zinc. The ore consists of sandstone, with the other minerals between the grains. The important ones are galena and sphalerite, but small amounts of barite, fluorite, calcite and pyrite also occur.

In rich ore almost all the original intergranular spaces are filled with galena or sphalerite, while in more average grade ore, the sandstone is spotted with patches of sulphides containing concentrations in the pore-spaces which reflect the original sedimentary structures of the sandstone. In the white sandstone dark galena shows up dramatically the current-bedding, lode-casts and other structures. Today there is very little pore space in the sandstones, and it is common to see the effects of pressure solution between sand grains and overgrowths on the grains that have filled in the remaining spaces.

The basement surface and the autochthonous sediments with the ore bodies dip gently to the west and are interrupted to some extent by faults, some of which pre-date the nappe thrusts.

The area of the mine was glaciated in the Quaternary, during which time a valley was cut down into the ore horizon, but the sub-outcrop of the ore formation beneath the present day lake sediment is only a few hundred metres long yet enough to yield the train of erratic boulders of ore that led to its discovery. Much of the area round the mine is covered by till and good exposures are rare.

Size and Grade

At a cut-off grade of 2% the Laisvall ore bodies total 80 Mt averaging 4·3% Pb, 0·6% Zn with 9 g/t silver and 30 g/t cadmium.

Geological Interpretations

Sulphides in the inter-grain spaces sealed off by pressure solution and over-growth, indicate that the ore was emplaced before the rocks were buried deep, and this most likely occurred when the nappes were overthrust during the Silurian. It has also been suggested that the change in stratigraphic position of the ore from the arkose in the east to the Upper Laisvall sandstone in the west, indicates that the ore was emplaced at a horizontal groundwater interface after the basement floor had been tilted (which must have been at some time after the Middle Cambrian). Most writers agree that the ore was emplaced by a mineralizing solution that permeated the sandstones, but there is less agreement about the mechanism of precipitation. Some prefer lead and zinc in a chloride-rich brine similar to the composition of the fluid inclusions found in the sphalerite meeting reduced sulphur in the formation; while others prefer to have lead and zinc sulphides brought in as soluble sulphide complexes that were precipitated by lowering of temperature, or mixing with less saline formation waters, or both.

The origin of such a mineralizing solution is subject to more speculation. The brine could have had its source in as yet undiscovered evaporite sequences in the main Caledonides sequence. At least the sulphur isotope ratios indicate a sea water source. It is suggested that the source of the metals could have been from hydrothermal activity in the interior of the Caledonides, and this is supported by the existence of some small lead veins in the tectonic basement 'windows'. Alternatively, it could have come from leaching of the alum shales, but the 'J'-type lead isotope ratios of the ore mismatch those of the shales. The metals could have come from leaching of the weathered top of the basement and its overlying arkoses, by formation brines passing along the unconformity.

Mining

Laisvall Mine is operated by Bolidens Gruvaktiebolag who mine 1·2 Mt/a of ore by room and pillar methods with large trackless equipment. Ore is hauled by an underground railway to a central shaft at the head of which is a modern semi-automated crushing, autogenous grinding and flotation plant, producing galena and sphalerite concentrates, which are trucked by road to the group smelter complex at Ronnskar on the Baltic coast. It is one of the largest and lowest grade lead-mining operations in the world, and succeeds despite the considerable water problem arising from its position under a lake.

Further Reading

LILJEQVIST, R. (1973), Caledonian Geology of the Laisvall Area, Southern Norrbotten, Swedish Lappland. *Sveriges Geol. Undrsokn.*, ser. C **691**, 43pp. [A good general description of the local geology with a good map.]

GRIP, E. (1954), Blymalmen vid Laisvall, dess Geologi och en Jamforelse med nagre utlandska Forekonster. *Geol. Foren. Forhandl.* **76**, 357–80. [The most comprehensive description but lacks the results of later researches. For those who do not wish to grapple with Swedish, a shorter paper in English was published as:

GRIP, E. (1960), The Lead Deposits of the Eastern Border of the Caledonides in Sweden. *Int. Geol. Cong. 21st ser.* Norden. Rep. **16**, pp. 149–59.]

RICKARD, P. T., MARINDER, N.-E. and WILLDEN, M. Y. (1977), Spatial relations at the Laisvall Deposit, Sweden. *Econ. Geol.* In press. [Contains a good factual summary but is mainly a genetic essay that takes into account a wide range of research.]

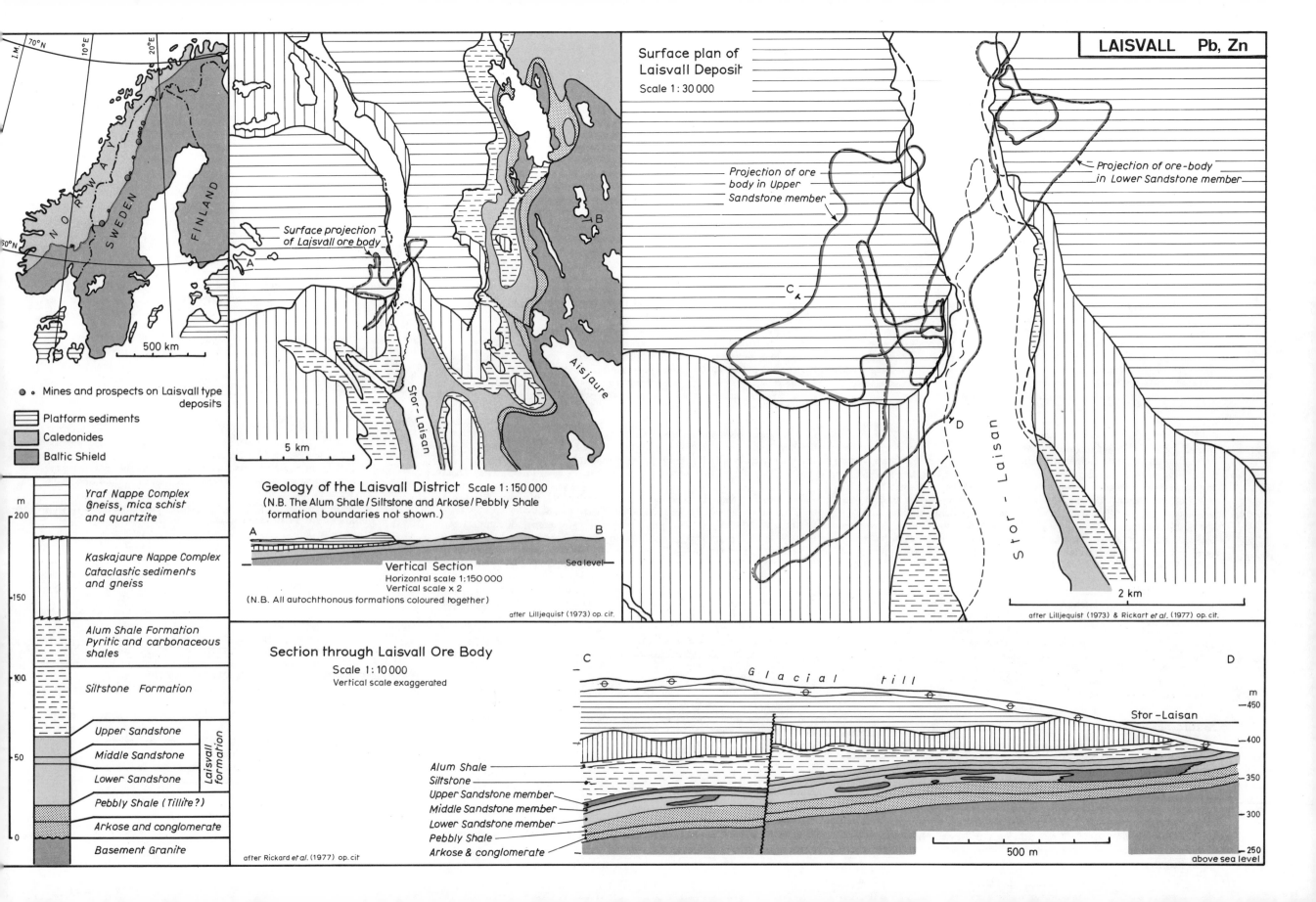

LAISVALL Pb, Zn

Mines and prospects on Laisvall type deposits

Platform sediments
Caledonides
Baltic Shield

Yraf Nappe Complex
Gneiss, mica schist and quartzite

Kaskajaure Nappe Complex
Cataclastic sediments and gneiss

Alum Shale Formation
Pyritic and carbonaceous shales

Siltstone Formation

Upper Sandstone
Middle Sandstone
Lower Sandstone

Laisvall formation

Pebbly Shale (Tillite?)

Arkose and conglomerate

Basement Granite

Surface projection of Laisvall ore body

Stor-Laisan

Aisjaure

Geology of the Laisvall District Scale 1:150 000
(N.B. The Alum Shale/Siltstone and Arkose/Pebbly Shale formation boundaries not shown.)

A B

Vertical Section
Horizontal scale 1:150 000
Vertical scale × 2
(N.B. All autochthonous formations coloured together)
Sea level

after Lilljequist (1973) op. cit.

Surface plan of Laisvall Deposit
Scale 1:30 000

Projection of ore body in Upper Sandstone member

Projection of ore-body in Lower Sandstone member

C

D

Stor - Laisan

2 km

after Lilljequist (1973) & Rickart et al. (1977) op. cit.

Section through Laisvall Ore Body

Scale 1:10 000
Vertical scale exaggerated

C

Glacial till

Stor-Laisan

D

Alum Shale
Siltstone
Upper Sandstone member
Middle Sandstone member
Lower Sandstone member
Pebbly Shale
Arkose & conglomerate

500 m

above sea level

after Rickard et al. (1977) op. cit.

The Picher Lead–Zinc Field – U.S.A.

The Picher, largest of the fields in the Tri-State district, is arguably typical of the zinc–lead deposits that have become known as the Mississippi Valley Type. It differs from other examples by the abundance of chert in the carbonate sequence that contains the ore, and in the halo of silicification surrounding the ore zones. Although there are several mineralized horizons, in each case the ore is clearly strata-bound.

Location

The mining field is centred on the town of Picher (latitude 36° 59′ N, longitude 94° 50′ W and 256 m altitude) in Ottawa County, Oklahoma, although the field stretches over into Kansas State.

Geographical Setting

The Picher field lies on the Osage Plains to the west of the Ozark uplands. The area is flattish rolling country around 260 m altitude, drained by several rivers which have cut modest valleys on their way eastwards to join the Mississippi. The area was originally prairie with scrubby woods in the valleys, but has been extensively farmed and ranched for 150 years. The area receives almost 1·5 m of rain during a year; has a wet spring, long warm summer, and receives light snow in winter. The area is moderately well populated, and is served by good roads and railways.

History

When the early French explorers penetrated into the Mid-West up the Mississippi in the early eighteenth century they found numerous Red Indian tribes who worked small lead deposits from limestones. European scale mining and smelting began almost at once. In 1821, when Missouri became a State of the Union, the influx of settlers soon discovered a trail of lead deposits across the state; the most spectacular finds being around Joplin (in the middle of the century). During the latter part of the century these were followed westwards by drilling under the cover of Pennsylvanian shales, and one such drill-hole penetrated ore in the Picher Field in 1901, although it aroused no interest at the time. Some oxidized zinc ore was turned up by a plough in the south of the Field about the same time, but no great finds were made. It is recorded that in 1914 a driller returning to Joplin after an unsuccessful summer season got his drill stuck in mud, and in an attempt to recoup some of his losses, decided to put down a wildcat hole. He struck rich ore in the main part of the field, and within four years there were 250 processing plants operating. Mining was complicated by the numerous small land holdings, and it was some decades before any large operations emerged. Nevertheless, by the Second World War a few large, modern mills had been built and larger companies (in particular the Eagle Picher) began to put mining on a firmer basis.

Geological Setting

The Picher Field lies on the boundary between the Ozark Dome and the Ouachita Basin. Pre-Cambrian rocks are exposed in the centre of the Dome, and are known from drill-holes to be between 400 m and 600 m beneath the field. Palaeozoic sediments dip off the Dome to the west and south. Several formations of Cambrian and Ordovician sediments are recognized in which dolomitic rocks predominate and, after an unconformity, comes the persistent Chattanooga Formation, a black fissile shale overlain by the main ore-bearing formation of the Tri-State district; the Boone. This includes crinoidal calcarenites, oolites, glauconitic and cherty limestones as well as massive chert beds, some of which are a cloudy, white type, known locally as 'Cotton Rock'. The sequence contains much evidence of thinning by solution, such as stylolitic limestones, and chert breccias derived from cherty limestones. A further unconformity separates the Boone from another series of carbonate-rich formations, and then above a more important unconformity, a change takes place to mixed estuarine beds of the Hale formation, which is taken as the base of the Pennsylvanian. Almost all of the Picher Field is covered by the Krebs Group shales, erosion having penetrated to the Mississippian rocks in one small area in the south.

The principal structural feature of the area is the gentle draping of the Palaeozoic sediments over the Ozark Dome. But there are a few other structural features of some importance, one of which is the Miami Trough, a graben-like structure which seems to reflect basement movements which continued throughout and long after the Palaeozoic sedimentation. Very little is known of the post-Pennsylvanian history of the area. The area was peneplaned and slightly rejuvenated in the Tertiary, so that the rivers have cut down exposing the limestones as cliffs.

Geology of the Zinc–Lead Deposits

Mineralization is sphalerite, poor in iron and relatively rich in cadmium, with galena, pyrite, marcasite and traces of other sulphides, with a gangue of quartz, pink dolomite and calcite. Oil and bituminous materials are found in the ore, and hydrogen sulphide is commonly found in mine workings. Ore occurs in a number of irregular bodies or zones confined within certain horizons of the Boone formation. The most productive of these is the Joplin Member or 'M' Bed, a thick-bedded and stylolitic, crinoidal calcarenite with chert nodules and some glauconite grains, which lies above the massive Grand Falls Chert and below another chert member, the 'L' Bed. In places the top of the 'M' Bed is an oolitic limestone known as the Short Creek Oolite. Other beds are mineralized, particularly the G–H bed, and some production has come from two other calcarenite beds, the 'H' and the 'E'. In the Grand Falls Chert, are several thin limestone beds close together and separated by chert, which are mineralized and known locally as the 'Sheet Ground'.

Within each mineralized horizon the pattern is much the same. The ore is irregular and very variable in texture, from cavities lined with large crystals to fine-grained compact varieties. There is much evidence of solution and collapse and the ore-bed is generally thinner in the actual ore zone than away from it. Surrounding each ore zone is a rock known locally as jasperoid, a fine intergrowth of quartz and cherty silica with many quartz-lined cavities and a few stringers of mineralization. Normally, the jasperoid merges gradually into a chert breccia and eventually into unaltered cherty limestone. Within the ore are zones of coarse-grained grey dolomite containing little or no mineralization.

The best areas of ore in the field lie on the intersection of a gentle anticlinal arch striking north-west, and the Miami graben. In detail the best runs of ore are on slight upfolded sections of the ore-bed. There is much evidence of solution and circular areas up to 30 m in diameter have collapsed down into solution caves in lower formations (known as 'pipe slumps') and caves filled with upper Mississippian and Pennsylvanian strata are found in the upper parts of the Boone.

Most of the field is preserved from weathering by the Krebs Group shale but, in the small sub-outcrop area in the south, some calamine ore was found in the early days of mining.

Size and Grade

The whole field has produced 200 Mt of ore containing about 3·2% Zn and 0·8% Pb, with a small quantity of ore left. There are probably large tonnages of material of sub-economic grade remaining, particularly in the 'sheet ground' in the Grand Falls Chert.

Geological Interpretations

Most writers on the Tri-State district have held that the deposits are formed by the accumulation of mineralization from some ore solution into spaces formed by the dissolution of limestone confined between more impermeable beds of chert and shale. This solution process probably also caused the replacement of some of the limestone by the grey dolomite and the introduction or redistribution of silica, to form the jasperoid. That it was a low-temperature process is evident from the study of fluid inclusions in sphalerite and dolomite that indicate temperatures no higher than 120°C and as low as 75°C. Fluid inclusions also indicate that the ore solution was highly saline, so eliminating the early idea that the ore solution was normal groundwater circulating through the Boone Formation from the west. Most modern writers favour an origin of the ore solutions from magmatic activity in the basement, which gained access via the Miami fault zone, but they do not explain why the fault zone is so poorly mineralized, nor why the lower carbonate sequences below the Boone are less mineralized. Others have suggested that the permeable, leached rocks of the Boone Formation were originally an oil pool containing a lot of sour gas, and that the ore solutions were formation brines derived from evaporites, that leached metals from large volumes of rock, and deposited them on encountering hydrogen sulphide in the gas pools. Certainly the Picher Field looks as if it could have been a suitable place for a series of oil pools. But there are many difficulties with this interpretation, because evaporites close to the Boone Formation stratigraphically are a long way away geographically, and the chemistry of such a process is not compatible with the wholesale redistribution of silica. No help comes from attempts to determine the age of the mineralization for the reason that the lead-isotope ratios in galena are of the anomalous 'J' type indicating Pre-Cambrian ages.

Mining

The whole field was mined by open stoping with irregular pillars for roof support. This was necessitated by the nature of the ore, and partly by the complicated pattern of landownership which controlled the mining rights. In any case, it is doubtful even today if larger mining units could be profitable on such a patchy low-grade field. In the later years most ore was raised up shafts to one of a number of custom mills, which carried out grinding and selective flotation of lead and zinc concentrates.

Further Reading

LYDEN, J. P. (1950), Aspects of Structure and Mineralization used as guides in the development of the Picher Field. *A.M.I.E. Trans.*, **187**, 1251–1259.

McKNIGHT, et C. and FISCHEL, R. P. (1970), Geology and Ore Deposits of the Picher Field, Oklahoma and Kansas. *U.S. Geol. Surv. Prof. Paper* 588. [Probably the definitive work on the area and a very good summary, well-illustrated with full bibliography.]

OHLE, E. L. (1959), Some Considerations in Determining the Origin of Ore Deposits of the Mississippi Valley Type. *Econ. Geol.*, **54**, 769–89. [Of the many, this is probably one of the most sensible genetic essays on the subject.]

BASTIN, E. S. *et al.* (1939), Contributions to a Knowledge of Lead–Zinc Deposits of the Mississippi Valley Region. *Geol. Soc. America, Special Paper* 24, 156pp. [One of the best collections of fact and opinion on the area.]

WEIDMAN, S. *et al.* (1932), *Miami–Picher Zinc–Lead District*. Univ. of Oklahoma Press. 177pp. [Interesting for its historical summary and description of the mining written in the heyday of the Field.]

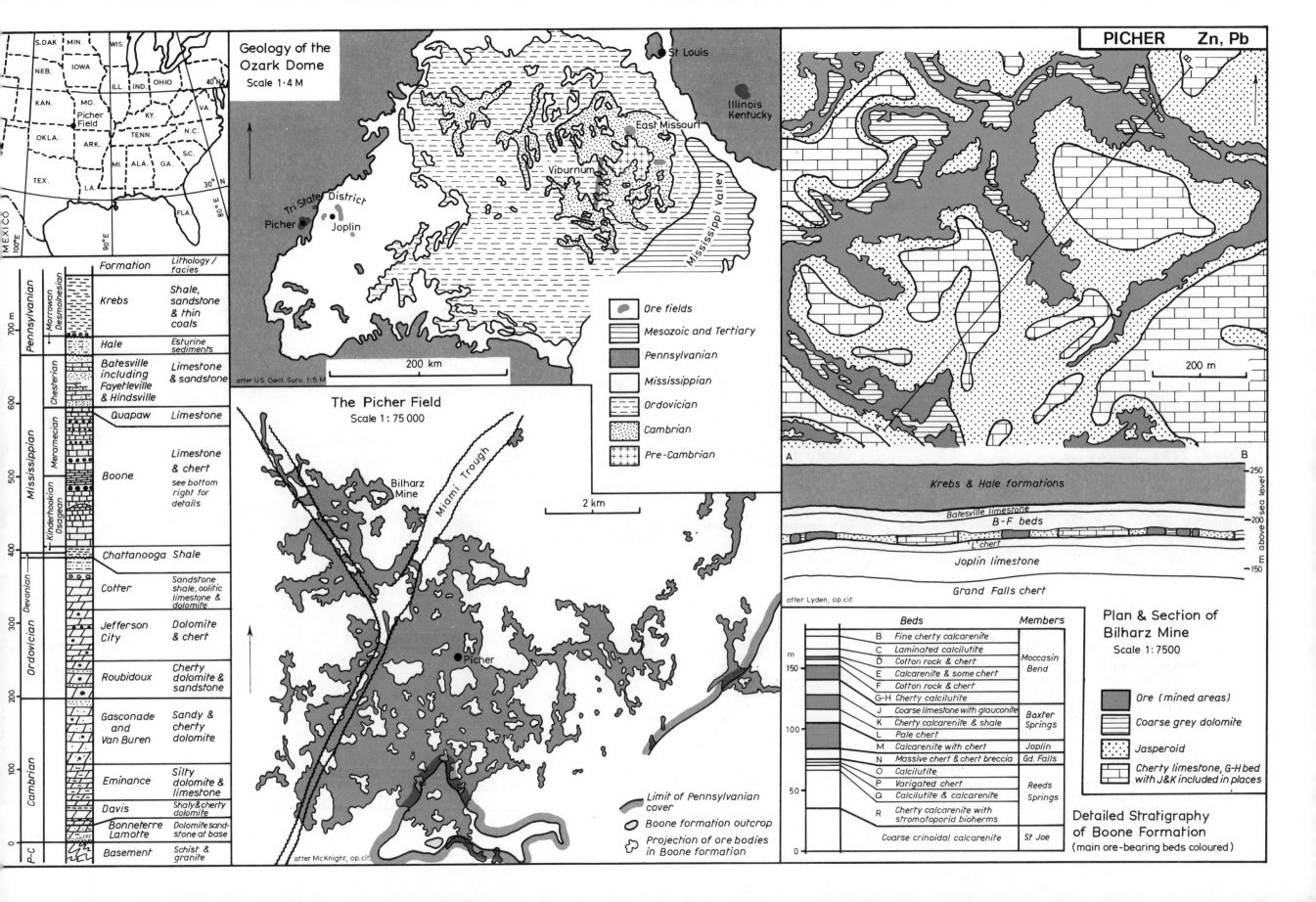

Geology of the Ozark Dome
Scale 1·4 M

after US Geol. Surv. 1:5 M

200 km

PICHER   Zn, Pb

200 m

- Ore fields
- Mesozoic and Tertiary
- Pennsylvanian
- Mississippian
- Ordovician
- Cambrian
- Pre-Cambrian

The Picher Field
Scale 1 : 75 000

Bilharz Mine

Miami Trough

Picher

2 km

after McKnight, op. cit.

Limit of Pennsylvanian cover

Boone formation outcrop

Projection of ore bodies in Boone formation

	Formation	Lithology / facies
Pennsylvanian — Morrowan Desmoinesian	Krebs	Shale, sandstone & thin coals
	Hale	Esturine sediments
Mississippian — Chesterian	Batesville including Fayetteville & Hindsville	Limestone & sandstone
Meramecian	Quapaw	Limestone
	Boone	Limestone & chert see bottom right for details
Kinderhookian Osagean		
Devonian	Chattanooga	Shale
Ordovician	Cotter	Sandstone shale, oolitic limestone & dolomite
	Jefferson City	Dolomite & chert
	Roubidoux	Cherty dolomite & sandstone
	Gasconade and Van Buren	Sandy & cherty dolomite
Cambrian	Eminance	Silty dolomite & limestone
	Davis	Shaly & cherty dolomite
	Bonneterre Lamotte	Dolomite sandstone at base
P-C	Basement	Schist & granite

Krebs & Hale formations

Batesville limestone

B-F beds

L' chert

Joplin limestone

Grand Falls chert

after Lyden, op. cit.

Plan & Section of Bilharz Mine
Scale 1 : 7500

- Ore (mined areas)
- Coarse grey dolomite
- Jasperoid
- Cherty limestone, G-H bed with J&K included in places

	Beds	Members
B	Fine cherty calcarenite	Moccasin Bend
C	Laminated calcilutite	
D	Cotton rock & chert	
E	Calcarenite & some chert	
F	Cotton rock & chert	
G-H	Cherty calcilutite	
J	Coarse limestone with glauconite	Baxter Springs
K	Cherty calcarenite & shale	
L	Pale chert	
M	Calcarenite with chert	Joplin
N	Massive chert & chert breccia	Gd. Falls
O	Calcilutite	
P	Varigated chert	Reeds Springs
Q	Calcilutite & calcarenite	
R	Cherty calcarenite with stromatoporoid bioherms	
	Coarse crinoidal calcarenite	St Joe

Detailed Stratigraphy of Boone Formation
(main ore-bearing beds coloured)

The Zinc, Lead and Barite Deposits of the Silvermines District – Ireland

This is the most varied of the mineralized districts in the Lower Carboniferous carbonate rocks of the Irish Midlands. Although the origin of the lead and zinc is unknown, it is clear that the minerals have accumulated in spaces in the rock created by a variety of diagenetic and tectonic processes. It is illustrative of a type of strata-bound lead–zinc deposit known from many localities, typically in the Mississippi Valley of America.

Location

The Silvermines district is 4 km long and 1 km wide, centred on latitude 52° 8′ N, longitude 8° 15′ W and about 160 m above sea level. The mineralized zones lie between the villages of Silvermines and Shallee, 8 km south-west of the town of Nenagh, County Tipperary, in the Republic of Ireland.

Geographical Setting

Wild moorlands rising above a rolling green country of sluggish rivers, lakes, peat bogs and meadows characterize the Irish Midlands, and Silvermines lies along an edge where such a moor descends from Shallee Hill (490 m) to the Kilmastulla Valley and Lough Derg. The landscape is lush and green, but the soils are thin or poorly drained and agriculture not very profitable, although Nenagh is a modest market town of 5000 people. The climate is very much influenced by the Atlantic; temperate, humid summers (12°–15°C) and winters mild and wet (5°C), rainfall totalling 1500 mm in a year. The district is served by a network of roads and the main railway from Dublin to Limerick passes by.

History

Outcropping mineralization south of Silvermines village was first recorded early in the seventeenth century, but little mining was done until the middle of the eighteenth century after the Geological Survey of Ireland reported various outcrops. Later in the century, and until the middle of the twentieth, lead, zinc and barite were worked south of Silvermines and Shallee villages and at Ballynoe. In the 1960s new companies came to the area and, with the help of geochemical and induced polarization surveys, and drilling, located and brought into production several new ore bodies in the poorly exposed ground between the villages.

Geological Setting

Folded and feebly metamorphosed clastic sediments of Lower Palaeozoic age underly the Irish Midlands, much eroded and covered by conglomerates, quartzites and shales of the Old Red Sandstone (Devonian). Beginning early in the Carboniferous, the whole area was covered by carbonate rocks of an arid, tropical shelf sea type. Among the varied lithologies found is the so-called 'reef' that consists of bank upon bank of micrites and calcarenites. Various types of mineralization, usually confined to one or a few horizons, occur characteristically in the dolomitized zones of the purer carbonate rocks, often in the so-called 'reef'.

There was some vulcanicity, notably from a large volcano at Pallas Green, south-east of Limerick. At the time when the much thicker carboniferous sequence in the south of Ireland was folded to form the northern flank of the Hercynian Mountains, the Midlands were gently folded and faulted; a south-westerly strike predominating. Although the evidence is fragmentary, the whole area was probably covered by Mesozoic sediments, but by the

Tertiary the carboniferous rocks were exposed and karst weathering developed; and during this time many dykes of dolerite were intruded, no doubt related to the great outpourings of basalt which dominate the Antrim Plateau in the north-east. During the Quaternary the area was covered by ice several times, so much so that the prevailing material at the surface over much of the area is glacial till or fluvioglacial sediment. In post-glacial times the flat low-lying topography and boreal climate caused large areas to be covered by thick peat bogs.

In spite of the till and bog cover, several mineral deposits are known, yielding lead, zinc, copper, barite with byproducts of silver, cadmium and mercury. The largest is a major lead–zinc deposit at Navan, 45 km northwest of Dublin, but Silvermines is interesting because of its variety.

Geology of the Silvermines District

The Silvermines fault zone runs east to west forming the northern boundary of an area of grey grits, shales of Silurian age with a cover of red sandstone, conglomerate and shale of the Devonian, Old Red Sandstone. North of the fault, and downfaulted against these rocks are the Lower Carboniferous carbonate rocks of the Kilmastulla syncline. The main mineralized horizon occurs 170 m above the base of the Carboniferous sequence, although mineralization also occurs further down and in the much faulted Old Red Sandstones. The rocks on the footwall of the main horizon exhibit a change from the sandy rocks at the transition from the Old Red Sandstone, through grey shaly rocks to nodular shaly limestones, a dolomitized calcarenite; and eventually to a richly fossiliferous, nodular limestone with whisps of shale, known locally as the 'Muddy Reef'. The main mineralized horizon is at the base of a unit of massive micrites and calcarenites of the 'reef' facies, which in the mineralized areas is largely converted to a dolomite breccia. In the central part of the district this is overlain by a very cherty limestone and a dark carbonaceous basinal limestone. The carboniferous rocks dip at about 20° to the north, steepening to 45° near the fault zone.

Geology of the Ore Deposits

The most important ore bodies consist of a series of zones where the ore horizon at the base of the dolomite breccia is thick enough and rich enough to be mined. In this horizon there is a pyrite–sphalerite–galena facies, a siderite and a barite facies. The two main zinc–lead bodies, the 'G' and 'B' are composed of beds of fairly massive pyrite displaying a variety of sedimentary and diagenetic textures with substantial amounts of sphalerite and galena. This is overlain by a more diffuse zone, in which the dolomite breccia is mineralized with sphalerite and galena. Siderite becomes more important than pyrite in the marginal parts of the 'B' ore body; up dip, it changes to massive bedded, nodular or brecciated barite, associated with a little jaspery hematite.

The mineable parts of the lead–zinc bodies average 6–9% zinc, 2–4% lead, with 15–40 g/t silver. The barite body contains 85–90% $BaSO_4$. The zinc–lead mineralization continues to the east where (south of Silvermines village) a small body occurs which was worked in the past and which differs from the others in that it had a capping of calamine ore (a mixture of zinc carbonates and silicates). Below the 'G' ore body is another, the lower 'G' ore body, which takes the form of a mineralized breccia in the Lower Dolomite. This ore differs from that of the main ore bodies in that it is richer in silver and is in a breccia that is clearly of much later formation. South of the main fault zone, sporadic mineralization is found in the Old Red Sandstone, particularly south of Shallee village where a series of NNW trending faults are mineralized with pyrite, sphalerite, galena and chalcopyrite. The faults of the main east–west zone are largely filled with barren clay gouge.

Size and Grade

The present-day workings of the zinc–lead ore bodies were started on reserves of 14 Mt of 2·8% Pb, 7·4% Zn and 27 g/t silver. The size of the Ballynoe barite body has never been published, but is thought to be over 5 Mt of 85% $BaSO_4$.

Geological Interpretations

It is probable that at least some of the mineralization in the main horizon was formed about the time of the sedimentation, particularly the barite, siderite and pyrite. The lead–zinc seems to have accumulated in rocks where sufficient permeability was created either by diagenetic dolomitization ('G' & 'B' ore bodies), tectonic brecciation (lower 'G'), or faulting (Shallee). It is also probable that the zinc and lead were transported in saline ground water, but opinion differs as to where it came from. Some have suggested diagenetic pore fluids of the carbonate rocks themselves, others by leaching of the Lower Palaeozoic shales; yet others suggest solutions associated with the Limerick volcanic complex. Some authors point to the main fault zone as a possible conduit for mineralizing solutions, others have suggested the permeable beds of the Old Red Sandstone may have played that role.

Mining

The zinc–lead ore is exploited by Mogul of Ireland Ltd. who operate an underground mine and flotation mill with a capacity of 2700 t/d delivering lead–silver and zinc concentrates by a spur, off the main railway to the Shannon estuary port of Foynes. The barite is exploited by Magcobar (Ireland) Ltd. who have an open pit mine producing 160 000 t of crude barite per year, which is shipped by the same route to a grinding plant at Foynes. Ballynoe is one of Europe's most important barite deposits, particularly because of its proximity to the oil-drilling operations of the north-west European continental shelf.

Further Reading

BARRETT, J. R. (1975), *Genesis of the Ballynoe Barite Deposit, Ireland and Other British Strata-bound Barite Deposits*. Ph.D. Thesis, University of London.

GRAHAM, R. D. F. (1970), *The Mogul Base Metal Deposits, Co. Tipperary, Ireland*. Ph.D. Thesis, University of Western Ontario. [These two theses are the most up-to-date descriptions of the district.]

MACDERMOT, C. V. and SEVASTOPULO, G. D. (1972), Upper Devonian and Lower Carboniferous Stratigraphical Setting of Irish Mineralization. *Geol. Surv. Ireland, Bull.* 1, 267–80. [A good summary of the host rocks of mineralization in the Irish Midlands.]

MORRISSEY, C. J., DAVIS, G. R. and STEED, G. M. (1971), Mineralization of the Lower Carboniferous of Central Ireland. *Trans. Inst. Min. Metall.*, 80, B173–B185. [A good summary of the mineralization itself with an evaluation of genetic ideas.]

RHODEN, N. H. (1959), Structure and Economic Mineralization of the Silvermines District, Co. Tipperary, Eire. *Trans. Inst. Min. Metall.*, 69, 67–94. [A timely and thorough description published just before renewed exploration began.]

WEBER, W. W. (1964), Modern Canadian Exploration Techniques reveal major Base Metal Occurrences at Silvermines, Co. Tipperary, Ireland. *Can. Min. J.*, 85, 54–7. [A record of the discoveries of the early sixties by the man who led the exploration team.]

WYNNE, A. B. (1861), *Mem. of the Geol. Surv. of Ireland*, sheet 134, Dublin, p.46. [A fascinating and informative account that makes one wonder why it took over a century to find the main ore bodies.]

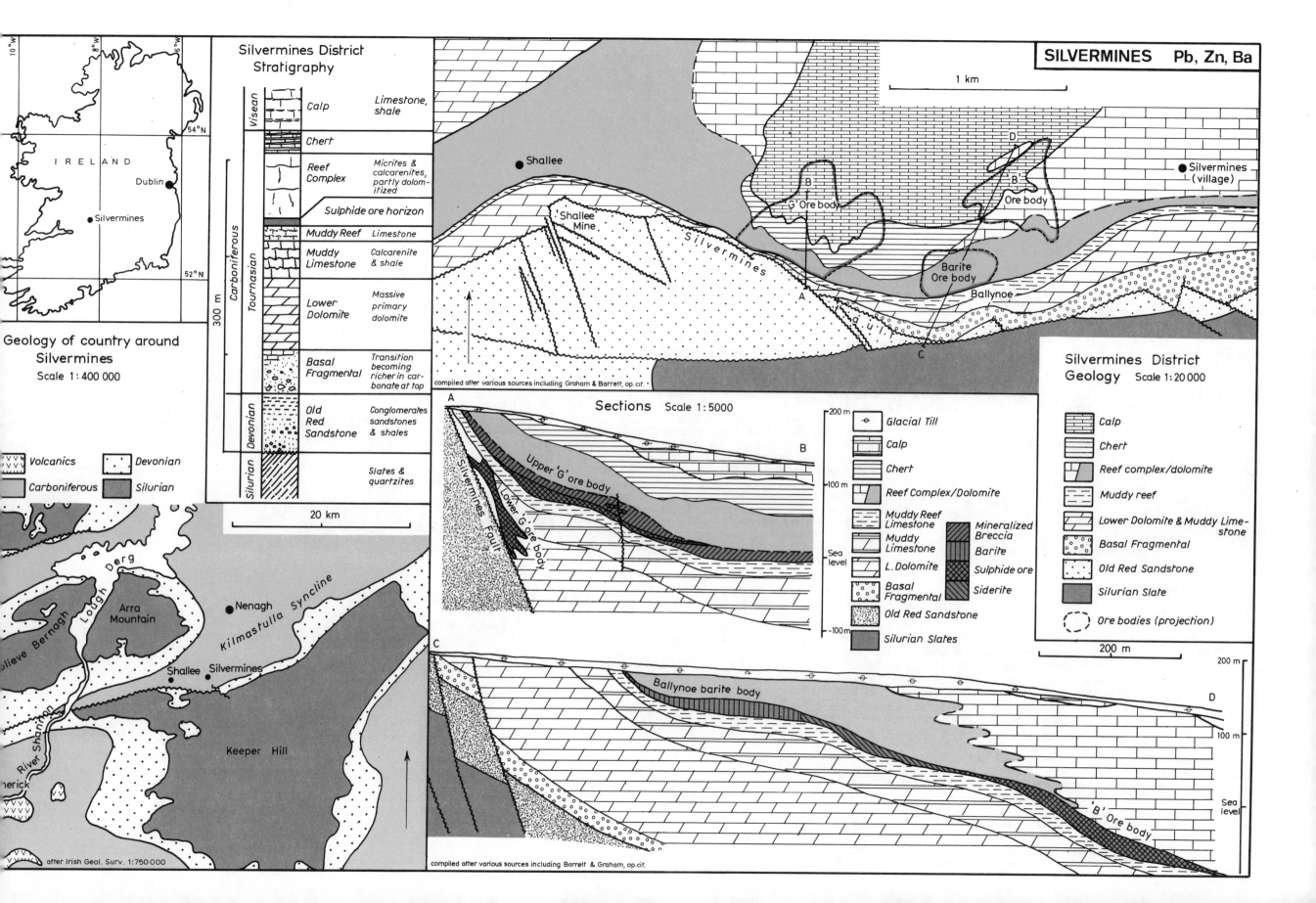

Geology of country around Silvermines
Scale 1 : 400 000

Silvermines District Stratigraphy

Carboniferous	Viséan		Calp	Limestone, shale
	Tournasian		Chert	
			Reef Complex	Micrites & calcarenites, partly dolomitized
			Sulphide ore horizon	
			Muddy Reef	Limestone
			Muddy Limestone	Calcarenite & shale
			Lower Dolomite	Massive primary dolomite
			Basal Fragmental	Transition becoming richer in carbonate at top
Devonian			Old Red Sandstone	Conglomerates sandstones & shales
Silurian				Slates & quartzites

Volcanics
Carboniferous
Devonian
Silurian

20 km

after Irish Geol. Surv. 1:750 000

SILVERMINES   Pb, Zn, Ba

1 km

compiled after various sources including Graham & Barrett, op.cit.

Sections   Scale 1:5000

Silvermines Fault
Upper 'G' ore body
Lower 'G' ore body

Ballynoe barite body

'B' Ore body

compiled after various sources including Barrett & Graham, op.cit.

Glacial Till
Calp
Chert
Reef Complex/Dolomite
Muddy Reef Limestone
Muddy Limestone
L. Dolomite
Basal Fragmental
Old Red Sandstone
Silurian Slates
Mineralized Breccia
Barite
Sulphide ore
Siderite

Silvermines District Geology   Scale 1:20 000

Calp
Chert
Reef complex/dolomite
Muddy reef
Lower Dolomite & Muddy Limestone
Basal Fragmental
Old Red Sandstone
Silurian Slate
Ore bodies (projection)

200 m

IRELAND
Dublin
Silvermines

Slieve Bernagh
Lough Derg
Arra Mountain
Nenagh
Kilmastulla Syncline
Shallee
Silvermines
River Shannon
Limerick
Keeper Hill

Shallee
Shallee Mine
Silvermines
'G' Ore body
Barite Ore body
Ballynoe
Silvermines (village)
'B' Ore body
fault

The Zinc–Lead Deposits of the Pine Point District – Canada

An example of a zinc–lead deposit in Palaeozoic carbonate rocks of which there are many in the central part of North America. It is particularly important to the understanding of this kind of mineralization because of its connection with petroleum and evaporite accumulations.

Location

Pine Point mining township is situated at latitude 60° 54′ N, longitude 114° 13′ W and 250 m altitude between the mouths of Peace and Buffalo Rivers on the southern shore of Great Slave Lake, in the MacKenzie District of Canada's North-west Territory.

Geographical Setting

The land is undulating glaciated, with many lakes, and is covered by thin coniferous boreal forest. It suffers a cold, dry continental climate with summer average temperatures of 15°C; −27°C in winter; and about 300 mm of precipitation as both snow and summer rain. Surface water is frozen for up to eight months in the year. Apart from the mine there is almost nothing at Pine Point and it is connected by road and rail to Hay River, a fishing and communications centre on the lake 105 km away. The well-known gold-mining town of Yellowknife is on the opposite side of the Lake.

History

Outcrops of limestone containing small amounts of galena on the shore of the Lake, reportedly known to the Indians, were claimed by prospectors *en route* from the Klondyke after the famous gold rush of 1896. Little was done with these claims until 1929 when they were taken under option by Cominco Ltd. The thick till cover, and the remote location hampered exploration for a long time until the advent of the induced polarization geophysical method, and the interest shown by the Canadian government in developing road and rail connections to the MacKenzie district. Several discoveries of ore were made in the early 1960s, and a mine was started in 1965 with a rail link to the then new Great Slave Lake Railway at Hay River.

Geological Setting

The Pine Point deposits occur in Middle Devonian rocks of the Elk Point Basin at its northern end. Gneisses, schists and granites of the Canadian Shield outcrop 100 km to the east, notable in this region for the Yellowknife gold veins that occur in Archean greenstones, and for a series of NE–SW fault zones, one of which (if projected below the Palaeozoic cover) lies underneath Pine point. The Elk Point Basin is large, extending from North Dakota in the U.S.A. to Great Slave Lake and continuing to the Arctic Ocean (although beyond the Lake it is known as the MacKenzie Basin). A few thin formations of Lower Palaeozoic rocks begin the sedimentary sequence in the Basin, but the main development was during the Devonian, continuing into the Mississippian. In the central and southern parts of the Basin this principal group is divided into the lower, Winnipegosis (carbonate) Formation, and the upper, Prairie (evaporite) Formation. The equivalent of the Winnipegosis in the Pine Point area is the Keg River dolomitic marl and limestone and a series of formations of variable facies. To the south, the Keg River is covered by the Muskeg Formation consisting of gypsum, anhydrite and shale, equivalent to the Prairie Formation. To the north occur limestone –shale sequences, and between the two is a very complex series of dolomitic

rocks known as the Presqu'ile Barrier Complex. The overlying rocks here lie on a palaeokarst surface and begin with green shales of the Watt Mountain Formation, the lateral equivalent of the Dawson Bay of the central part of the Basin. The rest of the Devonian sequence is largely of marine shales. The Presqu'ile Barrier Complex is known to continue for some distance to the south-west, underneath later cover. There were small amounts of sediment deposited during the early Mesozoic and a thick development of shales during the Cretaceous, after which time little happened until the Quaternary glaciation when the whole area was covered by ice several times, leaving the present topography and a great deal of till. However, weathering of the carbonate rocks has continued below the till so that sink holes are common.

In addition to the zinc–lead ores of Pine Point, the Presqu'ile Barrier is also host to the important Zame and Rainbow oil fields in northern Alberta.

Geology of the Pine Point Deposits

The district contains a number of ore bodies all confined to dolomitized rocks of the Pine Point Group, the local name for rocks of the Presqu'ile Barrier Complex. The base of the mineralized group is the Keg River Formation composed of dolomitic marls, thin limestone lenses and one persistent shale marker; the 'E' Shale. The top of the group is a somewhat irregular palaeokarst surface infilled by the waxy green shales of Watt Mountain Formation. Above this, and forming the base of the Slave Point Formation, is the Amco Shale, another persistent stratigraphic marker. The stratigraphy of the area is well-established, and the Pine Point Group is wholly contained within the Givetian Stage of the Middle Devonian. South of the mineralized area is the Muskeg Formation composed of tidal-flat anhydrites and, in the north, the Buffalo River, dark grey limestone–shale sequence occurs, passing down into bituminous limestones. Laterally between these two are a variety of coarse and fine dolomites containing enough undolomitized limestone residuals and relic textures to be understood as a reef complex. The rocks are richly fossiliferous with corals, stromatoporids, crinoids, brachiopods and gastropods. Dolomitization has yielded rocks that are mottled or zebra-textured, and some fairly persistent saccharoidal dolomitic horizons are found. The most spectacular rock is the so-called Presqu'ile dolomite, a very coarse cavernous dolomite which might originally have been a breccia, in which as much as 15% of the volume may be cavities lined with crystalline dolomite.

The ore minerals are sphalerite in crystalline and colloform varieties, galena, pyrite, marcasite, pyrrhotite with minor amounts of sulphur, gypsum and bitumen. There are two types of ore body. The largest, the so-called prismatic bodies, are cavity infillings of some vertical extent and show little relationship to the traces of bedding; the other, rather smaller, tabular bodies, are more conformable to the original bedding. The prismatic bodies are larger, richer and have a higher lead–zinc ratio. Some of the ore bodies are associated with solution collapse zones visible in the overlying drift.

Size and Grade

There are 40 known ore bodies in an area of 480 km², although some are too small to be worked economically. Those which had been evaluated up until 1971 totalled 40 Mt with an average of 6·0% zinc and 2·5% lead, but new reserves are continually being added.

Geological Interpretations

Oxygen and carbon isotope and fluid inclusion studies indicate a formation temperature of about 100°C and inclusions contain a rich chloride brine. Sulphur isotopes are consistent with derivation from sea water, and lead isotopes show model ages equivalent to simple development from the mantle in the Permian. Lead isotope interpretation is, however, notoriously difficult and contentious. Most authors accept that the ore was emplaced in spaces resulting from dolomitization, caused by magnesium-rich brines derived

from the Muskeg evaporites. Some have argued forcibly that the sulphur of the sulphide was formed by the reduction of sulphate, either as anhydrite or in solution, by hydrocarbons originally accumulated in the dolomites. Here, it is interesting to note that the high sulphur crude oil and sour gas of the Rainbow field occurs in an almost identical lithology to the ores of Pine Point. For the source of the metals there is far less agreement; some would have it leached from the marine shales of the MacKenzie Basin to the north by groundwater; others from the diagenesis of the carbonates in the Elk Point Group; yet others propose a deep source associated with the large fault zones in the underlying Pre-Cambrian Shield.

Mining

The deposits are mined by the Pine Point Mining Company managed by Cominco Ltd., who operate a series of open pits on the major ore bodies. Ore goes to a flotation mill capable of treating 3·5 Mt/a. Lead and zinc concentrates are transported via rail link to Hay River and the south.

Further Reading

BEALES, F. W. and JACKSON, S. A. (1966), Precipitation of Lead Zinc Ores in Carbonate Rocks as Illustrated by the Pine Point Ore Field, Canada. *Trans. Inst. Min. Metall.*, **75**, B278–B285. [The best account extant of the origin of the ore.]

CAMPBELL, N. (1967), Tectonic Reefs and Stratiform Lead–Zinc Deposits of the Pine Point Area, Canada. *Econ. Geol.* Monograph 3, pp.57–70. [The best account of the geology of the deposits.]

McCROSSAN, R. G. *et al.*, ed. (1964), *Geological History of Western Canada*. Alberta Society of Petroleum Geologists, Calgary, 232pp. [A large and comprehensive documentation containing a full description of the sedimentation in the Elk Point Basin.]

NORRIS, A. W. (1965), Stratigraphy of Middle Devonian and Older Palaeozoic Rocks of the Great Slave Lake Region, Northwest Territories. *Geol. Surv. Canada, Mem.*, **322**. [The standard work on the regional geology of the area.]

SKALL, H. (1975), Palaeo-environments of the Pine Point Lead–Zinc District *Econ. Geol.*, **70**, 22–47. [A full account of the facies variations in the Pine Point Group.]

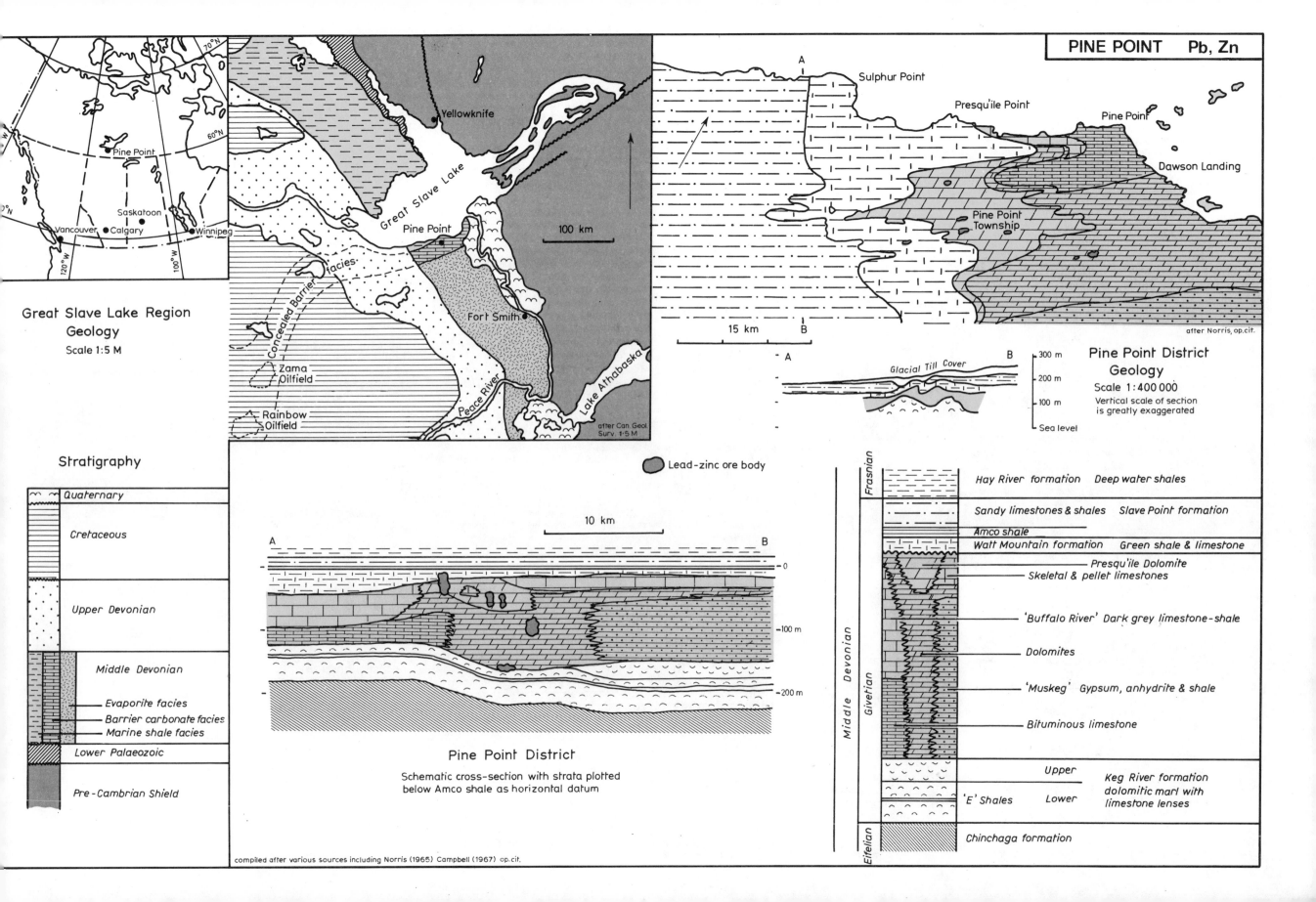

Great Slave Lake Region Geology

Scale 1:5 M

70° N

60° N

120° W   100° W

Vancouver   Calgary   Saskatoon   Winnipeg   Pine Point

Yellowknife

Great Slave Lake

Pine Point

100 km

Concealed Barrier facies

Fort Smith

Zama Oilfield

Rainbow Oilfield

Peace River

Lake Athabaska

after Can. Geol. Surv. 1:5 M

Sulphur Point

Presqu'ile Point

Pine Point

Dawson Landing

Pine Point Township

A

B

15 km

after Norris, op.cit.

A   B

Glacial Till Cover

300 m

200 m

100 m

Sea level

Pine Point District Geology

Scale 1:400 000

Vertical scale of section is greatly exaggerated

Lead-zinc ore body

Stratigraphy

Quaternary	
Cretaceous	
Upper Devonian	
Middle Devonian	
	Evaporite facies
	Barrier carbonate facies
	Marine shale facies
Lower Palaeozoic	
Pre-Cambrian Shield	

10 km

A   B

0

-100 m

-200 m

Pine Point District

Schematic cross-section with strata plotted below Amco shale as horizontal datum

compiled after various sources including Norris (1965) Campbell (1967) op.cit.

Frasnian		Hay River formation	Deep water shales
Middle Devonian	Givetian	Sandy limestones & shales	Slave Point formation
		Amco shale	
		Watt Mountain formation	Green shale & limestone
		Presqu'ile Dolomite	
		Skeletal & pellet limestones	
		'Buffalo River' Dark grey limestone-shale	
		Dolomites	
		'Muskeg' Gypsum, anhydrite & shale	
		Bituminous limestone	
	Upper	Keg River formation dolomitic marl with limestone lenses	
	Lower	'E' Shales	
Eifelian		Chinchaga formation	

The Sullivan Deposit – Canada

This is one of the largest known lead–zinc sulphide ore bodies. Despite having been discovered over eighty years ago and studied extensively by generations of geologists, its origins remain a mystery; being almost unique for its combination of discordant and concordant features, and its association with a monotonous sequence of muddy shallow water sediments. Its size and richness make it important not least to Cominco, which has grown to be one of the world's major mining companies primarily on the revenue of Sullivan ore.

Location

The deposit lies at latitude 49° 43′ N, longitude 160° 00′ W and at an altitude of 1400 m on the edge of the town of Kimberley in the Kootenay district of British Columbia.

Geographical Setting

Sullivan is in the Purcell Mountain Range which spans the U.S.A.–Canadian frontier just west of the highest part of the Rocky Mountains. The mine is only 15 km from the Rocky Mountain Trench, a major landmark which divides the steep rugged Rockies from the bold but gentler Purcells. The area is mountainous, rising up to 2600 m, interspersed with glacial valleys that fall below 1000 m. The climate is dry, being in the shadow of the Cordillera further west; annual precipitation is about 400 mm, about half as snow from November to April, and half as rain in summer. Winters are cold, −30°C not unusual; and summers hot, 38°C being common. Up to the tree line at 2000 m the land is good forest of red pine, larch and spruce, and was once a rich ground for hunting bear, caribou, deer, and trout in the rivers. The mine is on a mountain site at the head of a small tributary of the St. Mary River which flows eastwards to join the Kootenay River in the Rocky Mountain Trench.

The flattish floors of the glacial valleys form good agricultural land, particularly if irrigated, and timber is felled in the area. The mining town of Kimberley is quite large and is connected by road and rail to Cranbrook, 28 km to the SW and to the major industrial town of Trail.

History

Four prospectors from Cœur d'Alene in Idaho, Pat Sullivan, Ed Smith, John Cleaver and Walter Burchett, found the ore body at its outcrop in 1892 and staked claims on it named Hamlet and Shylock. They later sold out to a company who attempted to mine the ore in the early 1900s, but in those days zinc was an almost useless product and serious problems were encountered in getting a satisfactory lead output. However, in 1909, the company that eventually became Cominco began investigating the ore body and mining of the lead-rich parts of the body started in 1914, and over the years up to 1920 developed a concentrating plant capable of handling the complex mixed sulphide ore.

Geological Setting

The Sullivan deposit is situated in rocks of the famous Belt Super-Group, a sequence of late Pre-Cambrian sediments that outcrop from central Montana and Idaho over the frontier into British Columbia. The rocks are all fine-grained shallow-water clastic sediments devoid of fossils which are very difficult to correlate. In Canada the Belt is divided into the lower, Purcell and upper, Windermere Groups.

The massif formed of Belt rocks is bounded on the east by the Rocky Mountains, mainly composed of highly deformed Palaeozoic sediments; on the west by sediments, volcanics and granites of the Central Cordillera; to the south by the Idaho granite batholith; while in the south-west it disappears below the Columbia River Basalts. The massif contains two major mineral deposits; Sullivan and Cœur d'Alene in Idaho.

The Purcell Group is divided into a series of formations, of which the key one is the Aldridge in which the ore body occurs at the transition between its lower and middle members. One small granite cuts the Purcell rocks (dated at 800 Ma) which is regarded as about the age of deformation of the Purcell. The Purcell rocks have been metamorphosed to low greenschist facies, folded and faulted. Near the mine, the rocks form a broad anticlinorium and a series of north-easterly and northerly faults are important in that they define the block in which the ore body is found.

The Belt massif was covered at least in part by Palaeozoic sediments, some of which have been down-faulted into it, but it remained as a relatively rigid block during the folding of the Rocky Mountain area to the east.

The Purcell Mountains were covered by ice in the Quaternary, which receded slowly and irregularly, to leave a complex series of glacial deposits over the area.

Geology of the Sullivan Deposit

The deposit is a single complex ore body consisting of massive and banded sulphides with a gangue of minerals identical to those of the surrounding rocks. The main minerals are silver-bearing galena, dark sphalerite, pyrrhotite, pyrite and a number of others in smaller quantity, such as arsenopyrite, chalcopyrite, boulangerite, jamesonite, tetrahedrite and cassiterite.

The ore body is largely conformable with the host rocks and occurs at the boundary of the lower and middle members of the Aldridge Formation. The rocks in the footwall are thin-bedded impure quartzites with argillaceous partings, being conglomeratic at the immediate footwall of the ore zone. The hanging wall rocks are more arenaceous and thicker-bedded.

The ore body occurs in a faulted block of the Aldridge Formation rocks which includes diorites and related rocks, and the block is terminated on the north side by a major fault, the Kimberley Fault, that throws the Creston Formation against the Aldridge, a throw of at least 3000 m.

The ore body is in the form of a gentle anticline and is between 60 m and 90 m thick, and is known to underlie an area of 2·5 km². The upper part consists of rather massive ore with many offshoots into the walls, while down-dip the ore becomes banded with very fine lamina of sulphide and argillite, often showing very complicated folding, faulting and boudinage. There is a central iron-rich core where the dominant mineral is pyrrhotite surrounded by a zone of lead–silver rich ore with a lead–zinc ratio of 8 : 1. Beyond this, sphalerite becomes progressively more abundant so that at the margins the lead–zinc ratio becomes 1 : 6. The traces of arsenic tend to be concentrated in about the same zone as the lead-rich ore, and antimony is the converse. Tin, as cassiterite, is found as a patchy distribution around the central core and in a series of vein-type structures penetrating down into the footwall.

One of the remarkable features of Sullivan is the alteration of the wallrocks. Deep down below the ore zone the rocks are fractured, brecciated and indurated with tourmaline, giving the rock a chert-like nature; this is almost absent in the hanging wall where extensive albitization is found. Around the ore zone and in layers of rock within the ore zone, chloritization is common.

Size and Grade

Total recorded production and published reserves show over 170 Mt of ore with 9% combined lead and zinc; but since every year production continues, the published reserves show little sign of diminishing, it is fair to assume that the total size of the ore body is greater. The average grade is about 5% lead, 4% zinc and 55 g/t silver.

Geological Interpretations

There is still a wide divergence of opinion about the origin of Sullivan despite a long history of geological investigation. Isotope evidence suggests the deposit was formed 1340 Ma ago and that the sulphur was derived from deep crustal sources rather than from sea water. The age is consistent with formation at the time of sedimentation of the Aldridge Formation, from some deep-seated source; but all igneous rocks in the area are considerably younger. A part of the deposit has remarkable stratiform features but other parts show distinct cross-cutting relationships. The alteration and zoning is that normally found in deposits associated with igneous activity; and yet suitable parent intrusions are lacking and the formation of the finely banded ore by hydrothermal replacement is beyond the comprehension of many geologists. It has been suggested that it is a simple sedimentary deposit modified greatly during metamorphism, but this is an unsatisfactory compromise. It could possibly be an accumulation of sulphides from hot springs, but its occurrence in shallow water clastic sediments (a considerable way up a thick sequence, that once formed a continental edge) seems one of the least likely environments in which such activity might take place. Despite all, geologists who have worked on the mine seem to favour a hydrothermal replacement origin from some unknown deep-seated source.

Mining

Sullivan is mined by a division of Cominco Ltd. who have a substantial installation around the town of Kimberley. Ore is mined by several large-scale underground methods and brought to the surface via haulage adits. The ore is treated in a heavy media separation plant to remove waste rock, and then crushed and ground to remove the sulphides by selective flotation. The company has its own smelter in the area and produces 2·12 Mt of ore a year containing 9% combined lead and zinc with substantial amounts of silver.

Further Reading

FREEZE, A. C. (1966), On the Origin of the Sullivan Orebody, Kimberley B.C. In Tectonic History and Mineral Deposits of the Western Cordillera. Canadian Inst. Min. Met. Spec. 8, 263–94. [The most recent review of the origin and also contains a good summary of the relevant facts.]

RICE, H. N. A. (1937), Cranbrook Map Area, British Columbia Geol. Surv. Canada Mem. 207, 67pp. [General description of the area.]

Staff, Consolidated Mining and Smelting Company Ltd. (1924), The Development of the Sullivan Mine. Canadian Inst. Min. Met. Tr. 27, 401–65. [Fascinating historical account of the development of the mine.]

Staff, Consolidated Mining and Smelting Company Ltd. (1954), Geology (of the Sullivan Mine). In 'Cominco' – a Canadian enterprise. Canadian Min. J. 75 (5), 144–53. [A good account by the whole geological staff of the day.]

SWANSON, C. O. and GUNNING, H. C. (1945), Geology of the Sullivan Mine. Canadian Inst. Min. Met. Tr. 48, 645–67. [The standard work on the deposit but needs to be read in conjunction with Freeze, 1966 op. cit.]

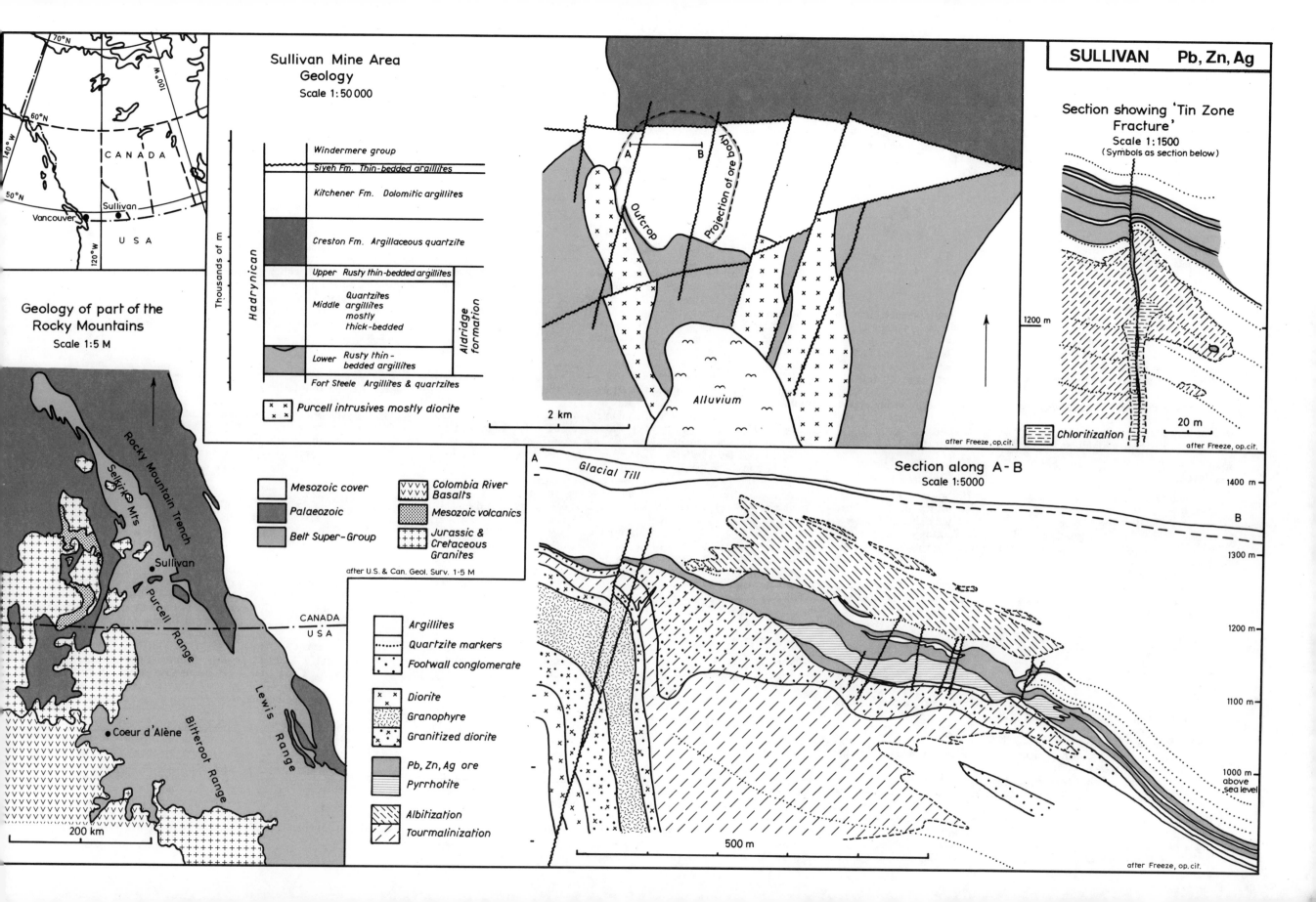

SULLIVAN   Pb, Zn, Ag

Sullivan Mine Area
Geology
Scale 1:50 000

Windermere group

Siyeh Fm. Thin-bedded argillites

Kitchener Fm. Dolomitic argillites

Creston Fm. Argillaceous quartzite

Upper Rusty thin-bedded argillites

Middle Quartzites argillites mostly thick-bedded

Lower Rusty thin-bedded argillites

Fort Steele Argillites & quartzites

Hadrynican

Aldridge formation

Thousands of m

x x / x x Purcell intrusives mostly diorite

2 km

after Freeze, op.cit.

Geology of part of the
Rocky Mountains
Scale 1:5 M

Selkirk Mts
Rocky Mountain Trench
Sullivan
Purcell Range
Coeur d'Alène
Bitteroot Range
Lewis Range

CANADA
USA

Mesozoic cover
Palaeozoic
Belt Super-Group

Colombia River Basalts
Mesozoic volcanics
Jurassic & Cretaceous Granites

after U.S. & Can. Geol. Surv. 1·5 M

200 km

Argillites
Quartzite markers
Footwall conglomerate
Diorite
Granophyre
Granitized diorite
Pb, Zn, Ag ore
Pyrrhotite
Albitization
Tourmalinization

Section showing 'Tin Zone
Fracture'
Scale 1:1500
(Symbols as section below)

1200 m

Chloritization

20 m

after Freeze, op.cit.

Section along A-B
Scale 1:5000

Glacial Till

A

B

1400 m

1300 m

1200 m

1100 m

1000 m
above
sea level

500 m

after Freeze, op.cit.

Outcrop
Projection of ore body
Alluvium

The Broken Hill Deposit – Australia

Among lead–zinc–silver deposits, Broken Hill is so large and important economically that it cannot be ignored, but is so complicated geologically, that interpretation of its origin and even its original geological environment is difficult. It is a very instructive example of the situation faced by geologists in trying to understand a deposit in which most of the evidence of its original nature has been destroyed by metamorphism and deformation.

Location

Broken Hill is situated at latitude 31° 58′ S, longitude 141° 28′ E and 300 m above sea level, in the extreme west of New South Wales, 440 km north-east of Adelaide.

Geographical Setting

Broken Hill lies on a rather flattish but relatively elevated area adjacent to the southern end of the Barrier Range. It is surrounded by large tracts covered by alluvium and sand. The climate is semi-arid and receives an average of 220 mm of rain a year, but very erratically, being twice as much in some years and considerably less in others. Temperatures as high as 46°C have been recorded and it has been known to freeze but, in general, it is hot for much of the year. The vegetation is sparse; scrub, a few trees and low-growing plants that grow profusely after rain. Some sheep ranching takes place in the area, but the mine is the main reason for the existence of the town of Broken Hill. It is connected by road and rail to Port Pirie, 350 km to the west on Spencer's Gulf, and to the cities of the east coast.

History

The original black gossanous outcrop of Broken Hill was noticed by Sturt's expedition in 1844, but no thought given to it for many years. Silver-rich lead ore was discovered some way away in 1876 and active prospecting began in the area. The outcrop was finally pegged in 1883 by Charles Rasp, an employee of a local sheep station and, after a year's exploratory mining, high-grade secondary silver was found 30 m below the outcrop. The first mine of the Broken Hill Proprietary Company started production in 1886 and exploration continued along the line of the outcrop and, by the early years of this century, several companies were active in the area. In 1936, the companies formed a Central Geological Survey and, despite the enormous amount of work done, the geology of the area is continually being reinterpreted.

Geological Setting

Much of the central part of Australia is covered by young sediments with exposed areas of underlying Pre-Cambrian Shield rocks. Broken Hill is situated in one such area called the Willyama. This area is bounded by younger sediments in the west and south-west. Two groups of metamorphic rocks are found; the upper Torowangee Group is of late Proterozoic age and consists in the main of low-grade metamorphosed clastic sediments; a major unconformity separates these from the Willyama metamorphic complex consisting of gneisses, amphibolites and schists.

The complex is divided into a north-western part, which consists of schists with some slates and a few small bodies of granite, and the main part composed almost entirely of gneisses, amphibolites and pegmatites. The structure is complex, several phases of folding having affected the area yielding a complex pattern of isoclinal and refolded folds, and a number of large faults and shear zones traverse the area. Radiometric dating suggests that the main deformation and metamorphism took place at 1·65 Ga before present, with a subordinate phase at 900 Ma simultaneous with the folding of the Torrowangee sediments. After this event the area became stabilized and eroded, and it is probable that the exposure and oxidation of the deposit took place some time during the Tertiary. A great deal of the area is covered by thin alluvium, most of which probably formed in a wetter climate than the present one.

Several types of mineralization are found in the area. The Broken Hill type of lenticular massive lead–zinc sulphide ore is the most important and occurs not only at Broken Hill but at the Pinnacles Mine and one or two other localities. Vein lead–zinc mineralization also occurs around the district, but these are generally small.

Geology of the Broken Hill Deposit

The deposit occurs in gneisses on the south-east side of the large Globe Vauxhall shear zone. Much of the host rock consists of banded sillimanite gneiss containing variable amounts of garnet, staurolite and other minerals. Within these can be distinguished a number of horizons of granular gneisses, amphibolites and the ore horizon. It is not easy to arrange these into a stratigraphy as a result of differing ideas on the structure. The ore horizon is about 30–60 m thick, and where the ore minerals are absent is characterized by the presence of bluish quartz, greenish feldspar, orange or pink garnet, gahnite and pyrrhotite. Sulphide bodies occur within this horizon, and several different ore bodies occur containing different proportions of the major ore minerals and different assemblages of gangue minerals.

The major ore minerals are silver-rich galena and iron-rich sphalerite, chalcopyrite and arsenopyrite and small traces of a large number of other sulphide minerals. Gangue minerals are quartz, calcite, rhodonite, manganhedenbergite and fluorite, in addition to the normal minerals of the ore horizon rock. Magnetite-quartz rock with some garnet, haematite, apatite and pyrrhotite (locally known as Banded Iron Formation) occur in persistent layers associated with the ore horizon. Zones rich in sericite are common largely in the numerous shear zones.

The structure and form of the ore bodies is very complicated. The series of ore lenses form tight anticlines and synclines on one flank of a large synclinorium, the ore lenses being on the scale of parasitic folds to the main structure. The crests and troughs of the folds plunge down to both the north-east and south-west, so that the ore bodies only outcrop at the central culmination. The structure is further complicated by a series of transverse shear zones that locally alter the direction of the fold-axes.

In the outcrop area the ore body was extensively oxidized, leaving a black manganese oxide-rich gossan on the surface, and considerable bodies of cerrusite, containing silver minerals, were mined in the early days from just below the gossan.

Size and Grade

Adding past production to present estimates of ore, it would seem that the total size of the deposit is well over 120 Mt, containing an average of 13% lead, 11% zinc and about 60 g/t of silver.

Geological Interpretations

The origin of the Broken Hill deposit has been a controversial subject for many years. For a long time it was (and by a few people still is) thought of as a hydrothermal replacement deposit. The idea was that some of the associated fault structures were conduits for solutions, presumed to be of igneous origin, that replaced certain favourable zones in the rock sequence after the main metamorphism. However, other geologists could not see how separate, and largely conformable, horizons of ore with contrasting mineralogies could be formed by such a process and were puzzled by the fate of the millions of tonnes of silicate rock that would be displaced. They held that the deposit was emplaced in sediments before deformation and folded with them. This idea became more tenable when it was shown that much of the evidence of such a deformation in the sulphide ore would be lost by 'annealing' after the event. It was also shown that the chemical composition of the gangue, when compared with that of the unmineralized 'ore horizon', was inconsistent with a replacement origin.

The nature of the host rocks has always posed a problem. The composition of the gneisses is consistent with an original sequence of clastic sediments, the more argillaceous rocks yielding the banded sillimanite gneisses, and the coarser-grained rocks, the more granular Potosi and Hangingwall ones. But, there are certain volcanic rocks that could equally well be precursors to the gneisses. The amphibolites seem most likely to have been basic igneous or volcanic rocks.

The deposit has been compared to those at Mt. Isa in Queensland and the MacArthur River, both broadly in the same group of Proterozoic rocks. The former is deformed but hardly metamorphosed and the latter relatively fresh. Broken Hill can be compared with the Sullivan deposit in Canada, and it may well be a series of lenticular deposits of the Sullivan type which have been metamorphosed and deformed. The only major difference is that the Broken Hill ore is associated with magnetite-quartz rocks that could well have originated as the type of jaspery exhalites commonly found capping the vulcanogenic sulphide deposits.

Broken Hill is instructive to exploration geologists; had it not been for its spectacular outcrop, who would have thought of looking for such a deposit in such a place?

Mining

Originally Broken Hill was mined by four companies, three of which were absorbed into the Conzinc Rio Tinto Group, with Broken Hill South remaining indepedent. Underground mining takes place along the length of the deposit, principally by horizontal cut-and-fill methods. Over 2·5 Mt of ore is produced per year. Ore is hosited via shafts to several selective flotation mills which send concentrates to the coast via the railway. Some of the ore is smelted and refined at Port Pirie.

Further Reading

GUSTAFSON, J. K., BURRELL, H. C. and GARRETTY, M. D. (1950), Geology of the Broken Hill Ore Deposit, Broken Hill, N.S.W. *Bull. Geol. Soc. Am.*, **61**, 1369–1414. [Perhaps the most complete description of the deposit, but based on work done in the late thirties, and many of the interpretations and particularly the genetic ideas lack the advantage of more recently available evidence.]

KING, H. F. and THOMPSON, B. P. (1953), The Geology of the Broken Hill District. In: *Geology of Australia Ore Deposits*, 1st edn (Edwards, A. B., ed.), pp.533–77, 5th Empire Min. Met. Congr., Melbourne.

KING, H. F., THOMPSON, B. P. and O'DRISCOLL, E. S. (1953), The Broken Hill Lode. In: *Geology of Australia Ore Deposits*, 1st edn (Edwards, A. B., ed.), pp.578–600, 5th Empire Min. Met. Congr., Melbourne. [These two papers set out the meta-stratiform interpretation. Several other useful papers on the deposit may be found in the same volume.]

LEWIS, B. R., FORWARD, P. S. and ROBERTS, J. B. (1965), Geology of the Broken Hill Lode, Reinterpreted. In: *Geology of Australia Ore Deposits*, 2nd edn (McAndrew, J., ed.), pp.319–32, 8th Commonwealth Min. Met. Congr., Melbourne.

CARRUTHERS, D. S. (1965), An Environmental View of Broken Hill Ore Occurrence. In: *Geology of Australia Ore Deposits*, 2nd edn (McAndrew, J., ed.), pp.339–51, 8th Commonwealth Min. Met. Congr., Melbourne. [These two papers are good reviews of the evidence and opinion of the day, putting alternative points of view on the origin.]

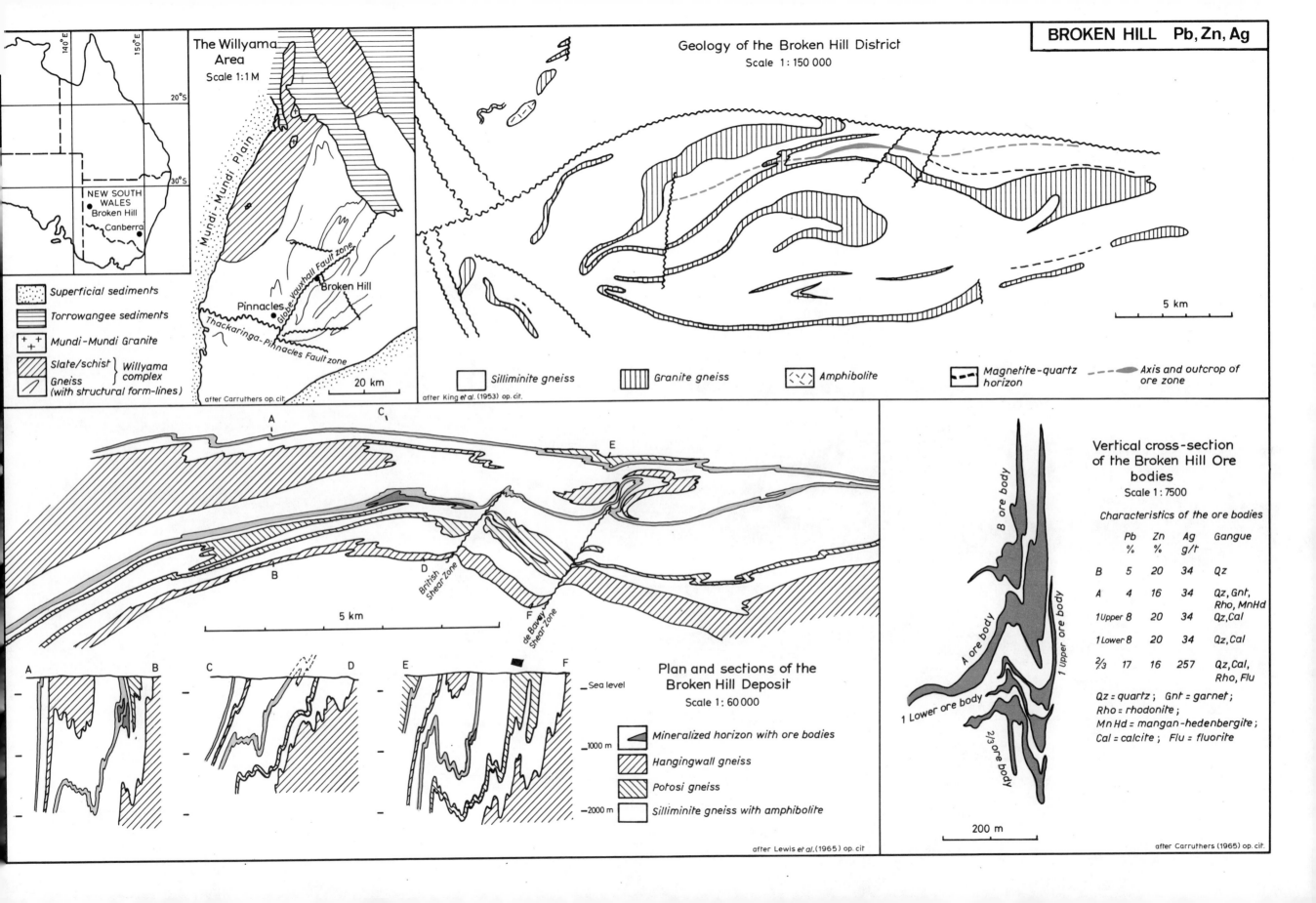

BROKEN HILL Pb, Zn, Ag

The Willyama Area
Scale 1:1 M

Mundi-Mundi Plain

Globe-Vauxhall Fault zone

Broken Hill

Pinnacles

Thackaringa-Pinnacles Fault zone

NEW SOUTH WALES
Broken Hill
Canberra

20°S
30°S
140°E
150°E

Superficial sediments

Torrowangee sediments

Mundi-Mundi Granite

Slate/schist } Willyama complex
Gneiss (with structural form-lines)

after Carruthers op. cit.

20 km

Geology of the Broken Hill District
Scale 1:150 000

5 km

Silliminite gneiss

Granite gneiss

Amphibolite

Magnetite-quartz horizon

Axis and outcrop of ore zone

after King et al. (1953) op. cit.

Vertical cross-section of the Broken Hill Ore bodies
Scale 1:7500

Characteristics of the ore bodies

B ore body
A ore body
1 Upper ore body
1 Lower ore body
2/3 ore body

	Pb %	Zn %	Ag g/t	Gangue
B	5	20	34	Qz
A	4	16	34	Qz, Gnt, Rho, MnHd
1 Upper	8	20	34	Qz, Cal
1 Lower	8	20	34	Qz, Cal
2/3	17	16	257	Qz, Cal, Rho, Flu

Qz = quartz; Gnt = garnet;
Rho = rhodonite;
Mn Hd = mangan-hedenbergite;
Cal = calcite ; Flu = fluorite

200 m

after Carruthers (1965) op. cit.

British Shear Zone

de Bavay Shear Zone

A B C D E F

Sea level
1000 m
2000 m

5 km

Plan and sections of the Broken Hill Deposit
Scale 1:60 000

Mineralized horizon with ore bodies

Hangingwall gneiss

Potosi gneiss

Silliminite gneiss with amphibolite

after Lewis et al. (1965) op. cit.

Section Three: Deposits Associated with Felsic Magmatic Environments

Introduction

A selection of mineral deposits will be dealt with in this section which are all associated with magmatic rocks or with geological environments in which magmatic processes played a part. For the most part the magmatic rocks concerned are felsic ones but, for convenience, certain deposits associated with more mafic rocks such as basalts, are included; the more important deposits associated with large bodies of mafic and ultramafic rocks will be described in Section V.

Many of the deposits included in this section are thought to have been formed by the crystallization of minerals from some aqueous solution generated directly or indirectly from magma. These so-called 'hydrothermal solutions' can escape to the land surface, or into the sea (and so produce mineral concentrations that have a syngenetic relationship to their host rocks), but they are also frequently channelled along fractures into spaces in host rocks and thereby precipitated epigenetically. The part played by the magma in generating hydrothermal solutions is not always clear. In some cases they can form by the unmixing or boiling of a hydrous phase from the magma itself. Most of the silicate minerals that crystallize from magma are anhydrous, and therefore any water present in the magma will concentrate as crystallization proceeds and a hydrothermal phase may develop as a late-stage phenomenon.

There is evidence in some deposits, based on the isotope ratios of oxygen and hydrogen, that at least a proportion of the water in hydrothermal solutions is meteoric in origin. There are two possible explanations of this; the first being that during its ascent through the crust a magma may encounter water-bearing strata and absorb groundwater. This will undoubtedly change the course of crystallization during the late stages, and this mechanism may account for the fact that some magmatic rocks are more mineralized than others and that the nature of the mineralization varies from one intrusion to another. The other possible mechanism is the circulation of meteoric water under the influence of a thermal gradient induced by the presence of the magma. It is possible in these circumstances for hot water to leach materials from large volumes of rock and deposit them in favourable structures at lower temperatures.

Mineralizing solutions probably vary in their composition and properties. Evidence from fluid inclusions shows that some of them are concentrated brines; and the study of the mineralogy of deposits and of associated alteration shows that some solutions contain other volatile materials such as carbon dioxide and fluorine. It is thought by some writers that metals are carried in hydrothermal solutions as polysulphide complexes, but other authors have suggested that they are carried as fluorides or chlorides. In most cases the available evidence shows that drop in temperature and pressure, or dilution by water of a different composition, are the main causes of precipitation, although in some cases it may be by reaction with some constituent of the host rock.

The detailed features of the mineral deposits that are formed by these processes depends to a large extent on the actual environment in which deposition takes place, so it is necessary first to review magmatic environments in general.

From the point of view of magmatic environments it is best to divide the earth's surface into a number of physiographic and tectonic domains somewhat different from those used in the discussion of sedimentary environments. Fundamentally there are oceans (areas of thin crust largely covered by water), and continents (areas of thick crust, only parts of which are covered by water). Below the water and thin layer of sediment in the oceans, is a layer 1 km to 3 km thick which has a seismic 'P'-wave velocity of about 5 km/s on top of a 4–5 km thick layer of denser material with a velocity of 6·6 km/s. The lower layer of continental crust has similar seismic characteristics to that of oceanic areas, but is generally 10 km–20 km thick and overlain by between 15 km–20 km of continental 'basement' with a 'P'-wave velocity of 6·1 km/s. This, in turn, is overlain in part by sedimentary or volcanic cover. Within oceanic areas are many islands, some of the most important and largest of which form long chains or arcs, and which have many of the geological features of continental areas but on a smaller scale.

In oceanic areas, magmatic activity takes place largely along the so-called mid-oceanic ridges where submarine volcanoes pour out large quantities of low-potash tholeitic magma, generally forming pillow lavas. Underneath sediment most of the ocean floor consists of this material. These lavas are often highly altered and permeable, and it is quite possible that magma feeding these eruptions can cause circulation of sea water through them. There are also chains of volcanic islands in the sea which usually include a wider range of rocks than is produced at the ridges. Rocks similar to those found on the ocean floor are found on a smaller scale in the marginal ocean basins that lie between island arcs and the main continental areas.

On the continents of today, most of the observable magmatic activity takes place along the edges, or associated with rift systems. There are three main types of continental edge, the passive, marked by degraded topography and accumulating off-shore sediment; the faulted, marked by large seismically active wrench faults; and the active ones marked by a 'cordillera' built up of volcanic rocks and immature sediments derived from them. Active continental edges are usually accompanied by an off-shore trench. The same configuration occurs along most of the island-arcs, and both are characterized by magmatic products of the calc-alkaline suit. The volcanic zones of cordillera and of island-arcs contain alkali-basalts and occasionally some tholeites, but are dominated by large thicknesses of andesite, dacite and rhyolite. Where deeply eroded, cordillera and island-arcs are found to have large masses of intrusive rock at deeper levels, often forming batholiths. These intrusions vary in composition from gabbro to granite, but their average composition is granodiorite.

The intrusive and volcanic rocks of these zones exhibit compositional polarity. Inland from the volcanic zones nearest the continental edge, there is a gradual decrease in the amount of magmatic rock, an increase in their potash–silica ratio, and progressive changes in the trace-element composition. In some island-arcs the compositional polarity is not so much spatial, as stratigraphic or temporal.

Both cordillera and island-arcs have a similar morphology and structure. Off-shore there is a trench usually containing a kilometre or more of sediment, further inland is a wedge of relatively immature sediment followed by a volcanic mountainland which gradually tails off to the interior. Beneath some of the younger zones of this kind are found the inclined plains of seismic epicentres, known as Benioff zones, that descend below the volcanic zone. These have been interpreted as zones along which oceanic crust is 'subducted' below the continent. As the oceanic crustal plate descends, it is heated by its contact with the hotter interior of the earth, and by friction. The sediment and volcanic layer of the upper oceanic crust carry water down, and this favours the partial melting of the oceanic crustal rocks, the underlying mantle and the overlying continental rocks. As the subduction proceeds, water is lost, rock is melted, and the composition of the material available for further melting changes (which presumably accounts for the compositional polarity). The subduction process generates bodies of calc-alkaline magma which rise as diapirs through the cordillera to form the volcanoes and plutons; the deeper the level from which the magma comes, the greater the chance it has to be contaminated by assimilation of continental crustal material on the way. At higher levels, the heat output of the magma may cause circulation of groundwater through previously formed piles of volcanic, pyroclastic and sedimentary rocks, and some of this water may be absorbed by the magma and alter the course of crystallization within it; explosive volcanic eruption being one of the more obvious results.

Within the continental areas, the most important magmatic activity is to be found associated with rift systems. Lines of small cones and fissure-eruptions producing the so-called 'flood basalts' are the commonest types, usually accompanied by a smaller proportion of very acid rocks (the so-called bi-modal suits): but where the activity continues for a long time, thereby producing large volcanoes, highly alkaline and undersaturated rocks may be produced. These rocks are thought to be generated by the partial melting of the upper mantle along zones of tension and high thermal flux accompanied by melting of upper crustal material.

The environments so far discussed can be observed today in relatively young crust, and by and large we can identify the same types in most of the tectonically mobile belts active during Phanerozoic times. Within these belts can be seen evidence of continental plate collision, where the oceanic crust has been totally subducted and two continental areas effectively welded together along a suture. Along these lines are found tectonically complex zones of metamorphic rocks (originally trench sediments) and so-called opholite complexes, which occur as thrust wedges in the metamorphic complexes; as obduction slabs, or as uplifted zones. If this interpretation is accepted, it means that the opholite complexes should represent sections of oceanic crust and it has been concluded, therefore, that such crust consists of the pelagic sediments and pillow lavas (known from ocean investigations), a lower dolerite dyke complex (feeders to the lavas) and an even lower layer of gabbro and peridotite intrusions. The study of uplifted and eroded parts of these suture zones provides evidence of the nature of the processes that take place during the subduction process.

As one progressively studies older and older rocks it becomes increasingly difficult to identify the same environments. This is not entirely due to depth of erosion because sedimentary and volcanic rocks originally formed at the surface are still preserved. In rocks that formed during the Proterozoic for instance, one finds very large mobile belts of continental scale of great structural complexity and including large areas of plutonic rocks, gneisses, migmatites and granulites, often associated with large layered basic intrusions. It is from these zones that good evidence comes, not only for the large-scale melting of felsic continental crust, but also of the transformation of supercrustal rocks to granitic-plutonic ones in the solid or semi-solid state; that is to say, granitization.

Further back in time in the Archean are found quite different rocks. The Archean shield areas are made up of large areas of granite and migmatite, with belts of supercrustal rocks known as 'greenstone belts' (because of their predominant degree of metamorphism). All that is preserved of these belts are synclinal 'keels', usually with a cuspate form surrounded by large circular granite plutons. The usual sequence of supercrustal rocks begins with tholeitic pillow-lavas containing small intrusions of gabbro and ultramafic rocks, both intrusive and extrusive. Some of the thicker volcanic sequences pass up into calc-alkaline suits with andisites, dacites and rhyolites, often dominated by pyroclastic rocks, displaying evidence of highly explosive eruption, and associated with immature sediments forming the characteristic lava–greywacke facies. At the top of the supercrustal pile are normally more mature sediments, usually separated from the lava–greywacke sequence by an important unconformity. Chemical sediments such as chert, or taconite, are very common throughout these rock groups. Plutonic rocks are of three kinds. First, there are large areas of gneisses and migmatites; then, round the edges of the greenstone belts, are found rounded plutons of granite normally foliated parallel to the contacts and displaying intrusive contacts towards the greenstones. (However, in rare cases, the greenstone may be seen to overlie gneisses.) The third kind displays smaller plutons of alkaline granite that cut the rocks of the greenstone belts. These usually have only narrow contact metamorphic aureoles, indicating high-level intrusion.

Greenstone belts have some similarities to more modern island-arcs, but some important differences. The supercrustal rocks seem to have formed on a thin granitic, primitive continental-type basement, which subsequently became extensively remobilized.

There are mineral deposits associated with all of the magmatic environments discussed above. One group dealt with in this section are the metal-sulphide deposits associated with various volcanic and volcano-sedimentary sequences. Most of these consist of bodies of massive, banded (or bedded) sulphides of iron, with smaller amounts of copper, lead, zinc and a host of minor elements. Many are associated with zones of disseminated mineralization, veinlet zones or stockworks, usually accompanied by alteration of the host rock. Many examples are found to have internal evidence of deposition of iron sulphide in some colloidal form and not a few contain so-called framboids and spherulites in their marginal zones, which look organic in origin. Usually, they are associated with silicious chemical sediments such as chert, jasper or umbers, often terminating, stratigraphically, the mineralized zone.

These deposits are found in several different kinds of volcanic environment. Quite a few are found on the association with explosive rhyolitic rocks, as for example, the Japanese 'Kuroko' deposits. Here the mineralization seems to have formed on the sea floor in pools between lava domes, the sulphide forming from volcanic exhalations preserved from the oxidizing effect of the sea by a layer of iron-oxide rich sediment, and later preserved from erosion by lava flows. These deposits seem to be characteristic of island-arc environments and are found throughout geological time (including the Archean). Closely related deposits occur in some of the greenstone belts, notably the Archean Abitibi Belt in Canada.

Somewhat similar deposits are found associated with rather less explosive sequences of rhyolites and covered by marine sediments, such as those of the Huelva province in Spain, and it may be that these were formed by a similar process in the quieter conditions of the marginal ocean basins (or at least on the side of an island-arc away from the trench). Yet others are found in association with basaltic rocks, usually tholeitic pillow lavas, and preserved by later generations of the same rock, or by pelagic sediments. It has been suggested that these form on the flanks of the mid-oceanic ridges, but they have never been seen in such an environment and only occur in what are interpreted as being thrust sheets, or obducted slabs, of original oceanic crust.

It now seems clear that the majority of the water involved in these mineralizing processes is meteoric, but the origins of metals and sulphur are less clear. The most favoured hypothesis is that the hot bodies of magma set up circulation systems of water in permeable, broken lavas or pyroclastics which leach out metals from the rocks, and these are then channelled into various structures that become feeders for mineralizing volcanic springs. Some of the minerals are deposited in altered rocks alongside the feeder-fractures to form the stockworks and disseminated zones; the rest flows out on to the sea floor to form the banded or bedded mineralization, but is only preserved if some subsequent process covers them over with lava or sediment.

Deposits formed by this mechanism are collectively known as volcanic exhalative, and at least one other type is essentially of the same origin. These are the so-called 'Algoma Type' iron deposits found in many of the Archean greenstone belts. They are sulphide, carbonate or oxide bedded iron-rich rocks associated with chert that occur in altered volcanic rocks. The cherts often contain evidence of life (some of the oldest signs of life on earth occur in these rocks), the remains of micro-organisms that abounded in the pools of warm water near the volcanic springs that provided the source of the mineralization. These iron-rich 'exhalites' quite frequently contain gold. Several important gold deposits may originally have formed from traces of gold of exhalative origin; although many of the examples of such gold deposits in rocks have been deformed and metamorphosed to such an extent that their exhalative origin is impossible to prove. The Homestake deposit is a good example.

Perhaps also of volcanic-exhalative origin are some mercury deposits like the famous one at Almaden in Spain. This is an impregnation of a permeable sandstone by mercury sulphide and metallic mercury, associated with some very altered tholeitic volcanic rocks. It would perhaps be more logical to call these 'volcanic sublimate' deposits. Such deposits are usually found in marginal ocean basin environments, or associated with terrestrial, intercratonic volcanic activity.

A somewhat specialized type of volcanogenic deposit is that of the Kiruna type. These are magnetite–apatite ores found above altered rhyolitic pyroclastics and preserved by overlying lavas. The discovery at El Laco in Chile of a lava flow of the same composition as the type example in the Swedish Pre-Cambrian confirms that these are volcanic but, so far, petrologists have not been able to suggest how such an ore-magma is generated.

The porphyry copper deposits are now well established as a type. They occur along the length of most of the younger cordillera and island-arcs, in calc-alkaline magmatic environments. Their essential feature is a zone of alteration, becoming more intensive towards the centre, that is associated with a stock of granodiorite or similar rock. When completely developed, the zones of alteration begin at the core with intense potash–metasomatism accompanied by some copper sulphide dissemination, surrounding which is an envelope of copper-iron sulphide mineralization in sericiltized rocks, surrounded in turn by pyritized and argilitized rocks which fade away into a wide zone of propilitization. Other types of alteration may be superimposed on these, such as kaolitization and anhydritization. In some examples these alteration features are confined to the intrusive; but generally they affect the wallrocks as well, which are, in many cases, volcanic rocks of similar composition to the intrusive stock. However, modification of the zonal pattern occurs when the wallrocks include sediments. The valuable minerals are copper sulphides, but many porphyry coppers contain subordinate (but useful) amounts of molybdenite or gold. On a regional scale the changes in mineral content parallel the compositional polarity of the magmatic rocks. The ore minerals occur in veinlets or disseminations for the most part in breccia zones, pipes, pebble dykes and similar structures, which are thought to be formed by sudden release of volatile material from the magma by boiling caused by sudden drop in pressure. Fracturing of the roof of the magma chamber is the probable cause of the pressure drop, and there is evidence that (at least in some cases) the porphyry copper stocks are the magma chambers feeding large strata-volcanoes, and thus the pulses of volatile release correspond to eruptions. It has become clear from the evidence of isotope studies that meteoric water plays a large part in the formation of the outer alteration zones. It is now thought that as the magma rises its heat begins to cause the circulation of water in the surrounding rocks, setting up a convection system. The zoning of alteration and mineralization is caused by the mixing of hydrothermal solution from the magma released by pressure drop, with the heated circulating groundwater. Although it has been suggested that the copper and other components of the mineralization are derived by wallrock leaching, most of the evidence points to an origin from the magma. The source of the metals in the magma may be either pelagic sediments, or oceanic lavas, which are partially melted during the subduction process. The occurrence of metals in sub-volcanic alteration zones may therefore depend on the original formation of submarine-volcanic, exhalative mineralization on an ocean ridge at some earlier date.

There are other kinds of mineral deposit found in high-level magmatic or sub-volcanic environments of the calc-alkaline suit. Along the American Cordillera, for example, are found many examples of vein gold–silver deposits and some tin deposits occur in South America in essentially similar environments.

More distantly related to these deposits are the hydrothermal veins associated with granites. Very often these are found in, or surrounding, cupolas at the top of large batholithic granite complexes, intruded into wedges of terrigenous sediment in fold belts. The typical hydrothermal vein deposits are fairly well-organized systems of veins in faults, filled with a great variety of oxides and sulphides of metals with quartz and carbonates. The minerals found are cassiterite, wolframite, uraninite, fluorite, gold and sulphides of copper, lead, zinc, silver and many more. The mineralization is usually zoned around the granite cupola, and wallrocks altered with the production of kaolinite, sericite, chlorite and tourmaline. Intense alteration of the granites can produce zones that are workable for their kaolinite content. Veins of this type are commonly crustified and have a range of textures that show evidence of the mineralization having been formed in a series of pulses of solution passing along fault planes. There is also evidence in the granites themselves of boiling of hydrothermal solution from the magma. The generation of hydrothermal solutions is usually thought to be due to the separation of a hydrous phase from the residual magma enriched in volatile and exotic constituents, but there is some evidence that at least a portion of the water is absorbed by the magma from the wallrocks, a circumstance which increases the probability of hydrous phase formation. It is perhaps significant that these deposits are often found in thick wedges of sediment, particularly those that form rapidly on active plate edges (or on passive ones that subsequently become activated) which are deposited wet and have little time to have the water expelled from them.

It has been suggested that granites do no more than supply the heat for the hydrothermal mineralizing process, and that the metals and other elements are leached from the surrounding sediments; but, although there does not seem to be much support for this interpretation in the case of the vein districts intimately associated with granite cupolas, it has some merit as a hypothesis for some other types of vein deposit in which the relationship to an intrusive is tenuous, or not observed. Two such groups of deposits are worth mentioning.

In thick sequences of terrigenous sediments, particularly those thought to have formed on passive continental margins, are found vein deposits. Carbonate-quartz veins are common, the carbonate commonly being manganiferous siderite. They are usually found in zones of tectonic dilation that are related to folding or faulting, and are frequently found in association with (but rarely in) large wrench faults or shear zones. Perhaps the best example of the type is the Cœur d'Alene district in the U.S.A. which, in addition to the veins of carbonate and quartz, contains sulphides of base metals, and above all, silver. In this district, intrusives are confined to one small part and, although some of the veins are closely associated with these granodiorite stocks, much of the richest mineralization is remote from them. The intrusives probably only represent evidence of a thermal source in the area. A mechanism that has been proposed for vein formation that may be applicable to districts like the Cœur d'Alene is as follows. It has been observed in some seismically active area that, immediately prior to an earthquake, springs dry up and then begin to flow again (sometimes in different places) after the shock. The explanation of this is that during the build-up of stress on a fault plane, the surrounding rocks are elastically dilated, and groundwater is absorbed into the spaces so formed. When the fault moves, producing the earthquake, the dilational fractures collapse, expelling the water along the fault plane. If the groundwater contains dissolved material and is warm or hot, leaching of the zone of dilation will occur, so generating a potential mineralizing solution. Repeated many times, this process can produce very large volumes of solution which can be channelled along faults into other more permanent dilation zones. This process, called 'seismic pumping' may account for the origin of several kinds of vein deposit, and although it does not require magmatic influence, it does seem likely that the solution and redeposition of minerals would take place more readily if the pumping occurred in an area subject to a high thermal gradient (related for instance to magmatic influence). Vein districts of several kinds including the Cœur d'Alene district could have been formed in this way, perhaps also the many uranium vein deposits that occur without close association to magmatic rocks.

The other major group of vein deposits that have only a tenuous connection with magmatic processes are those found in the Archean greenstone belts, carrying gold. Some of the richest gold deposits of the world (in grade if not in size) are of this type. They occur as veins, vein zones, mineralized shear zones, and a great variety of other forms containing native gold; sulphides such as pyrite, pyrrhotite, arsenopyrite and occasionally gold tellurides associated with quartz, carbonate and graphite. These deposits are often confined to particular lithologies in a given district, some occur in lavas, others occur in taconites or iron-rich sediments of exhalative origin. The actual gold veins are usually structurally controlled and are quite frequently associated with alteration that resembles retrograde metamorphism. Although frequently occurring in volcanic rocks and sometimes related spatially to intrusions, the relationship of these deposits to magmatic processes is more often than not, difficult to establish.

One possible mechanism for their formation is related to the metamorphism of the belts of rock in which they occur. During the tectonic events that formed the deeply downfolded synclinal 'keels' that are the characteristic form of the greenstone belts, the deeper and marginal parts of the belts were subjected to metamorphism of amphibolite facies. The mineralogical changes of the facies of metamorphism involve the expulsion of large volumes of water, and it is supposed that the water escapes through the overlying greenstone facies rocks, along fractures. Under the right conditions these metamorphic waters can leach and transport various materials from the rocks, in particular elements that fit less happily in the minerals of the amphibolite assemblage than in those of the greenschists. At lower temperatures, and in zones of tectonic dilation, these solutions may deposit their dissolved material and the water facilitates the retrograding of the surrounding rocks.

It may be postulated that vein and vein-like deposits are essentially hydrothermal and are formed from groundwater and formation water, that is either activated by compaction, tectonism, metamorphism or magmatic intrusion. The exotic materials carried in solution are derived from the leaching of sediments or volcanic rocks, or from magma by scavenging of the late-stage residual accumulations.

There are two other types of deposit associated with magmatic environments. The first one is the so-called contact replacement, or pyrometasomatic type of deposit. Many of these are found along the contacts of intrusions of the calc-alkaline suit (commonly components of large batholiths), particularly where high-level cupolas come into contact with carbonate rocks. Although such deposits contain a variety of minerals, they are most especially important today for scheelite. Their mode of formation seems quite simple. When the magma is intruded, the surrounding sediments are heated and contact metamorphic changes take place (in the case of carbonate host rocks producing the calc-silicate assemblage also known as 'skarn' or 'tactite'). As the magma crystallizes, the minor constituents tend to accumulate in the residual magma and some will diffuse into the contact zone, replacing minerals by others of modified composition; and these may include ore-minerals such as scheelite. Late-stage hydrothermal veins may be associated with these contact zones.

The final type of deposit associated with magmatic rocks not so far mentioned is the pegmatite type. These are found in (or associated with) many intrusive rocks. Typically they are zones of coarse-grained minerals, not much different from those of the associated rocks in composition, but very different in texture. Pegmatites are common 'hanging under the roof' of granite intrusions (or as pods in migmatites), but these are only of relatively minor economic importance, as sources of felspar or quartz crystal. Much more important are the large dykes or lenses of complex pegmatite that are found as peripheral intrusions in the wallrocks of large granites, especially the marginal granitic diapirs surrounding the Archean greenstone belts, and some of the larger granite bodies in the early Proterozoic metamorphic belts. The typical complex pegmatite is zoned, with an outer margin of a quartz-felspar mixture and a core of very complex mineralogy, that may include such exotic constituents as the lithum minerals, beryl, cassiterite, columbite-tantalite, and a host of others. The elements in such minerals are those found in small quantities in most granites, but in pegmatites they are concentrated. Many pegmatites display internal changes by replacement; commonly the partial replacement of potash felspar by albite.

Pegmatites are usually regarded as intrusions of residual magma. Large crystals do not normally grow in the rather viscous granitic magmas, and will only do so if the viscosity is lowered by the presence of water and other fluxing constituents. Since the minerals that crystallized from the magma are largely anhydrous, water will accumulate as crystallization proceeds, as will many other constituents that are not included in the crystal lattices of quartz and felspar. Given some tectonic disturbance towards the close of cooling, this residual can be intruded into the host rocks of the intrusive when it will crystallize to form a pegmatite, the water finally causing a number of hydrothermal-like changes towards the end. (Pegmatites often have quartz vein-like cores.) Pegmatites presumably form only where the magma contains water, but insufficient water to generate a distinct hydrothermal phase. So the relative production of pegmatites versus hydrothermal veins may be a matter of the presence of water in the wallrocks of the magma chamber and the ability of the magma to absorb it. The large size and richness in exotic constituents of the pegmatites associated with the marginal diapirs of greenstone belts, is probably due to the immense volume of sialic crystal rocks from which these exotic trace elements can be concentrated.

In summary, it may be said that the history of deposits of magmatic affiliation is also that of crustal water and the way it can circulate, become mineralized, and deposit its dissolved load. The examples which follow have been selected to display the range of mineral deposits that result.

Further Reading

CONDIE, K. C. (1976), *Plate Tectonics and Crustal Evolution*. Pergamon, 288pp. [Contains a very good modern review of magmatic processes and their relationship to global tectonic theories.]

The Helen Iron Deposit – Canada

Iron deposits in the Michipicoten area, of which this is the best known, were (largely because of their proximity to Lake Superior) the first to be developed in Canada. Although subsequently overshadowed by others, the Helen Mine remains in production. It is the best example of the so-called 'Algoma' type of iron deposit, found in various parts of the world, usually, like this one, in association with the acid volcanic rock sequences of the so-called Archean greenstone belts.

Location

The deposit lies at latitude 48° 02′ N, longitude 84° 45′ W and 384 m altitude, 12 km inland from the north-eastern shore of Lake Superior in the Algoma Province of Canada.

Geographical Setting

The land rises steeply from Lake Superior (183 m) over a rocky landscape down which the Magpie, Dove and Michipicoten rivers flow in a series of rapids and waterfalls. After a few kilometres the land levels off on to a rugged area of ridges, swamps and lake-filled valleys; the lower land being about 380 m and the hilltops exceeding 500 m above sea level. The mine is on the side of a prominent ridge 518 m high, formed by the outcrop of the Helen Formation. Originally the area was mixed boreal forest (largely of spruce and birch) in which deer, caribou and lake trout lived in abundance. But much of the forest was felled or burnt in a series of forest fires (following the building of the railways at the turn of the century). The climate is cold continental, with fair snowfalls in winter, and hot summers. The growing season is too short and the soils too thin for serious agriculture. Settlement in the area is confined to mining townships, lumber camps, and Indian reservations, but a substantial town has grown up around Michipicoten Harbour on the Lake. From here a railway climbs up and over the area to serve the mines, and joins the CPR further north. The high potential of the rivers, as they descend to the Lake, has been largely harnessed for electricity generation.

History

The first claim on the Helen property staked by Alois Goetz in 1898 was for gold, which at the time was being found in the area. However, even as early as 1866 soft brown iron-ore was being exploited down by the lake shore. In 1900, the outcropping brown ore at Helen was produced and continued to be so until the end of the First World War, when it was all but exhausted. The deeper zones, lower in grade but vastly larger, having been explored by diamond drilling, began to be mined, and production has continued to this day.

Geological Setting

The Superior Province of the Canadian Shield is an area that was tectonically stabilized at the end of a series of events about 2·5 Ga ago. It is a vast area, over half of which is a complex of granitic and gneissic rocks. Interspersed between these are the so-called greenstone belts, among which the Michipicoten belt is one of the smaller, being about 120 km long, and a 'Y' shape, with a maximum width of 50 km. It is only a fraction of the size of the great and richly mineralized Abitibi belt further east.

A general stratigraphy is recognizable, beginning with a lower group of basic and andesitic lavas, many showing pillow structure, passing up in places into rhyolite or dacite lava domes and pyroclastic rocks. Above this is the Helen Formation, composed of iron-rich sediments that can be separated into three facies; oxide, carbonate and sulphide. The dominating lithology of the first two of these is iron-rich chert.

Above the Helen Formation there is in places a second andesite lava sequence followed by the Dove Formation, consisting of conglomerate, conglomeratic greywacke, greywacke and argillite. Further basic and andesitic lavas cap the sediments. Intrusive rocks are found, many being small irregular bodies of meta-gabbro or meta-diorite which are probably co-magmatic with the volcanic sequences. There are dolerite dykes of much later age and small bosses of granite. The whole belt is surrounded by granitic rocks the contacts with which are never sharp, there being·a transitional zone of dykes and assimilation. One of the persistent features of the greenstone belts is that the surrounding granitic rocks seem to intrude the supercrustal rocks of the belt at the contact, whereas morphologically, the granitic rocks look like a basement to them.

The Michipicoten Belt was strongly folded roughly along a NE–SW axis and a strong planar fabric imposed. The volcanic rocks in particular show metamorphism of greenschist facies.

The Helen Formation is cut in places by quartz veins that carry gold, and a few mines have produced in the area. Quaternary glaciation was a major factor in forming the present landscape, but the hardness of some of the rocks and the ridges formed as a result has prevented widespread coverage with till.

Geology of the Helen Deposit

The deposit as mined is the lower part of the Helen Formation on one overturned limb of a fold that has been faulted into an isolated position. The deposit is tabular, dipping 75° to the south. Stratigraphically below the deposit (albeit on the hanging wall of the mine) are a series of rhyolitic and dacitic pyroclastic rocks with some recognizable zones of porphyritic rhyolite lava. This sequence is cut by bodies of meta-diorite. As the ore zone is approached, these volcanic rocks become highly altered with reduction of silica and iron and increase in carbonate.

The ore body contact is fairly sharp, beginning with a discontinuous, up to 9 m thick, bed of ferruginous chert that passes gradually into a massive, fine-grained, granular siderite rock between 30–100 m thick, which contains about 30–35% iron. This, which is the main ore, contains some pyritic shale bands, small amounts of fine silica and iron-silicate minerals. Above the main ore, sometimes separated by lenses of granular silica rock, is a coarser-grained rock which is a mixture of pyrite and siderite, the pyrite becoming more abundant at the top where it passes into a banded chert member. This is the thickest member of the formation, but is not ore. It consists of thin (up to 30 mm) bands of partially recrystallized chert with siderite, magnetite, iron silicates and narrow bands of graphitic schist. It contains about 30% iron at its base, progressively reducing to 10% at the top.

The footwall of the deposit is this chert member, but beyond it, higher up the succession, are andesites, mostly pillow lavas. The deposit is interrupted by a few faults and dolerite dykes. The original outcrop of the ore body, most of which was under Boyer Lake, was a brown hematite–goethite ore much higher in grade and purity than the underlying siderite ore. In the sandy material associated with it were fair amounts of granular pyrite and silica.

Size and Grade

The ore body is selectively mined to avoid the worst of the pyrite and chert, to produce ore containing about 33% iron, with about 8% silica, 1·5% sulphur, 0·01% phosphorus and 2% manganese. In the period after the First World War the mine was established on a reserve of 70 Mt, but although published figures are not available, the total amount of ore down to the bottom of the present mine is probably four times as large.

Geological Interpretations

The great thicknesses of cherty iron-rich sediments in Pre-Cambrian rocks has always been a puzzle to geologists; but largely as a result of the study of the many good examples in Canada and the adjoining parts of the United States of America, they can be divided into recognizable types. The so-called 'Algoma' type is characterized by association with volcanic rocks and is common in Archean terrains. The generally accepted model for the formation of this type is that they were sediments laid down in a shallow shelf sea at the dying stages of explosive volcanism, when fumurolic gases leached iron and silica from the volcanic rocks and released them into the sea. Thereafter, a sequence of chemical processes took place (as the fumurolic solutions gradually mixed with sea water) that formed the three facies of iron-rich sediment in a vertical sequence and, with some modification, in horizontal sense also. The environment, as shown by the rock sequence, is reminiscent of that found in present-day island-arcs, and the reason why the latter do not seem to contain this kind of iron deposit is a puzzle. It has been attributed to differences in the composition of the Archean sea water, the primitive state of organisms of the time, (organisms play an important role in the chemical sedimentation of both iron and silica) and to geochemical differences in volcanic rocks of the Archean compared with the present. Simple oxidation, hydration and leaching of silica is presumed to be the explanation of the brown ore at the outcrop.

Mining

The MacLeod Mine (formerly the Helen or New Helen Mine) is operated by the Algoma Steel Corporation, which also has an iron and steel-making complex at Sault Sainte Marie on the Canadian side of the outlet of Lake Superior. The mining is underground via several shafts that reach 600 m below surface. The ore at 33% iron is treated in a heavy media plant to remove low-density siliceous waste, and then roasted and sintered to produce a magnetite-rich material containing 50% iron. Together with ore from Steep Rock Lake Mine at the other end of Lake Superior, a portion of this roasted ore is now pelletized. The plant and mine has a capacity of 8000 t/d.

Further Reading

COLLINS, W. M., QUIRKE, T. T. and THOMSON, E. (1926), The Michipicoten Iron Ranges. *Geol. Surv. Can., Mem.*, **147**, 175pp. [Somewhat old, but the standard work on the area.]

GOODWIN, A. M. (1962), Structure, Stratigraphy and Origin of Iron Formations, Michipicoten Area, Algoma District, Ontario, Canada. *Geol. Soc. Am. Bull.*, **73**, 561–86. [Best description of the whole area.]

GOODWIN, A. M. (1973), Archean Volcanogenic Iron Formation of the Canadian Shield. In: UNESCO. *Genesis of Pre-Cambrian Iron and Manganese Deposits*. Proc. Kiev Symp. 1970 (Earth Sciences 9). [Covers similar ground to 1962 paper, but more philosophical.]

LANG, A. H., GOODWIN, A. M. *et al.* (1970), Economic Minerals of Canadian Shield. In: *Geology and Economic Minerals of Canada*. (Douglas, R. J. W. ed.), Geol. Surv. Canada, Econ. Geol. Rept. I, pp.153–226. [Contains a good summary of the problems of Pre-Cambrian iron ores and some details about the Algoma type, see pp.170–8.]

Staff, Algoma Ore Properties (1956), Algoma Ore Properties, Part II – Helen Mine Geology. *Can. Min. J.*, **77** (11), 80–7. [The best description of the mine itself.]

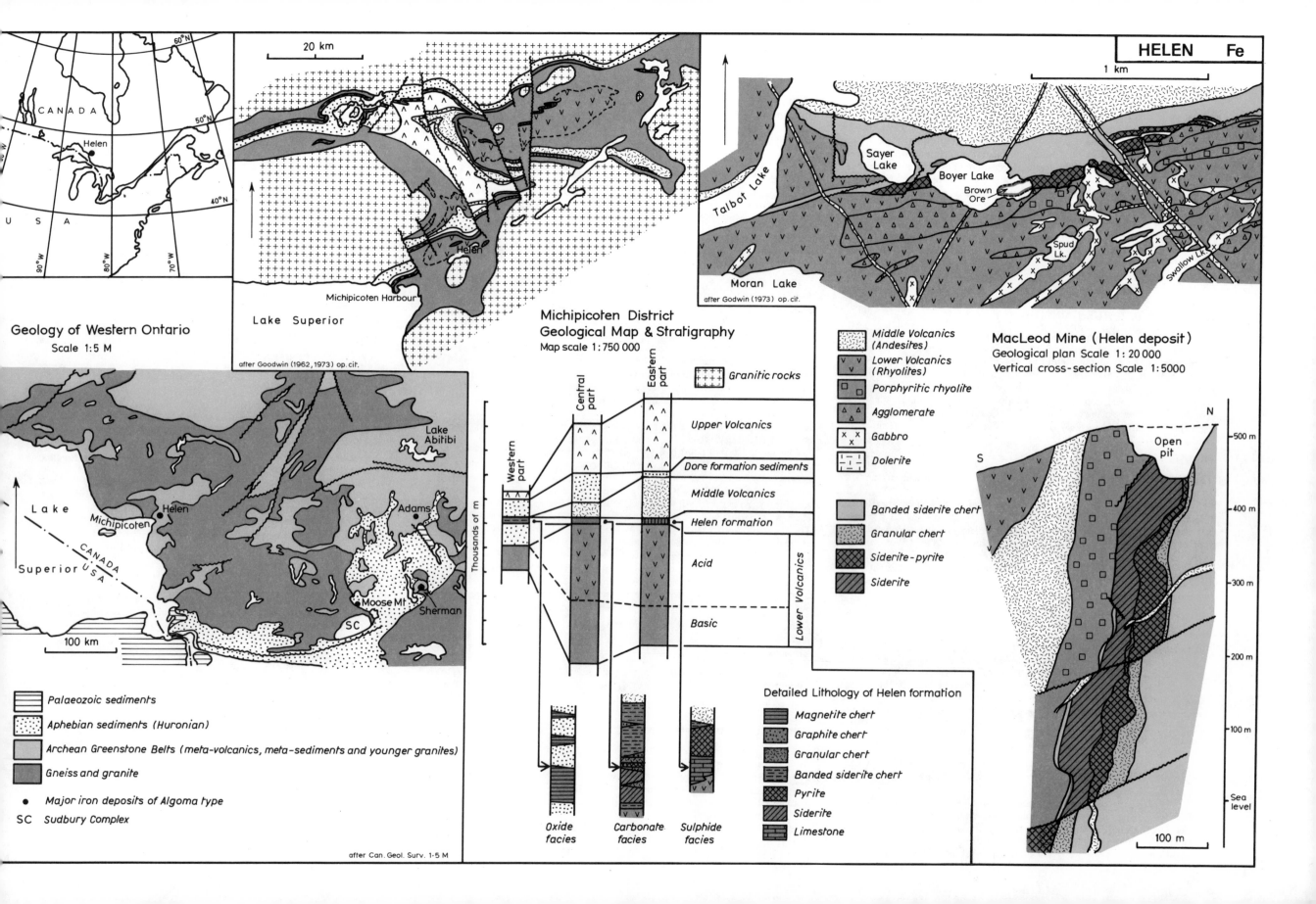

Geology of Western Ontario
Scale 1:5 M

after Goodwin (1962, 1973) op. cit.

Lake Superior

Michipicoten Harbour

20 km

CANADA

USA

Helen

60°N

50°N

40°N

90°W   80°W   70°W

HELEN   Fe

1 km

Talbot Lake

Sayer Lake

Boyer Lake

Brown Ore

Moran Lake

Spud Lk.

Swallow Lk.

after Godwin (1973) op. cit.

Michipicoten District
Geological Map & Stratigraphy
Map scale 1:750 000

Lake Superior

Helen

Michipicoten

CANADA
USA

Adams

Moose Mt

Sherman

SC

100 km

Palaeozoic sediments

Aphebian sediments (Huronian)

Archean Greenstone Belts (meta-volcanics, meta-sediments and younger granites)

Gneiss and granite

● Major iron deposits of Algoma type

SC Sudbury Complex

after Can. Geol. Surv. 1·5 M

Granitic rocks

Western part

Central part

Eastern part

Upper Volcanics

Dore formation sediments

Middle Volcanics

Helen formation

Acid

Basic

Lower Volcanics

Thousands of m

Oxide facies

Carbonate facies

Sulphide facies

Middle Volcanics (Andesites)

Lower Volcanics (Rhyolites)

Porphyritic rhyolite

Agglomerate

Gabbro

Dolerite

Banded siderite chert

Granular chert

Siderite-pyrite

Siderite

Detailed Lithology of Helen formation

Magnetite chert

Graphite chert

Granular chert

Banded siderite chert

Pyrite

Siderite

Limestone

MacLeod Mine (Helen deposit)
Geological plan Scale 1:20 000
Vertical cross-section Scale 1:5000

N

S

Open pit

500 m

400 m

300 m

200 m

100 m

Sea level

100 m

The Pyritic Deposits of the Tamasos Field – Cyprus

One of the difficulties that arises in the study of massive sulphide deposits is that they are frequently affected by post-depositional processes (see for example Noranda, Skorovas or Broken Hill). The interpretaton of the origin of these deposits depends very much on examples that are in a relatively fresh state; in the case of copper-pyrite deposits the best examples are those found on the flanks of the Trudos Mountains. There are other examples of this type of deposit, at Küre and Ergani in Turkey, and at Betts Cove at Halls Bay in Newfoundland.

Location

The Tamasos mining field extends over about 18 km² around the village of Mitsero (latitude 35° 02′ N, longitude 33° 08′ E and 400 m altitude), 26 km south-west of Nicosia.

Geographical Setting

The Tamasos Field lies in hilly country that separates the Trudos Mountain Range from the Plain of Mescoria. The mountains rise to a height of 1953 m while most of the plains are below 150 m. The plains are relatively dry but can be cultivated, particularly where irrigated; the mountains are largely forested with pines and oaks. The climate is Mediterranean with a wet, mild winter, and a hot, dry summer. In the mountains, snow remains for several months of the year. At the northern end of the mining district, there is a rail-head that is connected to Nicosia and the island's largest port, Famagusta.

History

The first written reference to the mining of these deposits is in Homer. The Iliad records copper mining from the Kingdom of Tamasos from which the field gets its name. The Skouriotissa mines were described by the Roman, Galen, in A.D. 166. From the end of Roman activity in the fourth century A.D. the mines became disused until 1882 when Limni was re-explored by an Englishman, John Pearce. From the time of the First World War onwards, there was an increase in exploration by American, British and Greek companies. The ancient mines of the Tamasos Field were explored in the 1930s by the Hellenic Mining Company, and new ore bodies, beneath the Roman workings, were found by drilling in 1951.

Geological Setting

There are three major units to the geology of the western part of Cyprus; the Trudos Complex, its sedimentary cover, and a thrust-sheet around Mamonia on the south-western flank of the mountains. The Trudos Complex has been divided into three parts. In the centre is the Trudos Plutonic Complex, composed of dunite, peridotite, with lesser amounts of gabbro. Surrounding this is an extensive area known as the Sheeted Complex, most of which consists of dykes of dolerite; dominantly low-potash tholeiites. Overlying the Sheeted Complex are pillow lavas, about 1400 m thick and cut by many dykes of dolerite and basalt which become less numerous towards the top of the sequence. The Basal and Lower Pillow Lavas are low potash tholeiites and the Upper group tend to be alkali-olivine basalts with some limbergites.

The sedimentary cover of the Trudos Complex begins with a thin and discontinuous unit of brown iron–manganese-rich shales known locally as 'umber', containing a micro-fauna of Campanjan age, and named the Perapedhi Formation. There follows a continuous group of carbonate sediments, the Lefkara Group, consisting of marl and cherty chalk ranging

in age from Maestrictian to Lower Miocene. Carbonate sediments continue with the Middle Miocene Pkhna Formation and, eventually, continental clastic sediments up to Recent age.

Copper-bearing pyrite deposits occur at several levels in the pillow lavas. The Skouriotissa ore body is in the Upper Pillow Lava Group and is capped by the Perapedhi Formation, but the ore bodies of the Tamasos Field are situated at the top of the Lower and Basal groups. The Trudos Plutonic Complex contains several deposits, in ultrabasic rocks, of chromite and asbestos.

Geology of the Tamasos Field

The Field is bounded on the north side by carbonate sediments of the Lefkara Group, at the base of which are found a few local patches of the Perapedhi umbers. Below these is an extensively faulted series of volcanic rocks in which all three units of the Trudos Pillow Lavas are represented. The Basal Group lavas are very altered and rusty at outcrop, and contain numerous zones of silification and some jaspers. Many dykes cut these rocks, corresponding in composition to both the Upper and Lower Pillow Lavas.

There are four ore bodies known in the field. Kokkinopezoula is the largest and occurs at the top of the highly altered Basal group and was probably covered, at one time, by the Lower Pillow Lavas. The other three, Kokkinoyia, Agrokipia A and B are situated on the boundary between the Lower and Upper Pillow Lavas.

Two main types of mineralization occur. The first is massive sulphide ore containing a large amount of pyrite with lesser amounts of chalcopyrite, sphalerite and trace amounts of other sulphides. The amount of copper can vary from a trace to over 5%, that of zinc from nothing to 11%. Some of this ore is dense but in places it is quite porous (up to 50% porosity in places) and friable, and frequently has a conglomeratic texture. The dense massive ore is composed of well-formed crystals, but the porous ore is a mixture of badly corroded crystals and collomorphic pyrite.

The second type of ore is a stockwork cutting altered lava. The lava is usually argillized and, to some extent, chloritized and silicified on the borders of the veinlets. The veinlets themselves contain pyrite and small amounts of other sulphides. In the deeper parts of the stockwork zones, which may persist for a hundred metres below the massive ore, the veinlets are less numerous and the rock as a whole may only contain 20% sulphide. The rocks associated with the stockwork usually contain 5 or 10% of pyrite in small veinlets, or in pockets between the lava pillows.

In cases where the ore body is capped by lavas there is often a narrow layer of ocherous limonite on the contact. Where the ore outcrops on the present land surface, there are siliceous limonite and jarosite gossans which may contain pockets of earthy material rich in gold and silver. Some of the outcropping ore bodies have an enriched layer containing chalocite (and other similar minerals) below the gossan.

Size and Grade

The size of the ore bodies is between 4 Mt and 6 Mt containing on average about 30%. Of this, about a tenth is massive copper–zinc-rich grading about 5% Cu.

Geological Interpretations

Recent studies have shown fairly conclusively that these deposits were formed during the waning stages of the volcanic activity just below the sea floor. The process envisaged is the fracturing of the lava, and the passage of sulphur and iron-rich fumarolic gas (forming the stockworks) and the gradual alteration, corrosion and removal of the lava forming a hollow on the sea floor that became filled with sulphide. The formation of the very porous ores is disputed but the most likely explanation seems to be that it was formed by leaching, after formation. It is thought that part of the oxidized pyrite resulting from this leaching is present as the ocherous capping. This

latter process is quite distinct from the formation of gossan which took place on the land surface, after the uplift and erosion of the area during and since the Upper Miocene.

Marine geological and geophysical research has shown that oceanic crust has a three-layer structure, and the Trudos Complex may be a part of an original ocean floor that was either domed upwards, or thrust up into its present position.

The consequences of this idea is that the pyritic deposits must have formed off the centre of a mid-oceanic ridge (in this case in the now lost Tethys Ocean). Only in special circumstances, where the mineralization was preserved below a later lava flow, are ore bodies preserved.

Mining

The Tamasos Field was taken under licence by the Hellenic Mining Company in 1937, since which time the four ore bodies have been discovered and mined. The Kokkinopezoula and Agrokipia A bodies were worked by open-pits, the others are worked by underground cut and fill methods. Much of the ore was, and is, simply crushed and shipped for sulphuric acid manufacture. The richer copper–zinc ores are now treated by selective flotation to recover sulphide concentrates.

Further Reading

BEAR, L. M. (1963), The Mineral Resources and Mining Industry of Cyprus. Cyprus Geol. Surv. Bull. 1. [The most comprehensive descriptive work on the island, much of which is devoted to the Trudos pyritic deposits.]

BEAR, L. M. (1960), The Geology and Mineral Resources of the Akaki–Lythrodondha Area. Cyprus Geol. Surv. Mem., 3. [A regional monograph that includes the Tamasos Field.]

GASS, I. G. and MASSON-SMITH, D. (1963), The Geology and Gravity Anomalies of the Trudos Massive, Cyprus. Phil. Trans. Roy. Soc. Lon. Ser. A 1060, 255, 417–67. [Probably the most significant work on the nature of the Trudos Complex.]

CONSTANTINOU, G. (1972), The Geology and Genesis of the Sulphide Ores of Cyprus. Thesis submitted to University of London Ph.D. 275pp. [A modern comprehensive work which includes a full description of the Tamasos Field but is difficult to get hold of.]

CONSTANTINOU, G. (1973), Geology, Geochemistry and Genesis of Cyprus Sulphate Deposits. Econ. Geol., 68, 843–58. [A shorter version of his thesis, more readily available.]

STRONG, D. F., ed. (1976), Metallurgy and Plate Tectonics. Geol. Ass. Can. Spec. Paper No. 14, 66opp. [This very good symposium volume includes some papers relevant to the understanding of these deposits and the comparison with others; see particularly a paper by Sawkins, F. J., pp.221–42.]

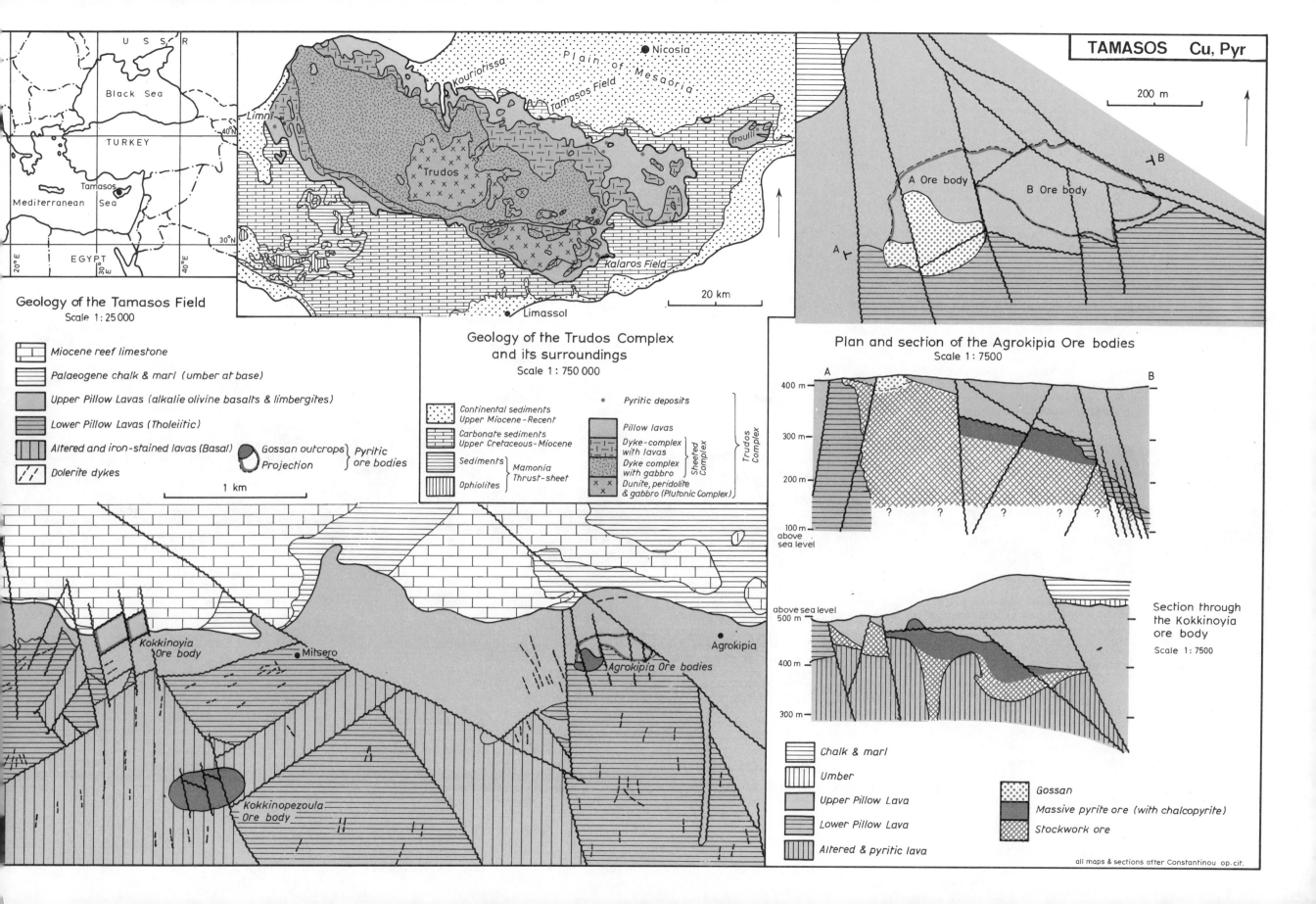

TAMASOS Cu, Pyr

200 m

Geology of the Tamasos Field
Scale 1 : 25 000

Black Sea

U.S.S.R.

TURKEY

Tamasos

Mediterranean Sea

EGYPT

Limni

Kouriotissa

Plain of Mesaoria

Nicosia

Tamasos Field

Troulli

Trudos

Kalaros Field

Limassol

20 km

Legend — Geology of the Tamasos Field

- Miocene reef limestone
- Palaeogene chalk & marl (umber at base)
- Upper Pillow Lavas (alkalic olivine basalts & limbergites)
- Lower Pillow Lavas (Tholeiitic)
- Altered and iron-stained lavas (Basal)
- Gossan outcrops } Pyritic ore bodies
  Projection
- Dolerite dykes

1 km

A Ore body   B Ore body

A   B

Geology of the Trudos Complex
and its surroundings
Scale 1 : 750 000

- Continental sediments Upper Miocene - Recent
- Carbonate sediments Upper Cretaceous - Miocene
- Sediments } Mamonia Thrust-sheet
- Ophiolites
- Pyritic deposits
- Pillow lavas
- Dyke-complex with lavas } Sheeted Complex
- Dyke complex with gabbro
- Dunite, peridolite & gabbro (Plutonic Complex) } Trudos Complex

Plan and section of the Agrokipia Ore bodies
Scale 1 : 7500

A   B

400 m

300 m

200 m

100 m above sea level

? ? ? ? ?

Section through the Kokkinoyia ore body
Scale 1 : 7500

above sea level
500 m

400 m

300 m

Kokkinoyia Ore body

Mitsero

Agrokipia

Agrokipia Ore bodies

Kokkinopezoula Ore body

- Chalk & marl
- Umber
- Upper Pillow Lava
- Lower Pillow Lava
- Altered & pyritic lava
- Gossan
- Massive pyrite ore (with chalcopyrite)
- Stockwork ore

all maps & sections after Constantinou op.cit.

The Skorovas Pyritic Deposit – Norway

Although not very large, this is a good example of the type of base metal-bearing pyritic deposit that occurs along the Caledonian–Appalachian fold belt from the north of Norway, through the British Isles, Newfoundland, New Brunswick and on into the north of New England. Skorovas is instructive when compared with other deposits in volcanic sequences, such as those in Cyprus and Japan.

Location

The deposit is located at latitude 64° 38′ N, longitude 13° 04′ E and 700 m altitude in the Nord Trøndelag district of Norway, 40 km north-east of the town of Grøng.

Geographical Setting

Skorovas outcrops on a bare hillside overlooking a small glacial lake in a rather wild and semi-mountainous part of Norway. There are several mountains around the district that rise well over 1000 m, and the land drains down to the Sanddøla River, to the Nansen and the Atlantic. The climate is wet and cool, for the most part frozen and snow-covered in winter and rarely averaging more than 5°C in summer. The district is not only affected by the cold air-mass over central Scandinavia and Russia, but is also close enough to the sea to be influenced by 'westerly' weather. The annual rainfall is over 1·5 m. The vegetation is principally scrub birch and moss, with a few pines and taller birches on the lower ground. There is a mining village on the lakeside, but little else. This is connected by road to the larger village and railway station of Flåtådal, 25 km to the west. The Swedish frontier is only 25 km to the east.

History

A weathered rusty outcrop was reported from the area in 1873 but it aroused little interest at the time. The occurrence was investigated in 1904 and again in 1910 when the Elektrokemisk Company took an interest in the whole Grøng area. The first systematic work on the area was carried out by Foslie in the late thirties, when a series of drill holes was put down which revealed a sizeable body of ore, largely pyrite, but with small amounts of zinc and copper. Elektrokemisk finally opened a mine in 1952, which exploited and marketed crushed pyrite for sulphuric acid making. Later, however, because of marketing difficulties, a mill was built in order to produce zinc and copper concentrates.

Geological Setting

The Caledonian belt of Scandinavia is largely composed of a series of flat thrust sheets or nappes of Lower Palaeozoic sediments and volcanic rocks. Groups of these nappes are geographically separated by culminations of the underlying Pre-Cambrian basement. Although the palaeontological control is poor, it seems that in general these thrust sheets tend to contain rocks in reverse order of stratigraphic age and the higher units tend to be of higher-grade metamorphism than the lower ones.

Skorovas lies in a series of nappes north of the Grøng basement high and south of a series of basement windows exposed on Børgefjell and Fjallfjallet. The host rocks of the pyritic deposits of this district are the upper unit of the Seve–Koli Nappe Group. The lower part of this group that is exposed to the east (largely in Sweden) called the Seve Nappe, consists of low-grade meta-sediments with a few irregular masses of ultrabasic rocks. The Koli Nappe Complex has been divided into a series of individual nappes, the uppermost of which, the Gjersvik, consists of a group of metamorphosed basaltic, andesitic and dacitic volcanic rocks associated with bodies of gabbro, diorite and granodioritic rocks, and a sequence of calcareous sediments. The magmatic rocks are separated from the sediments by an important polymictic conglomerate containing pebbles of all the magmatic rock types of the Gjersvik Group, many already metamorphosed or altered at the time of the conglomerate formation. The Koli Nappe Complex is overlain by a further one, the Helgelad, of much higher grade of metamorphism. The Koli rocks are deformed into a series of flat-lying recumbent folds, and again buckled into a series of more open folds apparently related to the uplift of the basement highs. These two phases of folding combined with the rugged glacial topography tend to make the geology rather complicated.

Such evidence as there is suggests that the rocks in the Grøng area range in age from the Middle or Lower Cambrian to the Lower Silurian, and the evidence of conglomerates suggests that some deformation took place during this period. From outside the district there is general evidence that the main thrusting and nappe formation took place during the Silurian and early Devonian.

Almost nothing is known of the history of this area from this time until the Quaternary when the area was buried deep under the Baltic ice-cap, the melt-waters of which considerably modified the topography and left the valleys full of till.

There are five economic-sized pyritic deposits in the Grøng area, all located in the basalt–andesite sequence, and usually associated with the more acid intercalations. Only Skorovas and Gjersvik occur in the Gjersvik nappe, the others, Joma and the two ore bodies at Stekenjokk in Sweden, are situated in the lower nappes of the Koli Group. Apart from the ore bodies themselves, horizons of banded sulphide or iron oxide mineralization are widespread in the volcanic sequence of the area. Small occurrences of copper–nickel–iron sulphide with some platinum are known in the ultrabasic rocks of the Seve Nappe.

Geology of the Skorovas Deposit

The deposit is situated on the contact between basic pillow lavas of the Gjersvik Group and one of the many intercalations of dacite. The structure of the area consists of a series of small recumbent folds on the overturned limb of a very much larger fold of similar style. The Hardalsvalin polymictic conglomerates lie underneath the volcanics, although they are stratigraphically higher. The ore itself consists of a complicated series of folded layers or lenses of sulphide ore, forming a flat lying body 600 m long and 200 m wide and with an average thickness of about 30 m.

The principal mineral is pyrite, with smaller amounts of sphalerite, chalcopyrite, magnetite and trace quantities of galena, arsenopyrite and tennantite. Silver and cadmium are present. The ore is medium- to fine-grained and strongly banded and, although the structure is complex, there is a general stratigraphic sequence to the mineralization. At the stratigraphic top (usually on the footwall of the ore body) are jaspers and banded iron oxide, carbonate or silicate rocks. Next is normally found a banded pyrite-magnetite rock, and then a pyrite-sphalerite ore which passes into a chalcopyrite-bearing zone.

The ore body seems to lie on an horizon within the volcanic sequence that is elsewhere marked by thin zones similar to the topmost part of the ore body. These can be traced for some distance round the district. The banded pyrite-magnetite, iron-silicate rocks are locally known as 'vasskis'.

Size and Grade

The total reserves of the Skorovas deposit are between 8 and 9 Mt containing 35% S, about 2% Zn and 1% Cu. About 10 g/t silver is present and some cadmium is also found.

Geological Interpretation

The form, texture and structure of the ore body makes it difficult to uphold the earlier view that this deposit was formed by replacement, associated with the nearby intrusions of gabbro, diorite and granodiorite. Rather, the present view is that it was formed at the original site of deposition of the volcanics before deformation and thrusting. The associated rocks are low-potash thoelitic basalts, basaltic andesites, andesites and dacites, many showing evidence of sub-aqueous deposition and exhibiting the characteristic alteration associated with the hydrothermal circulation of sea water. The accompanying intrusions are gabbros, diorite and granodiorite (of a special type known as trondhjemite). Some of the intrusions have the so-called 'bi-modal' compositions. The assemblage of magmatic rocks is held to be characteristic of an early stage in island-arc development just off the edge of a continental mass. Skorovas has therefore been interpreted as a volcanic-exhalative deposit formed under water on a chain of volcanic islands that once existed in an ocean which has since vanished, by the collision and welding, in Palaeozoic times, of the Baltic and North American cratons. During this process it is supposed that slices of the original island-arc rocks were thrust on to the Baltic Shield in the form of a series of nappes.

Mining

The mine is operated by Elkem Spigerverket A/S and is an underground mine, using cut-and-fill methods with pillars. From 1952 to 1975 the ore was crushed and screened to various sizes at the mine, and shipped by aerial ropeway and rail to various places in Europe for sulphuric acid manufacture. However, today there is a selective flotation plant on site which produces copper and zinc concentrates and pyrite fines.

Further Reading

HALLS, C. et al. (1977), Geological Setting of the Skorovas Orebody within the Allochthonous Volcanic Stratigraphy of the Gjersvik Nappe, Central Norway. Trans. Inst. Min. Metall. (Sec. B: Appl. Earth Sci.), pp.128–51. [A modern and comprehensive work on the deposit and its setting. Also contains a good bibliography.]

OFTEDAHL, C. (1958), Oversikt over Grøngfeltets Skjerp og Malmforekonster. Norges Geologiske Undersokelse 202. [A description of the whole Grøng district in Norwegian, but with an English summary. Contains the first description of the volcanic-exhalative metallogenic theory.]

GJELSVIK, T. (1960), The Skorovas Pyrite Deposit, Grøng Area, Norway. Rep. 21st Int. Geol. Congr. Part 16, pp.54–66. [A good description of the ore body itself with maps and sections.]

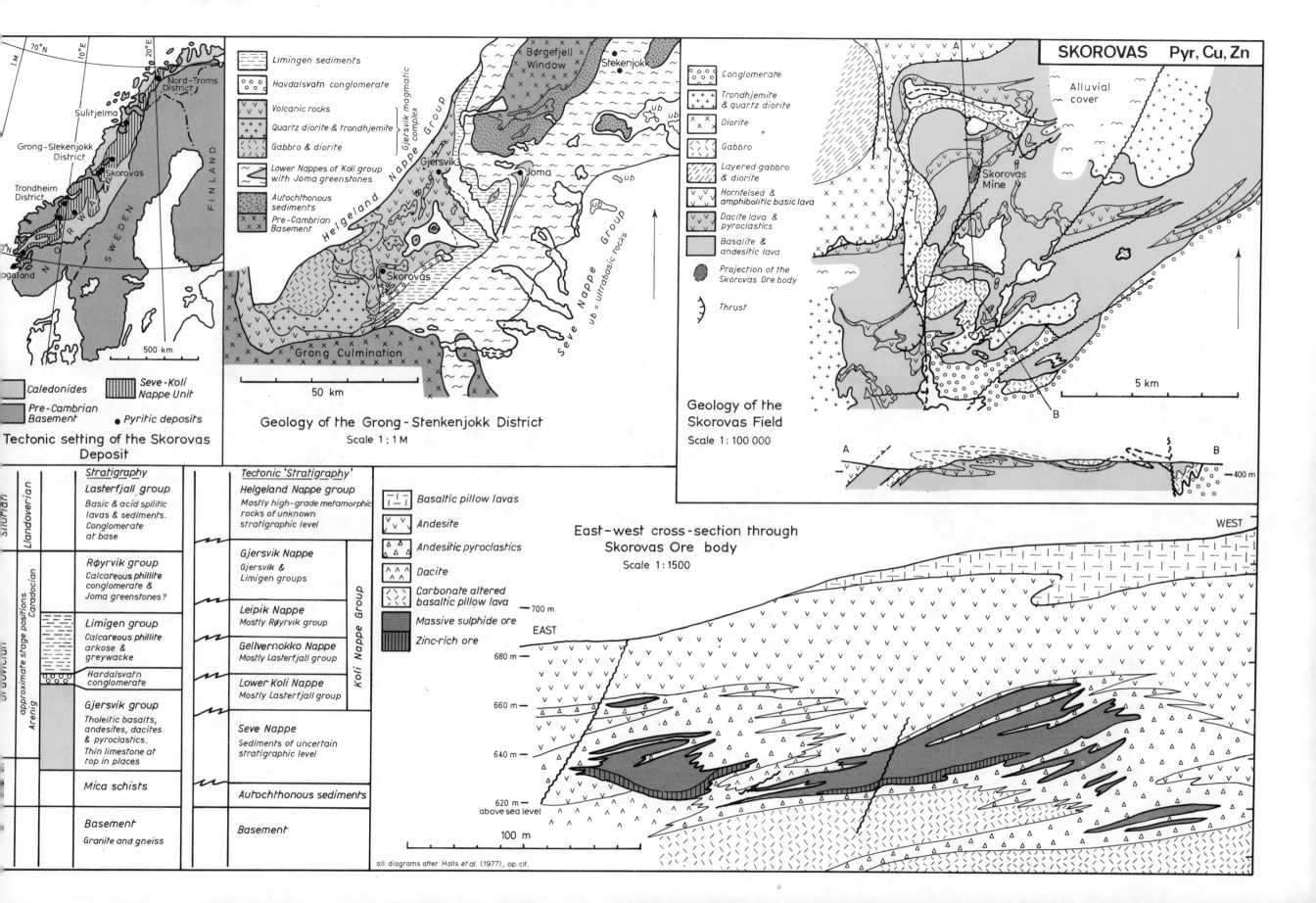

The Rio Tinto Deposits – Spain

'For the coastlands shall wait for me, the ships of Tharshish first bring your sons from far, their silver and gold with them.' (Isaiah LX, 9). Though it is a matter of dispute, some biblical scholars hold that Tharshish refers to the mining district, then in Carthaginian hands, now known as Rio Tinto.

Location

The township of Minas de Rio Tinto is situated at latitude 37° 42′ N, longitude 6° 35′ W and at 350 m above sea level in the Spanish province of Huelva, some 56 km from the port of the same name, and 60 km north-west of the city of Sevilla.

Geographical Setting

The Rio Tinto flows through valleys cut into the raised plateau that forms the ground to the south of the mines. To the north is the higher ground of the Sierra Morena. Originally the area was covered by Mediterranean oak scrub, although around the mines most of this was felled for fuel, or destroyed by atmospheric pollution. The climate is Mediterranean with a little Atlantic influence, 600 mm rain per year, mostly in the temperate winter (10°C), the summers being warmer with midday temperatures in the high twenties and low thirties.

Small villages are numerous in the area, the largest of which is the old Roman town of Neva. The mines are connected to the port of Huelva by a railway as well as by road.

History

Rio Tinto has probably been mined for over 3000 years. Signs of early mining are not confined to the biblical references, but it was the Romans who first developed the district on a grand scale. Mechanized mining began in 1878 with the arrival of a British company, eventually passing into Spanish hands in 1955. In the 1960s, the arrival of modern exploration and open-pit mining techniques, led to the discovery of the San Antonio ore body and the mining of the Cerro Colorado low-grade ore zone.

Geological Setting

The Huelva pyrite province is a belt 50 km wide that stretches from near Sevilla through Portugal almost to the Atlantic coast, some 250 km in all. It comprises a series of Upper Palaeozoic rocks, folded along an east–west axis, bounded to the north by older, more crystalline rocks of the Sierra Morena and covered by coastal plain sediments of Tertiary and Quaternary age to the south.

The oldest rocks exposed in the pyrite belt are sediments of Devonian age. Overlying these are volcanic rocks that are divided into a lower group, dominated by basaltic and andesitic, and pillow lavas; and an upper group, which includes rhyolites and dacites accompanied by a very variable series of coarse- and fine-grained pyroclastics.

The volcanic rocks are generally terminated by fine ashes and overlain by sediments; shales, quartzites including turbidites of Lower Carboniferous age. The whole sequence was folded in the late Carboniferous, and all but the most resistant lavas developed cleavage.

Nothing is known of the Mesozoic history of the area and the next recorded event is erosion down to a peneplain in the Tertiary, which was subsequently rejuvenated to form the present dissected plateau.

The dominant type of mineralization is lenticular massive pyrite bodies. Some are associated with zones of disseminated or stockwork mineralization.

Copper usually accompanies the pyrite with some more complex types that include zinc, lead, gold and silver.

Along the belt the mineralized horizons are often marked by chert, jasper and silicious manganese mineralization, the latter occasionally oxidized to a workable quality.

Geology of the Rio Tinto Deposits

The mineralization at Rio Tinto is associated with the boundary between acid volcanic rocks and Carboniferous sediments. Flow-banded and brecciated rhyolites occur, forming domes and flows with large quantities of volcanic breccia and tuffs. The transition to the sediments is marked by a zone of fine ash mixed with sediment often accompanied by jasper or chert. The volcanics form a broad anticline, but the sediments and uppermost tuffs show folds of smaller amplitude, and a strong, almost vertical, axial plane cleavage is seen in most rocks. In the zone of mineralization, the volcanic rocks are chloritized, but beyond they are sericitized or only feebly altered.

The mineralization comprises massive sulphide ore bodies on the tuff-sediment boundary, the stockwork mineralization below, and the supergene products of the two.

Massive sulphide bodies. These bodies are composed almost entirely of dense fine-grained pyrite that often shows colloform and framboidal textures under the microscope, usually modified by crystal overgrowths. Chalcopyrite and a few other sulphide minerals are found in small quantities, and substantial parts of the bodies contain 1% copper.

The Planes–San Antonio bodies are slightly different. The upper part of Planes is a relatively rich chalcopyrite pyrite body, but as it continues down the plunge (San Antonio) it becomes a strongly layered and banded body, with considerable amounts of galena and sphalerite, interbedded with pyroclastic debris, showing slump structures and other sedimentary features.

The stockworks. Beneath, and alongside the massive ore bodies are zones where heavily chloritized volcanic rock is traversed by numerous veins of pyrite chalcopyrite and quartz. Such rocks often contain between 20 and 80% sulphide and between 1 and 2% copper. These rich stockworks occur especially under the ore bodies along the north side of the hill and beneath the Planes body. Further away from the massive bodies less intense stockwork mineralization is found, typically zones of narrow veinlets that may be selectively mined with between 0.6 and 0.8% copper.

Supergene products. The massive pyrite bodies and the rich stockworks were originally marked at outcrop by thick gossans, and a very large area of gossan occurs on the Cerro Colorado although no massive ore is present beneath. The gossans are an open-textured mixture of irregular and concretionary masses of silicious hematite cemented by limonite. Near the base of the gossans, yellow and black earthy layers are found that contain jarosite, and very finely divided gold and silver.

Beneath the gossans the volcanic rocks are leached for between 10–40 m, below which is a zone of disseminated chalcocite containing between 1 and 2% copper.

A feebly bedded material looking rather like gossan occurs in the valleys below the main hill. It contains no other metals and contains a Tertiary flora.

Size and Grade

Up to 1960 the area had probably produced 115 Mt and had at that time over 50 Mt reserves of pyrite containing an average of 48% sulphur and perhaps between 0.5 and 1% copper. The present mine on the Cerro Colorado was started on the basis of 40 Mt of stockwork-ore with 0.8% copper and 18 Mt of gossan with 2.4 g gold and 42 g silver per tonne. The reserves of both pyrite and stockwork copper ore are probably much greater, and discoveries are still being made.

The San Antonio body contains 5 Mt with 42% sulphur, 1.4% copper, 0.9% lead and 1.6% zinc.

Geological Interpretations

It is now generally agreed that most of this mineralization was deposited at about the end of the volcanic episode in or just under the mud on the sea floor. After a violent phase of explosive activity, fumarolic activity continued up fractures which are today represented by some of the stockwork veins, emplacing sulphide in the fractured volcanic rocks, and forming more continuous masses of sulphide in trenches on the sea floor spreading out laterally to form a sheet of pyritic sediment over a wide area. After the deposition of the sedimentary cover, the area was deformed, yielding the present configuration of ore bodies and probably causing the redistribution of some sulphides into new sets of fractures. The area was probably exhumed during the Tertiary, at which time all but the synclinal parts of the folded pyrite sheet was weathered away, forming the gossans and associated supergene products. As the Tertiary peneplain was rejuvenated, parts of the gossans were eroded and redeposited on the lower ground around the hill.

Mining

The Union Explosivo Rio Tinto operates an open-pit on the San Dionisio body and an underground mine on the San Antonio, has underground developments in stockwork ore, and operates an open-pit, working gossan and stockwork ore on the Cerro Colorado. Pyrite ore is shipped as it is for sulphuric acid manufacture, while stockwork ore is treated in flotation plants to recover a sulphide concentrate that goes to the companies' own smelter and refinery at Huelva. A substantial amount of copper is recovered by leaching low-grade ore with acid mine-water, and precipitation by scrap iron.

Further Reading

GARCIO PALOMERO, F. (1974), Caracteres Estratigraficos del Anticlinal de Rio Tinto. *Studia Geol.* (Salamanca), **8**, 93–124.

GARCIA PALOMERO, F. (1975), Estudio Geologico de la Masa Piritica de San Antonio (Rio Tinto) *Jornadas Minero Metalurgicas Bilbao*, **1**, 175–91. [Two very good descriptions by the chief mine geologist of Union Explosivo Rio Tinto.]

RAMBAUD-PEREZ, F. (1969), El Sinclinal Carbonifero de Rio Tinto (Huelva) y Mineralizcicciones Asociadas. *Mems. Inst. Geol. Min, Esp.*, **71**, 229pp. [An earlier but interesting description of the whole district.]

SCHERMERHORN, L. J. G. (1971), An Outline Stratigraphy of the Iberian Pyrite Belt. *Bol. Geol. Min.*, **82**, 239–68.

STRAUSS, G. K. and MADEL, J. (1970), Geology of the Massive Sulphide Deposits in the Spanish–Portuguese Pyrite Belt. *Geol.* **63**, 191–211. [Two somewhat complementary accounts of the whole province, mostly not about Rio Tinto itself.]

WILLIAMS, D. (1933), The Geology of the Rio Tinto Mines. *Trans. Inst. Min. Metall.*, **43**, 593–640. [Still remarkably good reading despite later reinterpretations of the geology by the mine geologist of the day, who later became professor of Mining Geology at the Royal School of Mines, London.]

WILLIAMS, D., STANTON, R. L. and RAMBAUD, F. (1975), The Planes–San Antonio pyritic deposit of Rio Tinto, Spain: its nature, environment and genesis. *Trans. Inst. Min. Metall.*, **84**, B73–B82. [An up-to-date account of the recent discoveries.]

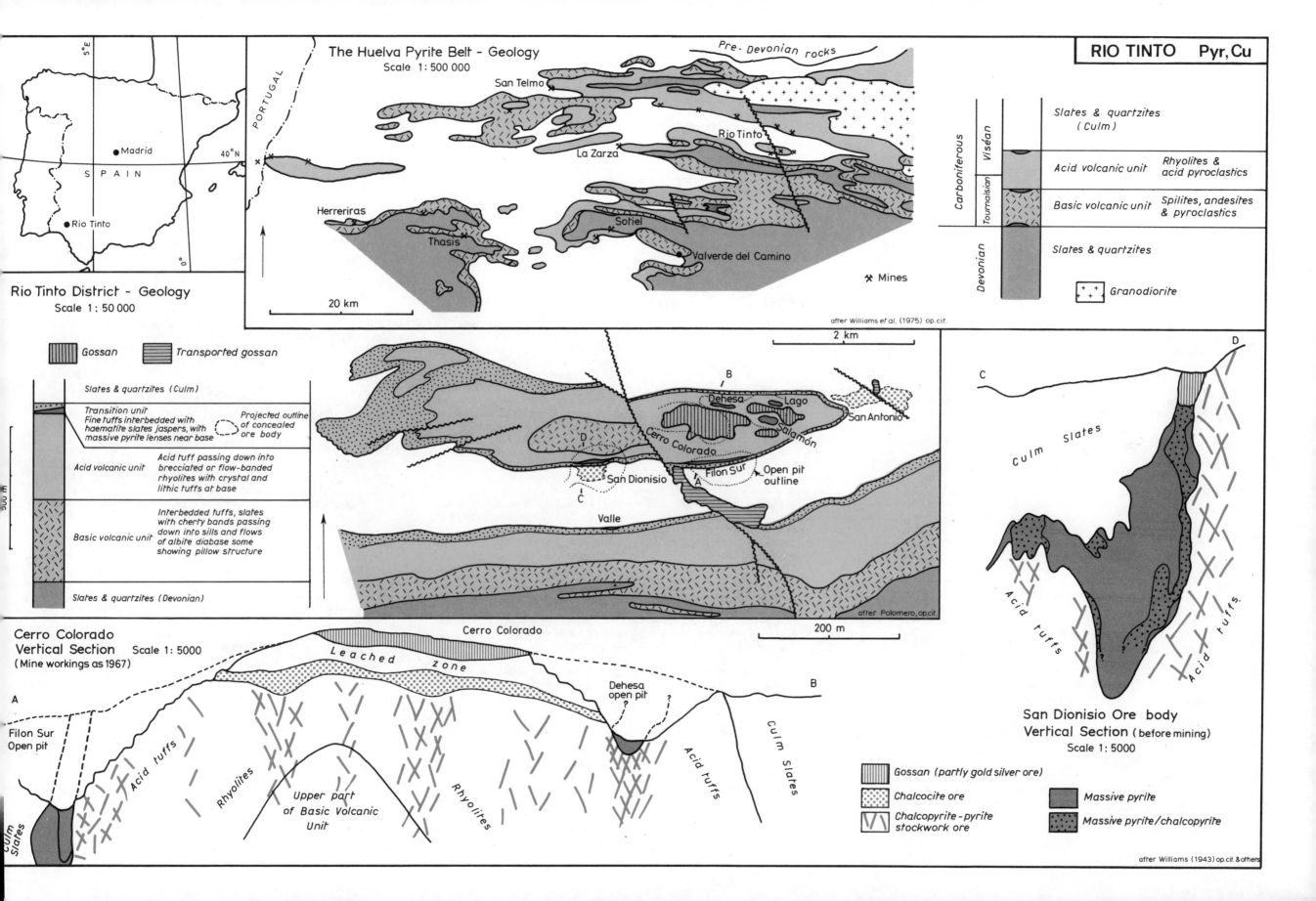

RIO TINTO Pyr, Cu

Rio Tinto District – Geology
Scale 1 : 50 000

The Huelva Pyrite Belt – Geology
Scale 1 : 500 000

San Telmo

Rio Tinto

La Zarza

Herreriras

Thasis

Sotiel

Valverde del Camino

Pre-Devonian rocks

PORTUGAL

SPAIN

Madrid

Rio Tinto

40° N

20 km

✵ Mines

after Williams et al. (1975) op.cit.

		Slates & quartzites (Culm)
Carboniferous	Viséan	Acid volcanic unit — Rhyolites & acid pyroclastics
	Tournaisian	Basic volcanic unit — Spilites, andesites & pyroclastics
Devonian		Slates & quartzites

Granodiorite

Gossan Transported gossan

Slates & quartzites (Culm)

Transition unit
Fine tuffs interbedded with
haematite slates jaspers, with
massive pyrite lenses near base

Projected outline
of concealed
ore body

Acid volcanic unit
Acid tuff passing down into
brecciated or flow-banded
rhyolites with crystal and
lithic tuffs at base

Basic volcanic unit
Interbedded tuffs, slates
with cherty bands passing
down into sills and flows
of albite diabase some
showing pillow structure

Slates & quartzites (Devonian)

2 km

B

Dehesa   Lago
Cerro Colorado   Salomón
San Antonio
D
San Dionisio   Filon Sur   Open pit
A   outline
C
Valle

200 m

after Polomero, op.cit.

C   D

Culm Slates

Acid tuffs   Acid tuffs

San Dionisio Ore body
Vertical Section (before mining)
Scale 1 : 5000

Cerro Colorado
Vertical Section   Scale 1 : 5000
(Mine workings as 1967)

Cerro Colorado

Leached   zone

Dehesa open pit

A   B

Filon Sur
Open pit

Acid tuffs

Rhyolites

Upper part
of Basic Volcanic
Unit

Rhyolites

Acid tuffs

Culm Slates

Culm Slates

Gossan (partly gold silver ore)

Chalcocite ore

Chalcopyrite-pyrite
stockwork ore

Massive pyrite

Massive pyrite/chalcopyrite

after Williams (1943) op.cit. & others

The Noranda Field – Canada

The complex of sulphide ore bodies worked at the Horne Mine, Noranda, are famous because their development sired the growth of one of the world's great mining companies. Although not the clearest example geologically, Noranda is the best known of a distinct type of deposit which contains iron sulphide with considerable amounts of copper, gold and zinc, associated with rhyolitic volcanic rocks. It makes an interesting contrast to the Kuroko, Rio Tinto and Cyprus types of pyritic sulphide deposits. Similar deposits occur not only in Archean terrains as does Noranda, but also in younger rocks such as the famous Skelefteå district in Sweden.

Location

The Noranda deposit (Horne Mine) outcrops at latitude 48° 15′ N, longitude 79° 01′ W, at 300 m altitude on the shores of Lake Tremoy in Timiskaming County, western Quebec.

Geographical Setting

Noranda lies in an area of flattish country, underlain by glacial clays and sands interrupted by steep rocky ridges where the harder volcanic rocks outcrop. Elevations are about 300 m above sea level with peaks of the ridges rising to about 460 m. The area was originally forested with spruce, pine and birch, but much has been felled, and for a long time regrowth was inhibited by smelter gasses. The climate is the cool continental one of most of the Canadian Shield, but there are sufficient frost-free days and hours of sunshine in summer to allow farming on the lower clay-belt lands. The twin towns of Noranda and Rouyn have grown up largely as mining towns, and are connected to the Great Lakes region by road, rail and air.

History

In the early years of the twentieth century trails, and later railways, came to the area and this stimulated much prospecting. In 1906, there was a gold rush to Kirkland Lake and Larder Lake to the west of Noranda, and soon discoveries of gold were being made all along the southern limit of the Abitibi region. Ed Horne, an experienced prospector, became interested in some gossan outcrops on the shores of Lake Tremoy which he visited in 1911 and 1914, but it was not until 1920 that he found backing to stake claims over the outcrops that were later revealed by drilling to be the uppermost of the Noranda ore bodies. Two years later the Noranda company was formed and, after extensive evaluation, a smelter was blown-in in 1927 and production began in 1928. The mine was developed down over 2000 m from surface and became more or less exhausted by 1976.

Geological Setting

Noranda is situated in the centre of the southern limb of the great Abitibi fold belt. This is a zone of tight folds and fault zones in Archean volcanic and sedimentary rocks that stretches for 200 km through the mining districts of Kirkland Lake, Larder Lake, Noranda, Cadillac, Malartic and the Val d'Or. There are considerable problems in understanding the geology of the area because of the complexity of the folding, the problem of correlating ancient meta-volcanic rocks, and the widespread cover of glacial deposits.

Regarded by most people as the oldest rocks in the area are the pale quartz-rich mica-schists of the Pontiac group. These are thought to underlie the meta-volcanic rocks of the area, a series of basic, intermediate and acid rocks of great complexity. The basaltic and andesitic rocks frequently show pillow structure, while the acid rocks are largely fragmental. Some geologists recognize two groups of meta-volcanic rocks; the Malartic and Blake River, separated by conglomerates and greywackes of the Kewagama Group,

although it has been suggested that the latter are a facies of the Pontiac.

A few sediments, albeit of locally derived volcanic material, occur in the volcanic sequences, but above an unconformity there are much better-developed clastic sediments called the Cadillac Group.

The Archean supercrustal rocks are intruded by acid plutonic rocks, granodiorites in the main, but in the Noranda area there are more alkaline granitic rocks and small bodies of syenite. Most of these plutonic rocks have radiometric ages of about 2·5 Ga.

There is a cover over part of the area of conglomerate and conglomeratic greywacke belonging to the Hudsonian Super-Group (Cobalt Group), thought by some to be tillites. Apart from some younger Pre-Cambrian dolerite dykes and the reactivation of faults, there is little record of the geological history of the area until the Quaternary when the whole area was glaciated; after the last ice retreated it was covered for a time by a large lake that left parts of the area covered by varved clays, fluvi-glacial sands and eskers.

The rich mineralization of the region includes the massive sulphide deposits of the Noranda-type gold quartz veins, and gold deposits in alteration zones.

Geology of the Noranda Ore Bodies

Over thirty ore bodies occur in a steeply dipping series of rhyolitic rocks in a block bounded by fault zones. The host rocks of the ore are an interfingering sequence of rhyolite flows, brecciated rhyolite, tuffs and a few quartz porphyry intrusions. The sequence is cut up by a very complex network of dykes of diorite and quartz diorite, variable in grain size, which are off-shoots of the mass of rocks lying on the north side of the rhyolitic sequence. These intrusions cut the mineralized zone into pieces and sometimes completely surround bodies of sulphide ore. The ore zone is terminated on the southern side by a fault, the other side of which is a sequence of andesitic lavas.

There are several types of ore; massive pyrite (with and without copper, zinc or gold) is one, but massive pyrrhotite, chalcopyrite and magnetite ore with variable amounts of gold and silver form the main ore bodies. There are also zones of altered rhyolite rich in silica that contain some gold (3–6 g/t) that serve in the mine both as a gold ore and as a flux for the copper smelters. Relatively low-grade copper-bearing 'flux ore' is also mined, and some gold is derived from zones of chloritization containing a complex paragensis of gold tellurides.

In addition to the dioritic intrusions, the ore zone is also interrupted by faults and by some dolerite dykes.

Size and Grade

During the life of the Horne Mine the Noranda ore bodies have yielded 58 Mt of ore averaging 2·4% copper and 5·5 g/t gold with minor, but useful, amounts of silver and other metals. The pyrrhotite–chalcopyrite ore, however, was much higher in grade with about 7% copper and the pyritic ore averaged about 0·7% copper and 6 g/t gold. In addition, fluxing ore of about 5 Mt with a low content of copper was extracted from neighbouring ore zones.

Geological Interpretations

The Noranda deposit exemplifies several of the problems that plague the understanding of ore genesis. Cross-cutting and replacement relationships seem to show that the ore was accumulated later than the rhyolites, the diorite dykes and, in some cases, later even than the dolerite dykes. However in other deposits in the southern Abitibi belt, there is clear evidence (and at Noranda there is some rather equivocal evidence) of the possibility that the ore accumulation was related to the alteration of the volcanic rocks following their deposition. The dolerite dykes have a radiometric age of more than 1 Ga younger than that of the Lake Default granite, regarded as at least as

old as the deformation and metamorphism of the Blake River volcanics. So there is something of a conflict. Broadly there are two schools of thought: one holds that the mineralization was emplaced by wholesale replacement of the volcanic rocks by solutions of unknown origin at, or shortly after, the intrusion of the dolerite dykes. Such an interpretation might be consistent with certain features of Noranda, but not of other similar deposits in the area. The other interpretation is that the mineralization accumulated in permeable, fragmental rocks with some replacement, at the closing stages of the rhyolitic volcanism at a time when a new phase of andesitic volcanism began to affect the area. The diorite dykes are regarded by some as feeders to the andesite flows that originally overlay the rhyolite, and now, by coincidence, lie against them on the other side of the Andesite Fault.

Mining

The Horne Mine was the original property of Noranda Mines which has since grown into a large mining group. In the early days there was some surface 'millhole' working, but most of the ore was extracted by cut-and-fill stoping underground. The mine which eventually reached a depth of 2440 m below surface had at its maximum a capacity of 5000 t/d. The richest of the ore was smelted directly, the leaner ores were concentrated by flotation before smelting. Pyrrhotite concentrates were produced and used to some extent for cementing granulated slag as a back-fill material; and pyrite concentrates (after cyanidation to remove gold) were used to produce sulphuric acid, particularly when the demand was high during the uranium boom of the 1950s. The mine had its own reverberatory furnaces and convertors for copper production which, despite the exhaustion of the mine, remain in use as a smelter for other ores in the area. Copper from the mine was refined by the Group's own electrolytic plant situated in eastern Quebec, where hydro-electric power is readily available.

Further Reading

PRICE, P. (1948), Horne Mine. In: *Structural Geology of Canadian Ore Deposits*, pp.763–72. Canadian Inst. of Min. and Metall., ed., Montreal. [Concise account of the geology of the mine by the chief geologist of the day. The volume also contains many descriptions of other mines in the area.]

PRICE, P. (1949), Noranda Mines Ltd. In: *Geology of Quebec*. (Dresser, J. A. and Denis, T. C., Quebec Dept. of Mines Geol. Report No. 20, V. 2, pp.338–61. [A fuller description in a very good volume that also contains descriptions of other mines in the area and an account of the general geology of Quebec.]

WILSON, M. E. (1941), Noranda District, Quebec. *Can. Geol. Surv. Mem.*, **229**, 162pp. [The standard work on the general geology of the area.]

WILSON, M. E. (1962), Rouyn–Beauchastle Map Areas, Quebec. *Can. Geol. Surv. Mem.*, **315**, 140pp. [Covers the area to the south of Noranda and contains a much more up-to-date geological description of the rocks.]

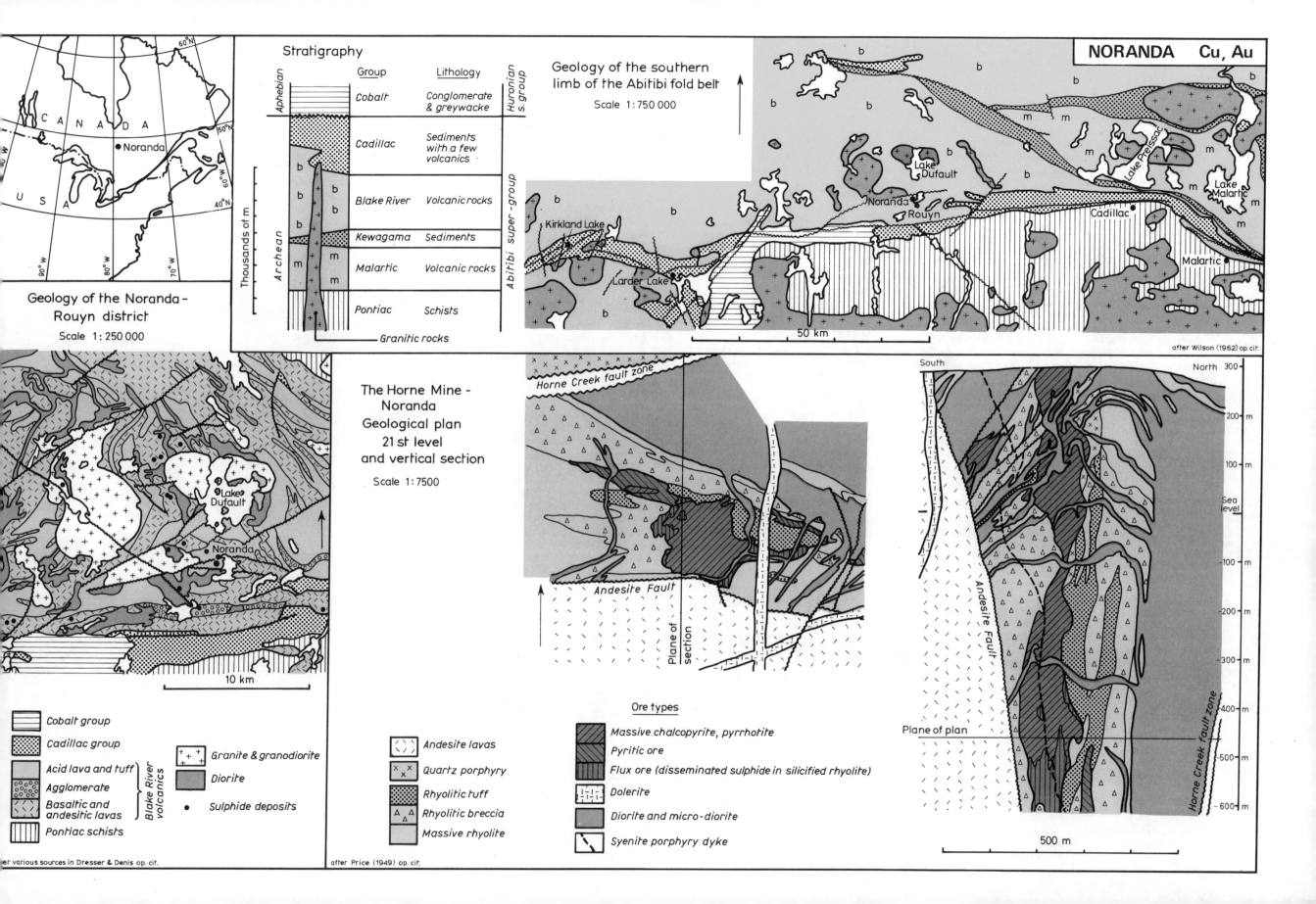

NORANDA Cu, Au

Geology of the Noranda–Rouyn district
Scale 1 : 250 000

Stratigraphy

	Group	Lithology	
Aphebian	Cobalt	Conglomerate & greywacke	Huronian s. group
Archean	Cadillac	Sediments with a few volcanics	Abitibi super-group
	Blake River	Volcanic rocks	
	Kewagama	Sediments	
	Malartic	Volcanic rocks	
	Pontiac	Schists	
	Granitic rocks		

Thousands of m

Geology of the southern limb of the Abitibi fold belt
Scale 1 : 750 000

Lake Preissac
Lake Dufault
Noranda
Rouyn
Cadillac
Lake Malartic
Kirkland Lake
Larder Lake
Malartic

50 km

after Wilson (1962) op. cit.

The Horne Mine – Noranda
Geological plan
21 st level
and vertical section

Scale 1 : 7500

Horne Creek fault zone
Andesite Fault
Plane of section

Lake Dufault
Noranda

10 km

Cobalt group
Cadillac group
Acid lava and tuff
Agglomerate
Basaltic and andesitic lavas
Pontiac schists

Blake River volcanics

Granite & granodiorite
Diorite
Sulphide deposits

after various sources in Dresser & Denis op. cit.

Andesite lavas
Quartz porphyry
Rhyolitic tuff
Rhyolitic breccia
Massive rhyolite

after Price (1949) op. cit.

Ore types

Massive chalcopyrite, pyrrhotite
Pyritic ore
Flux ore (disseminated sulphide in silicified rhyolite)
Dolerite
Diorite and micro-diorite
Syenite porphyry dyke

South
North 300 m
200 m
100 m
Sea level
-100 m
-200 m
-300 m
-400 m
-500 m
-600 m

Andesite Fault
Plane of plan
Horne Creek fault zone

500 m

CANADA
USA
Noranda

The Deposits of the Kosaka District – Japan

Kosaka is a district of the so-called 'kuroko' (black ore) deposits. These rich accumulations of pyritic, lead–zinc–copper–silver ore are associated with submarine lava domes of Cainozoic age and, as they are not complicated by later folding or metamorphism, they have yielded much valuable information about the association between sulphide mineralization and volcanism, that has proved critical to understanding of many other deposits in similar but more complicated environments.

Location

The district is situated at latitude 40° 20′ N, longitude 140° 41′ E and at an altitude of 560 m, 15 km from the town of Odate in the north-eastern part of Akita Prefecture on Honshu Island.

Geographical Setting

The mining township of Kosaka lies in the thickly wooded hills of northern Honshu, on the south flank of a mountainous, volcanic terrain which includes the high peaks and crater lakes of the Towada National Park. The region is fairly extreme climatically, cold ($-5°C$) winters with snow on the north winds, hot ($28°C$) wet summers and, although less violent than further south in the country, storms bring 1·5 m of rain in a year. Kosaka is connected by road and narrow-gauge railway to Odate and the coast.

History

It is recorded that a farmworker on the land of the feudal lord Namsu found the original outcrop, in 1862, of what is now known as the Motoyama Ore Body and that the owner developed a silver mine on the gossan and black ore. In 1884, a company was formed to develop the mines which continue to this day. By 1896, the mines were recorded as having an annual production of 6000 kg silver and 80 t of copper. Several blind ore bodies were found by underground exploration and drilling, some as recently as 1962.

Geological Setting

The district lies in what is known as the 'Green Tuff Region' covering a large part of north-western Honshu and the western part of Hokkaido. The early geological history of the area is revealed in sporadic inlyers of Palaeozoic rocks that are better exposed in the Kitahami mountains to the east. Here sequences of sediments, folded during the Permian and intruded by granite, are overlain by sediments and volcanic rocks of Cretaceous age. The majority of the rocks in the area around Kosaka are volcanic, and are of Miocene to recent age. The widespread occurrence of greenish pyroclastic rocks and immature sediments derived from them, gives the region its name. There is much evidence that the region was downwarped during the late Oligocene and that violent submarine volcanic activity continued throughout most of the Miocene. Great outpourings of basalt were followed by andesites and dacites. Evidence of several such cycles may be seen in places. Late in the Neogene the region emerged and subaerial pyroclastics covered much of the area. Such activity has continued almost to the present day, and the region still suffers earthquakes.

Geology of the Kosaka District

In the north of the district a small exposure reveals a basement of Permian phyllites and quartzites, overlain by the largely volcanic Kosaka formation. Original irregularities in the basement were in-filled with conglomerates and overlain by tuff-breccias. Between the conglomerates and the tuff-breccias nine lava domes were intruded which consist of dacite (known locally as the Motoyama). These are highly altered and brecciated towards the top of each dome. The mineralization occurs associated with these breccias and is often capped by the Kosaka–Tetsuschiei member, a red cherty rock resembling jasper. Away from the lava domes the upper members of the Kosaka formation become mixed with sediment which contains a warm marine microfauna of Burdigalian age, corresponding approximately to the Burdigalian of European terminology.

The overlying Harukizawa and Yagaratai formations are mixed sequences of basalt flows, dacite and andesite lava and pyroclastics, that tend to fill the troughs between the Kosaka lava domes and thus do not obscure the mineralized zones.

The younger Tobe and Noguchi dacite and tuff formations tend to cover the mineralized zones, and practically all the rocks mentioned so far are covered by the pumice and ash falls of the Quaternary, Towada formation. These younger formations show a progressive increase in the proportion of sub-aerial deposits, in contrast to the totally sub-marine character of the Kosaka formation.

Geology of the Ore Deposits

Six clusters of sulphide ore bodies have been discovered in the district, which have the form of irregular lenses varying from 100–200 m in length and in thickness, from a few metres to 50 m. Some mineralization occurs in the upper, altered, parts of the lava domes, but most is in, or just above, the volcanic breccias. The ore bodies are strikingly zoned in texture and mineralogy. The most complete sequence is displayed by Uchinotai-nishi ore body, which has been divided into four zones. The lowermost zone is of veinlet ore in volcanic breccia, which, though altered, still retains its original texture. The veinlets contain pyrite and chalcopyrite, the quantity of which increases upwards and eventually changes to a zone of massive pyrite and chalcopyrite. The third zone is the true kuroko (black ore) consisting of banded (and probably bedded) sulphides including pyrite, silver-rich galena, sphalerite, tetrahedrite and some barite. The banding in this zone is usually conformable with the overlying sediment. A barite-rich ore with some sulphide occurs as a fourth zone, and all are normally capped by jasper. The other ore bodies of the district show a similar arrangement, but some have, in addition, a fringe zone of anhydrite or gypsum-rich rock, containing a little sulphide, associated with intensely silicious rocks. There are two known mineralized stockworks with a complex mineralogy but with no associated massive ore.

Size and Grade

The ore bodies in the district are said to total about 10 Mt, averaging about 2% Cu, 1·7% Pb, 5% Zn, 14% S (as pyrite) with significant quantities of silver.

Geological Interpretations

Most authors agree that these deposits were formed under submarine conditions, by the action of hot, aqueous volcanic fluids. The sequence of events envisaged begins with the eruption of dacite lava which was fragmented on contact with sea water to form the tuff-breccia. Then the lava domes were pushed up under the pyroclastic pile, and terminated with violent steam eruptions and brecciation of the lava. Thereafter followed a period of submarine fumarole activity that caused the alteration and mineralization of the breccias on, or just under, the sea floor. Some of the sulphide was locally eroded and redeposited as a sediment; the jasper representing the reaction layer between it and the sea water. The anhydrite has been interpreted as a reaction product between the sulphide-bearing fluids and oxygenated sea water.

Opinion varies about the source of the metals; some authors hold that they are products of magnetic differentiation, others that they are leached from the fractured lava or pyroclastic rocks in groundwater convection cells set up by the volcanic heat source.

Mining

The district began life as a silver producer, mostly from an open-pit on the outcrop of the now largely exhausted Motoyama ore body. Other ore bodies have been, and are, exploited by underground methods by the Dowa Mining Company. The ore zones are to some extent selectively mined and treated by flotation at a rate of over 500 000 t/a. Some developments have taken place towards a process for direct smelting of the mixed sulphide ore to obviate the expensive and difficult selective flotation of such fine-grained ores.

Further Information

HOROKOSHI, E. and SATO, T. (1970), Volcanic Activity and Ore Deposition in the Kosaka Mine In: *Volcanism and Ore Genesis.* (Tatsumi, T., ed.), University of Tokyo Press, 448pp. [A reasoned and full description of the district in English, containing a full bibliography of Japanese language literature on the district.]

MATSUKUMA, T. and HORIKOSHI, E. (1970), Kuroko Deposits in Japan – A Review. In: *Volcanism and Ore Genesis.* (Tatsumi, T., ed.), University of Tokyo Press, 448pp.

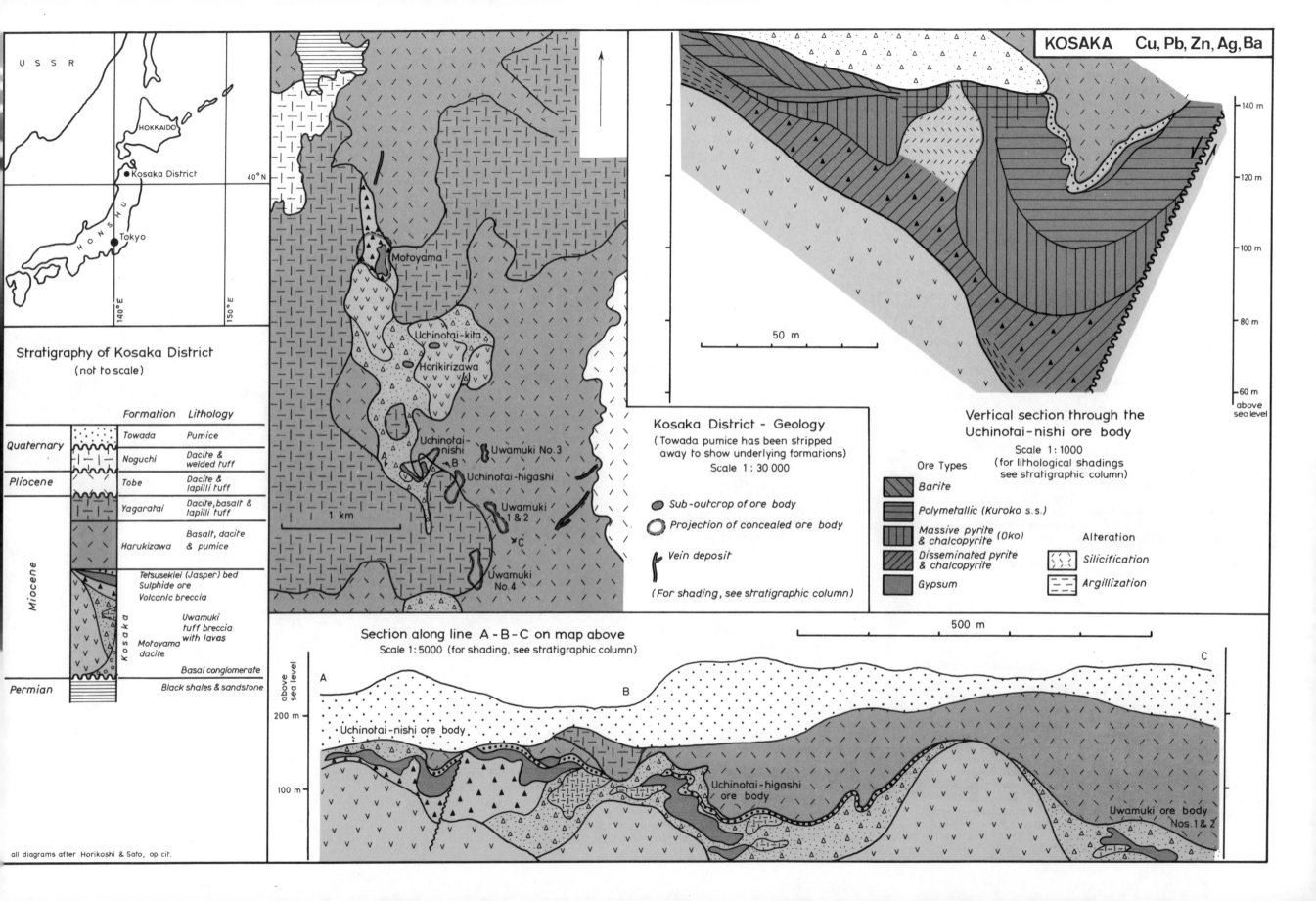

KOSAKA   Cu, Pb, Zn, Ag, Ba

USSR
HOKKAIDO
Kosaka District
40°N
HONSHU
Tokyo
140°E
150°E

Stratigraphy of Kosaka District
(not to scale)

	Formation	Lithology
Quaternary	Towada	Pumice
Pliocene	Noguchi	Dacite & welded tuff
	Tobe	Dacite & lapilli tuff
	Yagaratai	Dacite, basalt & lapilli tuff
Miocene	Harukizawa	Basalt, dacite & pumice
		Tetsusekiei (Jasper) bed Sulphide ore Volcanic breccia
		Uwamuki tuff breccia with lavas
		Motoyama dacite
		Basal conglomerate
Permian		Black shales & sandstone

Kosaka

Kosaka District - Geology
(Towada pumice has been stripped away to show underlying formations)
Scale 1 : 30 000

Motoyama

Uchinotai-kita

Horikirizawa

Uchinotai-nishi
B
Uchinotai-higashi
A
Uwamuki No.3
Uwamuki 1 & 2
C
Uwamuki No.4

1 km

Sub-outcrop of ore body

Projection of concealed ore body

Vein deposit

(For shading, see stratigraphic column)

Vertical section through the Uchinotai-nishi ore body
Scale 1:1000
(for lithological shadings see stratigraphic column)

50 m

— 140 m
— 120 m
— 100 m
— 80 m
— 60 m
above sea level

Ore Types

Barite

Polymetallic (Kuroko s.s.)

Massive pyrite & chalcopyrite (Oko)

Disseminated pyrite & chalcopyrite

Gypsum

Alteration

Silicification

Argillization

Section along line A - B - C on map above
Scale 1:5000 (for shading, see stratigraphic column)

500 m

A
B
C

above sea level
200 m
100 m

Uchinotai-nishi ore body

Uchinotai-higashi ore body

Uwamuki ore body Nos.1 & 2

all diagrams after Horikoshi & Sato, op.cit.

The Almaden Mercury Deposit – Spain

Almaden is the largest mercury deposit in the world. It consists of a series of strata-bound ore bodies situated in a rather ordinary sequence of clastic sediments that include a few volcanics. For several centuries the deposit has been so important in the economy of the Spanish state, that it has been regarded as a currency reserve and is to this day managed by the Ministry of Finance.

Location

Almaden is situated at latitude 38° 47′ N, longitude 4° 50′ W and 500 m altitude in the province of Cuidad Real, south-west of Madrid.

Geographical Setting

The deposit is in an area of open country about 500 m in altitude, bounded by the more hilly country of the Sierra Morena to the south and the lower ground of the valley of the river Guadiana. Prominent rocky outcrops of quartzite alternate with rolling hills underlain by shales. The climate is warm and semi-continental with a wet winter and long dry summer. Average winter temperatures are 5°C, in the summer 25°C or more, and 500 mm rainfall. Grain is grown extensively in the area with some rough hill pasture and more intense small-scale cultivation on some of the lower lying and wetter land. The rivers, which are seasonal, are dammed for irrigation in several areas. Almaden is a mining town, rather isolated, and 90 km from the provincial capital of Cuidad Real.

History

It is almost certain that Almaden is the mining district of Sisapo mentioned by Pliny, although Roman activities are not obvious in the area. The name Almaden suggests a Moorish origin (Al Maden means mine or ore in Arabic) but there is little evidence except for some potshards. (References by the Moorish chronicler El Rasis to a mercury mine almost certainly refers to one at Usago in Badajoz.) The mine became important after the re-establishment of the Catholic Kingdom in 1645, since which time it has been operated by the King or State. In 1755, the mine was totally destroyed by fire (quite spectacular in a mercury mine) causing lower levels to be flooded with liquid mercury. After the fire, new shafts were sunk, some of which are in use today and the mine developed to one side of and below the zone destroyed by fire.

Geological Setting

The deposit occurs in an extensive tract of the Palaeozoic rocks which underlie large areas of the Spanish 'Meseta'. The oldest rocks are non-fossiliferous, and may be late Pre-Cambrian in age, and consist of 6000–7000 m of greywacke and slate. These were folded and eroded in the so-called 'Sardic' tectonic event in late Cambrian or early Ordovician times. On these were deposited about 4000 m of siltstones, sandstones, thin carbonate horizons and occasional volcanic intercalations, ranging in age from late Ordovician to mid-Devonian. As the tectonic events of the Hercynian began, Carboniferous rocks of more variable facies were deposited in basins to the south, while the region round Almaden was folded into a series of anticlines and synclines trending NW–SE. In the Middle Carboniferous (302 Ma) the great Pedroches granite batholith was emplaced in the core of the Sierra Morena, and now outcrops 30 km south of Almaden.

This part of the Meseta was eroded down and eventually thinly covered by sediments during the Tertiary, but even these were largely removed by late Tertiary and Quaternary erosion. The Tertiary cover begins to outcrop 80 km east of Almaden. The landforms of the region indicate a much wetter climate than today, probably dating to the immediate post-glacial period.

Mercury is found over a wide area round Almaden, but always in a particular horizon of sandstone of Llandovery age.

Geology of the Almaden Deposit

The ore formation is a sequence of orthoquartzites interbedded with ripple-marked siltstones, generally dipping steeply to the north, or near-vertical, forming the southern limb of a syncline. Just south of the mine is a prominent ridge formed of the outcrop of the Canteras Quartzite, above which are argillites and a further quartzite. These are succeeded by the Footwall Argillites that contain a poorly defined upper-Ordovician fauna. The immediate footwall of the ore formation is sometimes marked by a rather altered basalt flow.

The ore formation consists of three (sometimes two) quartzite members, separated by slaty siltstones. The quartzites are fairly clean medium- to fine-grained. In many places the immediate hangingwall of the ore formation is black slate that contains monograptids of mid-Llandoverian age, and thus the ore formation itself probably represents the base of the Silurian.

Above the ore formation (or the graptolitic slates) is a sequence of pyroclastic rocks known locally as the 'Frailesca', followed by a sequence of interbedded volcanics and sediments, that pass eventually into an alternation of siltstones and impure quartzites generally regarded as Devonian in age.

There are three main ore horizons corresponding to the three sandstone members of the ore formation, known from bottom to top as the San Pedro, San Francisco and St. Nicholas. Within the sandstone units the mineralization is in the form of lenses probably corresponding to the more porous parts.

The ore zones are between 3–5 m thick and are worked down to the 14th level of the mine; a depth of some 385 m below surface, and extend along strike for 300–400 m. The ore is quartzite with minor amounts of pyrite, kaolinite and organic matter with the principal ore mineral cinnabar and small amounts of native amalgam. The cinnabar fills the pore spaces between the sand grains, microfractures in the rock and in individual sand grains. The content of mercury varies between under 0·6% (less than which it is not regarded as ore) up to 20%, and its distribution tends to reflect variations in grain size of the quartzite and sedimentary structures.

There are some minor flexures in the ore horizon and a few faults, mostly orientated north-south, that disrupt the ore to a slight extent.

Size and Grade

The mine is worked today at a grade of between 2 and 3% mercury and has yielded during its recorded history 250 000 t of mercury, won from about 5 Mt of ore. It seems probable that almost as much is still present as reserves.

Geological Interpretations

The association of the cinnabar with the detail of sedimentary structure and grain size variation, as well as its restricted stratigraphic distribution, has been taken to indicate an early age of deposition. The association with volcanic rocks has generally been interpreted as genetic. It is suggested that the mercury was emplaced in the more porous parts of the sandstones (probably corresponding to palaeo-channelways) from volcanic sublimates. Some redistribution of the mineralization during the folding of the rocks is thought probable. The puzzle of Almaden is this: of all the enormous number of occurrences in the world of volcanism associated with this type of rock sequence, why here?

Mining

Almaden is in the centre of a 25 km radius circular concession that is the exclusive province of the Almaden Mining Division of the Spanish Ministry of Finance, and the mercury produced is regarded to some extent as a currency reserve. Ore is mined from cut-and-fill stopes above concrete-lined drives and hoisted in vertical shafts. In the nearby plant, it is crushed and roasted at 750°C in special vertical kilns, fired by oil. Fans draw off the mercury and sulphur vapour, the former being recovered in air-cooled condensers, washed chemically and filtered. The workforce of the mine is rather special because each miner is, for reasons of health, only allowed to work 8 days in every 32. Most have smallholdings, small farms or businesses in the town as well. Almaden produces about 60 000 flasks of mercury per year (2070 t) being about a quarter of the world's production.

Further Reading

SAUPE, F. (1973), La Géologie du Gîsement de Mercure d'Almaden. *Sciences de la Terre, Mem.*, **29**, 342pp. [A full modern thesis on the deposit.]

ALMELA, A. *et al.* (1962), Estudio Geologico de la Region de Almaden. *Bol. Inst. Geol. Min. Esp.*, **73**, 193–327. [The most up-to-date description of the general geology of the area.]

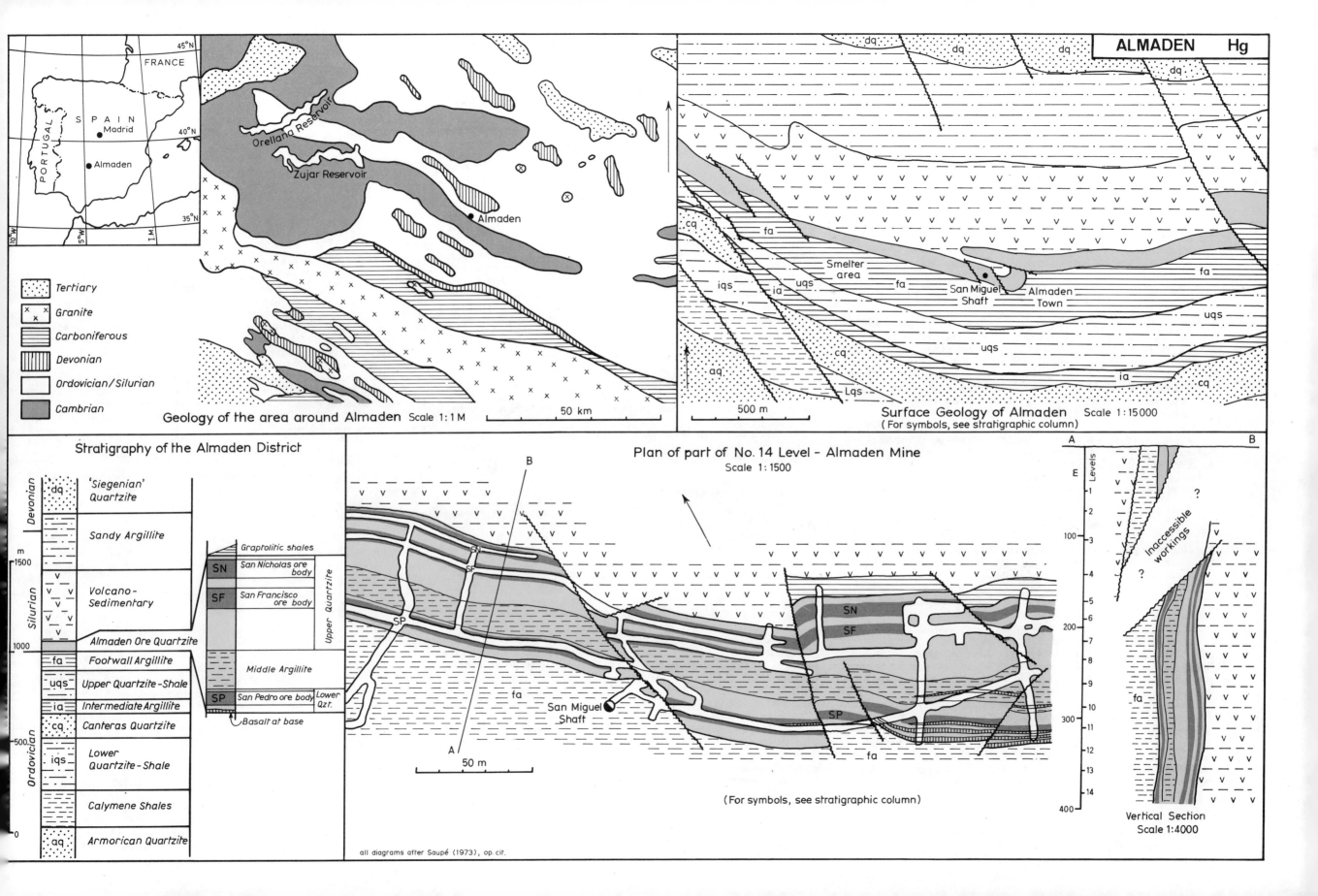

ALMADEN Hg

Legend (Geology of the area around Almaden):
- Tertiary
- Granite
- Carboniferous
- Devonian
- Ordovician/Silurian
- Cambrian

Geology of the area around Almaden Scale 1:1 M   50 km

Inset map: FRANCE / SPAIN / PORTUGAL; Madrid; Almaden; 45°N, 40°N, 35°N; 10°W, 5°W; 1 M

Orellana Reservoir; Zujar Reservoir; Almaden

Surface Geology of Almaden Scale 1:15 000
(For symbols, see stratigraphic column)
500 m

Labels: dq, cq, fa, iqs, uqs, ia, aq, Lqs, Smelter area, San Miguel Shaft, Almaden Town

Stratigraphy of the Almaden District

System	Symbol	Unit
Devonian	dq	'Siegenian' Quartzite
		Sandy Argillite
Silurian	v	Volcano-Sedimentary
		Almaden Ore Quartzite
	fa	Footwall Argillite
	uqs	Upper Quartzite-Shale
	ia	Intermediate Argillite
Ordovician	cq	Canteras Quartzite
	iqs	Lower Quartzite-Shale
		Calymene Shales
	aq	Armorican Quartzite

Column markers: m 1500, 1000, 500, 0

Right sub-column:
- Graptolitic shales
- SN San Nicholas ore body — Upper Quartzite
- SF San Francisco ore body
- Middle Argillite
- SP San Pedro ore body — Lower Qzt.
- Basalt at base

Plan of part of No. 14 Level - Almaden Mine
Scale 1:1500

Labels: B, A, SN, SF, SP, fa, San Miguel Shaft
50 m

(For symbols, see stratigraphic column)

Vertical Section
Scale 1:4000

Labels: A, B, Levels 1–14, E, 100, 200, 300, 400
Inaccessible workings; ?; SN; SF; fa

all diagrams after Saupé (1973), op. cit.

The Deposits of the MacIntyre–Hollinger Field – Canada

This is one of the best examples of the type of gold-quartz vein deposit that occurs frequently in Pre-Cambrian metavolcanic terrains, particularly the so-called Archean greenstone belts. It lies at the western end of the Porcupine District which has, since 1910, produced 1·6 million kilos of gold; a production surpassed only by the Witwatersrand.

Location

The deposits lie on the eastern outskirts of the town of Timmins (latitude 48° 28′ N, longitude 81° 19′ W and 304 m altitude) in Cochrane Province of Ontario. They form the western end of the so-called Porcupine Mining District.

Geographical Setting

The area is rather flat, about 300 m above sea level with ridges of hills on outcrops of the harder rocks up to 410 m altitude. To the north is an extensive area of land covered by glacial clays and thick clay soils, and to the south there are many areas covered by fluviglacial sands. The area gets its name from the Porcupine River that drains the area through several lakes and swamps. The mines are situated round a group of lakes much used for dumping tailings. The area suffers a continental climate, averaging 18°C with spells up to 32° in summer; and −18°C in winter, with cold spells down to −30°C or lower. Heavy snow falls with strong north-west winds in October to November, and thunderstorms are common in summer. The area was once forested with spruce and birch, much of which has been removed for mine timber or burned in forest fires.

The chief town is Timmins, which has grown to a population of over 30 000. It is largely a mining town and is served by rail and road, and has an airport.

History

Between 1896 and 1899, gold was reported along the lake shores by two geologists; Burwash and Parks of the Ontario Bureau of Mines. But no serious prospecting was done until the railway was driven north from the Cobalt District. A famous prospector, Reuben d'Aigle, visited the area in 1904 but found little. Then George Bannerman found good gold showings on Pearl Lake in the summer of 1909 and the rush was on. John Watson found a vast gold-rich outcrop on the property that became the Dome Mine at the eastern end of the area. In the rush of 1909–10 came John Miller and Alex Gillies (who are remembered by two lakes named after them) with Sandy MacIntyre and Benny Hollinger who staked claims on the two mine properties later named after them. It was the heyday of sharp eyes and a pan. Hollinger, then only in his teens, found the first outcrop of his mine just a few metres from a trench dug by d'Aigle five years earlier. Mines went into production almost at once, but were set back in 1911 by a disastrous forest fire. From 1912 onwards, the mines steadily developed and three, the Dome, Hollinger and MacIntyre, went into great depth, each producing gold worth hundreds of millions of dollars. Hollinger Mine became exhausted in 1968 and closed, but though the reserves are depleted and costs have risen, the field still produces.

Geological Setting

The Hollinger–MacIntyre field is at the western end of the Abitibi Greenstone Belt. This belt is surrounded by granitic rocks, and is composed of a thick sequence of volcanic and sedimentary rocks deposited in the Archean era.

Stratigraphic correlation is difficult, because of folding and faulting, and because the volcanic rocks are arranged round a number of centres with little to help correlate between them. Three main facies of rocks are found deposited in any one centre in approximately the following order. The lower suit is of massive or pillow-structured basalts and andesites, with a fair number of intrusions of peridotite, gabbro and diorite. Above this is a sequence also of volcanic rocks, but with progressively more and more dacites and rhyolites and becoming increasingly fragmental; intrusions of quartz porphyry being often found. Both interbedded with the volcanic rocks, and lying as a thick unit above, are sediments. Some of these are very immature and ill-sorted clastic rocks composed entirely of volcanic material. Others are more regular sequences of conglomerate, greywacke and shale. Over large parts of the belt there is an important unconformity below the topmost sequences of sediments; but it is not necessarily the same age in all parts. All the rocks have been folded, often more than once, and have a strong planar fabric. There are several large shear or fault zones that almost parallel the trend of the major fold structures. The rocks were metamorphosed, mostly to greenschist facies, but in some areas, more intensely.

In some places round the edge of the belt, the surrounding granitic rocks seem to form a basement to the supercrustal rocks, but in most places the contact is intrusive and a large number of plutons of granite and granodiorite intrude the volcanic and sedimentary rocks. The age of the deformation and granitic activity is about 2·4 Ga.

The belt is covered in part by sediments belonging to the Huronian Super-Group, particularly in the south, but little more is known of the geological history of the area except that it is cut by a few swarms of younger dolerite dykes, and is partly covered by the products of the Quaternary glaciation. These include extensive areas of varved clays, fluviglacial sands, eskers and the like.

The Abitibi Belt is one of the world's most richly mineralized provinces, with vein gold, base-metal sulphide deposits and a variety of other types, mostly associated with the volcanic sequences.

Geology of the Deposits

The deposits occur in one of the thickest sequences of the Abitibi volcanic rocks and in an area with a large proportion of acid rocks. The lower basic lavas, here known as the Tisdale Group, contain a number of persistent and recognizable units that has enabled local correlation to be made and the structure elucidated (which has been of importance because some are favourable for mineralization). The upper part of the volcanic sequence is largely acid pyroclastic rocks and immature sediment, known as the Krist Formation. There are two overlying sequences of sediments; the Hoyle Group which is argillacious, and above an important unconformity, the Timiskaming conglomerates, greywackes and argillites. The rocks are folded along a north-easterly axis and cut by a sub-parallel fault zone, the Hollinger Main Fault. Elongated bodies of quartz porphyry are found in the volcanic sequence, the largest of which, the Pearl Lake, occupies the position of the Hollinger Main Fault in the MacIntyre Mine; whereas in the Hollinger Mine smaller ones penetrate the axial planes of the syncline and anticline on either side of the Main Fault.

The principle mineralization consists of quartz veins, or zones of quartz veining. They contain quartz, ankerite, some albite tourmaline and sericite, with native gold, scheelite, arsenopyrite, pyrite, chalcopyrite, galena, sphalerite and small amounts of gold tellurides. Alteration of the volcanic rocks on the sides of the veins is general, with feldspars converted to carbonate and a dissemination of pyrite and arsenopyrite. Gold occurs in the quartz-ankerite veins and in the altered wallrocks. The veins occur in a variety of situations, along the contacts of certain specific lava-flows, and frequently congregating around the ends of tongues of porphyry or along the porphyry contacts.

In the MacIntyre Mine there is a different kind of mineralization. A zone of heavy shearing and alteration in the main Pearl Lake porphyry is shot through with veinlets containing sulphides associated with anhydrite, which contains good copper values and some gold.

Size and Grade

Grade information is difficult to give because of changing economic and mining conditions, and the notorious variability of grade in this type of deposit. One estimate for the whole district is 180 Mt, averaging 7·5 g/t gold and 1·5 g/t silver. In its life Hollinger Mine milled 60 Mt of ore, yielding over 600 000 kg of gold, 130 000 kg of silver and 300 tonnes of scheelite (as a byproduct during the Second World War and Korean war). MacIntyre Mine has produced about two-thirds of Hollinger's production and still has some reserves, plus about 5 Mt of porphyry ore containing 1% copper and 6 g/t of gold.

Geological Interpretations

For a long time the field was thought to be quite a simple case of high-to-medium-temperature hydrothermal mineralization derived either directly from the porphyries, or the granite plutons. But many people have found this unsatisfactory, particularly because of the stratigraphic control of many of the ore zones, and lack of zoning. Some have suggested that the importance of the porphyries is structural, behaving as resistant masses causing local dilation of certain rocks around them in response to the folding or fault movements. Others have suggested that, with space created by dilation, metamorphic water leached metals and other elements from the volcanic rocks below, as they underwent more intense metamorphism, and deposited them in the dilation zones. This association of rocks, structure and minerals is a common one in Pre-Cambrian greenstone belts throughout the world, and in some younger, similar environments. There are several other such fields in the Abitibi belt itself.

Mining

The two adjoining mines, MacIntyre and Hollinger, both operated extensive underground workings via numerous shafts down to 2000 m below surface. Most of the working was by horizontal cut-and-fill stoping using waste rock, glacial sand or mill tailings, as fill. In the early days, shrinkage and square-set methods were used. Long hole sub-level methods were introduced to work the copper ore body at MacIntyre. When working, Hollinger had a 4000 t/d cyanidation plant, and MacIntyre runs a more complex mill that combines gravity, flotation and cyanidation with a capacity of 3000 t/d, two-thirds of which is the copper–porphyry ore.

Further Reading

GOODWIN, A, M, and RIDLER, R. H. (1920), The Abitibi Orogenic Belt. In: *Symposium on Basin and Geosynclines of the Canadian Shield* (Baer, A. J., ed.), Geol. Surv. Canada, Paper 70–40, pp.1–44. [Best general description of the whole belt.]

GOODWIN, A. M. (1965), Mineralized Volcanic Complexes in the Porcupine–Kirkland Lake, Noranda Region, Canada. *Econ. Geol.*, **60**, 955–71. [Good account of the mineralization in the richest part of the Abitibi Belt.]

DUNBAR, W. R. (1948), Structural Relations of the Porcupine Ore Deposits.

JONES, W. A. (1948), Hollinger Mine.

FURSTE, B. D. (1948), McIntyre Mine. All in: *Structural Geology of Canadian Ore Deposits*. Can. Inst. Min. Metall., Montreal (pp. 442–56, 464–81, 482–96). [Best and most authoritative works on the area.]

BURROWS, A. G. (1924), *The Porcupine Gold Area*. Ann. Rept. Ont. Dept. of Mines, Vol. 33, pt.2. [Written in the district's heyday, with lots of background information not found in more recent works.]

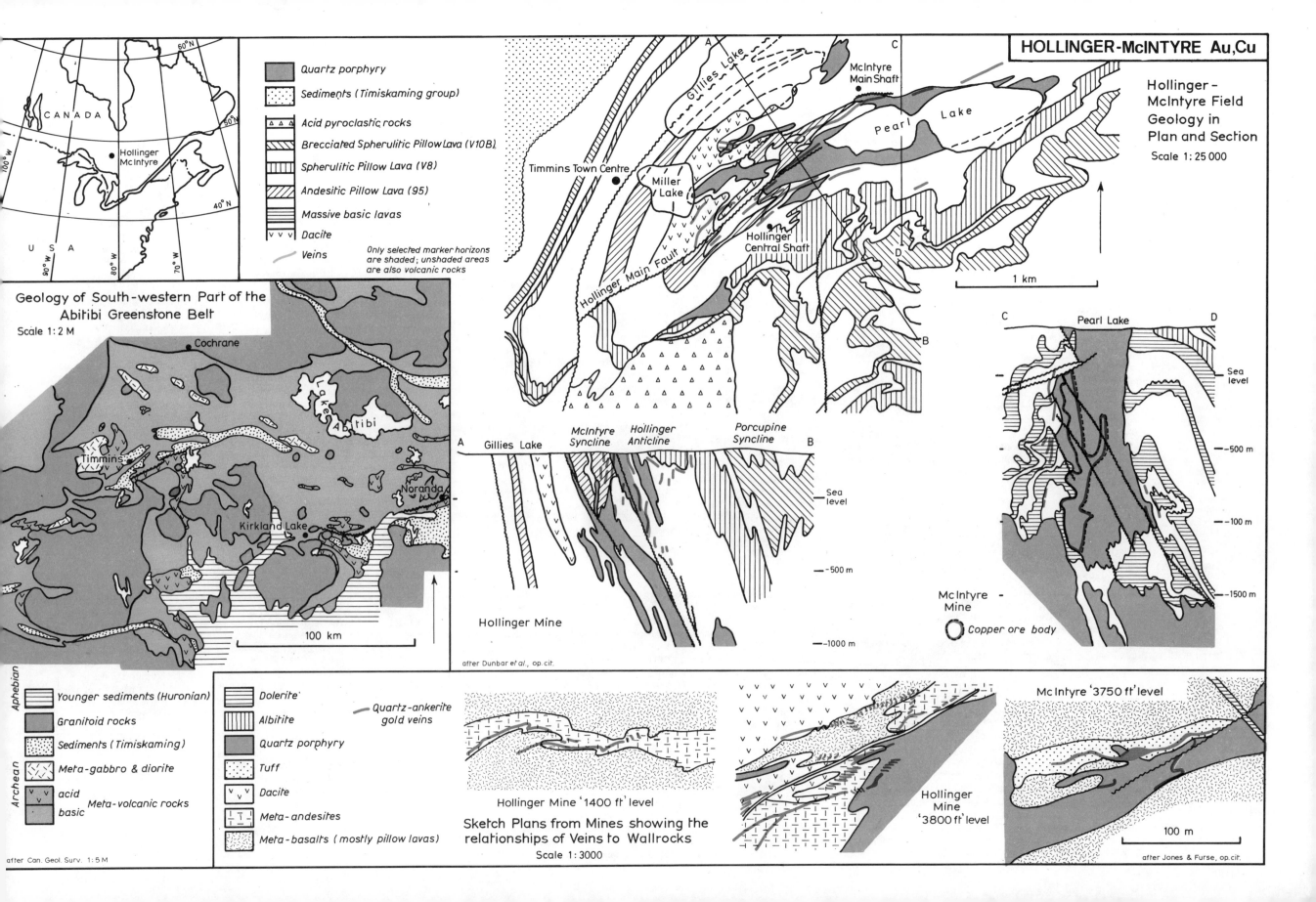

HOLLINGER-McINTYRE Au,Cu

Hollinger-McIntyre Field Geology in Plan and Section

Scale 1:25 000

Legend (top):
- Quartz porphyry
- Sediments (Timiskaming group)
- Acid pyroclastic rocks
- Brecciated Spherulitic Pillow Lava (V10B)
- Spherulitic Pillow Lava (V8)
- Andesitic Pillow Lava (95)
- Massive basic lavas
- Dacite
- Veins

Only selected marker horizons are shaded; unshaded areas are also volcanic rocks

Plan labels:
Gillies Lake, McIntyre Main Shaft, Pearl Lake, Timmins Town Centre, Miller Lake, Hollinger Central Shaft, Hollinger Main Fault

1 km

Section C–D:
Pearl Lake, Sea level, −500 m, −100 m, −1500 m

McIntyre Mine
○ Copper ore body

Section A–B:
Gillies Lake, McIntyre Syncline, Hollinger Anticline, Porcupine Syncline
Sea level, −500 m, −1000 m
Hollinger Mine

after Dunbar et al., op.cit.

Geology of South-western Part of the Abitibi Greenstone Belt

Scale 1:2 M

Cochrane, Lake Abitibi, Timmins, Noranda, Kirkland Lake

100 km

after Can. Geol. Surv. 1:5 M

Legend (bottom left):

Aphebian
- Younger sediments (Huronian)

Archean
- Granitoid rocks
- Sediments (Timiskaming)
- Meta-gabbro & diorite
- acid / basic Meta-volcanic rocks

- Dolerite
- Albitite
- Quartz porphyry
- Tuff
- Dacite
- Meta-andesites
- Meta-basalts (mostly pillow lavas)
- Quartz-ankerite gold veins

Sketch Plans from Mines showing the relationships of Veins to Wallrocks

Scale 1:3000

Hollinger Mine '1400 ft' level

Hollinger Mine '3800 ft' level

McIntyre '3750 ft' level

100 m

after Jones & Furse, op.cit.

(inset top left):
CANADA, U S A, Hollinger McIntyre
60°N, 50°N, 40°N, 100°W, 90°W, 80°W, 70°W

The Homestake Gold Deposit – U.S.A.

One of the world's largest gold mines, certainly the largest in America, Homestake has produced over one million kilos of gold, and continues to produce it at the rate of 18 000 kg per year from workings well over 2000 m below surface.

It is a typical example of a type of gold deposit that is associated with deformed and metamorphosed iron-rich sediments. There are many other examples; notably Morro Velho in Brazil.

Location

The mine is almost in the town of Lead at latitude 44° 22′ N, longitude 103° 46′ W and 1675 m altitude in Lawrence County, South Dakota in the northern part of the Black Hills.

Geographical Setting

The Black Hills area of South Dakota is elliptical, 200 km long and 100 km wide, consisting of forested hills rimmed by a hog's-back ridge, that rises from the great plains of the American Mid-West. The land rises to 2200 m but is well dissected by a network of steep-sided valleys that drain outward to the Cheyenne River. The climate is continental, with mid-day temperatures in summer up to 38°C and winters averaging − 7°C. An average of 740 mm precipitation falls during the year, half as spring rains, and the rest as snow in winter. Settlement in the area is confined to the few small towns based on the mines, ranching and lumbering townships, and Indian reservations. The small towns of Lead, Deadwood, Central City, Sturgis, Whitewood and Spearfish all depend to a large extent on the mine. Although the mine employs only 1500 people, some 28 000 either directly or indirectly gain their means of livelihood from it. The town of Lead is connected by the main railway network to the industrial east.

History

Until 1875, the Black Hills were regarded as a reservation for the Sioux Indians who had little interest in mining gold, but in that year exploration under government protection began, and the following year Frederick and Moses Manuel discovered and staked the main outcrop of the deposit. Production began on several claims, which by 1878 had been amalgamated into a single holding. By 1885, it was realized that the mine was substantial and a more organized development was started. Moreover, the size of the deposit, unlike so many other gold mines, has exceeded even the wildest dreams of its discoverers.

Geological Setting

The core of the Black Hills consists of a foliated and metamorphosed series of sediments, dominantly arenaceous with subordinate amounts of argillaceous and carbonate. There are many bodies of amphibolite, and in the south a complex of granitic intrusions is found which includes very coarse-grained granites and pegmatites. These rocks probably belong to the Aphebian period and are overlain by the Middle Cambrian Deadwood formation which forms a rim round the basement core and a series of hill-top outlyers. The Deadwood has a base of conglomerate and passes up into carbonates and glauconitic shales. Limestones surround the Deadwood outcrops, ranging in age from Ordovician to Mississippian (including the Pahaspa Formation, famous for its numerous caves). The remaining part of the Palaeozoic section is sandy and shaly. As a topographic feature, the Black Hills are delimited by a hog's-back ridge of the Lakota Formation, a cross-bedded sandstone which forms the base of the thick Mesozoic sequence that underlays the surrounding plains; with Cretaceous marine shales being the most important unit. A whole variety of Tertiary and Quarternary continental deposits cover the area in part and throughout the Black Hills there are thick alluvial accumulations in the valleys, obscuring much of the outcrop. Early in the Tertiary, the northern part of the Hills was invaded by stocks, laccoliths, dykes, sills and flows of rhyolite, quartz porphyry and related rocks.

Besides its occurrence in the Pre-Cambrian metamorphic rocks at Homestake, gold occurs widely in the northern part of the Hills in quartz veins, in the Deadwood conglomerates, in breccia zones in the rhyolites, and as placers. In addition, the granitic southern part of the area contains a variety of pegmatite mineral occurrences, and the surrounding Palaeozoic sediments contain a few small lead–zinc deposits.

Geology of the Lead District

In the northern lobe of the Black Hills the Pre-Cambrian rocks have been divided into six formations. The lowest, the Poorman, the base of which is not seen, is a monotonous sequence of grey phyllites and it is at its top that the important Homestake formation occurs. This thin but persistent unit is composed of a rather unusual quartz-sideroplesite rock that becomes rich in cummingtonite where metamorphism was of higher grade. It is in this rock that the gold occurs for the most part. The overlying Ellison Formation consists of mixed schists and includes further but less persistent bands of cummingtonite schist, and a number of quartzites. The upper three formations contain no further gold-bearing horizons. The North-Western Phyllite is partly cut out by an unconformity on top of which are schists of the Flag Rock Formation. This contains several quartzitic horizons thought to have been cherts, and an iron-rich chert horizon that forms prominent rusty outcrops locally called the Iron Dyke and useful in revealing the structure. Then the lithology returns to grey phyllites of the Grizzly Formation. Over the whole area metamorphism increases in grade progressively from south-west to north-east. The basement rocks are cut by numerous dykes of rhyolite and quartz porphyry often, as in the mine itself, occupying the axial planes of folds, or faults.

The area is dotted with hill-top outlyers of the Deadwood Formation capped by rhyolite. The variable thicknesses of conglomerate at its base often contain gold.

The Homestake Gold Mineralization

North of Lead the Homestake Formation forms a series of tight folds orientated just west of north. The main structure is the Lead Syncline which is flanked by smaller, tighter anticlines: the Poorman to the west, Independence and Pierce to the east, between which are the smaller DeSmet and Caledonian synclines. Each of these tight structures is mineralized, and the ore bodies are narrow and plunge down the crests of the folds. In the mineralized zone the cummingtonite has been converted to chlorite, and the rock is invaded by arsenopyrite and pyrrhotite or pyrite, with some magnetite. Free gold occurs, as well as gold locked in the sulphide minerals. Some gold is found again with sulphides, in the rhyolites and quartz porphyries, particularly in brecciated zones; but in the numerous veins and pods of quartz, little gold occurs. Gold also occurs in the Deadwood conglomerates, partly as free gold between pebbles, and partly in zones of oxidized sulphide impregnation. Some of the gold in the prophyry breccias occurs as tellurides.

Size and Grade

Down to present working levels Homestake contains about 120 Mt of ore, containing about 10 g/t gold and 2 g/t silver, of which about 15 Mt still remains as reserves. However, 2070 m below the surface the deposit still persists. The reserves are somewhat sensitive to cut-off grade, and to the price of gold and mining costs.

Geological Interpretations

Gold is in, and associated with, fold crests in the Homestake Formation and is also related to rhyolite and porphyry intrusions. Since detrital gold occurs in the Deadwood conglomerate, it must be Pre-Cambrian. The rhyolites are part of a swarm that cut the Mesozoic cover of the Black Hills and are, therefore, no older than Lower Tertiary. Arguments have raged over the relative importance of these two influences, but the general conclusion is that the greater part of the gold was emplaced with the sulphides after the folding of the Homestake Formation and was modified in its distribution and perhaps its quantity during the Tertiary magmatic activity. Some believe that the gold was in the Homestake Formation at least before its most recent folding, and that its concentration into fold crests resulted from deformation. The Homestake Formation probably started as an iron-rich sediment of the cherty carbonate facies (a lithology that is often associated with gold in Pre-Cambrian areas as widely separated as Brazil, Australia and India). Some have suggested that gold may be formed by the precipitating action of such a lithology on hydrothermal solutions containing gold. The ultimate origin of the gold, however, remains a mystery.

Mining

The Homestake Mining Company mines the deposit still, as it has since 1877. It now has two deep shafts down to the lower levels and a large number of active stopes using cut and fill, shrinkage, sublevel and square-set stoping. Ore is hoisted to a mill with a capacity of 4000 t/d, recovering gold by gravity concentration, cyanization and the charcoal-in-pulp methods, depending on the size fraction. The bullion is parted in a refinery, to produce gold and silver. Homestake, like the Cœur d'Alene district, has a special place in the minds of Americans, being one of the great precious metal mines that has survived into the modern era to support the dollar.

Further Reading

DARTON, N. H. and PAIGE, S. (1925), *Central Black Hills Folio– South Dakota*. U.S. Geol. Surv. Folio No.219. [Beautifully presented atlas of maps with descriptions and illustrations of the regional setting.]

PAIGE, S. (1924), Geology of the Region Around Lead, South Dakota. *U.S. Geol. Surv. Bull.*, **765**. 58pp. [A more analytic work based on the same fieldwork as the Folio with many plans of the upper levels of the mine, and a thorough description of the mineralography.]

NOBLE, J. A. and HARDER, J. D. (1948), Stratigraphy and Metamorphism in a part of the Black Hills and the Homestake Mine, Lead, South Dakota. *Bull. Geol. Soc. Am.*, **59**, 941–76.

NOBLE, J. A., HARDER, J. D. and SLAUGHTER (1949), Structure of a Part of the Northern Black Hills and the Homestake Mine, Lead, South Dakota. *Bull. Geol. Soc. Am.*, **60**, 321–52. [A pair of papers containing a full description of the geology of the area.]

HOSTED, J. O. and WRIGHT, L. B. (1923), Geology of the Homestake Orebodies and the Lead Area of South Dakota. Part I: *Eng. Min. Jr.*, **115**, (18) 793–99; Part II: *Eng. Min. J.*, **115**, (19) 836–43.

McLAUGHLIN, D. H. (1931), Homestake Mine – Ore Genesis and Structure. *Eng. Min. World*, **2**, (10), 640–5. [Papers by mine geologists of the day, the last of which is one of a series of papers covering many aspects of the mine.]

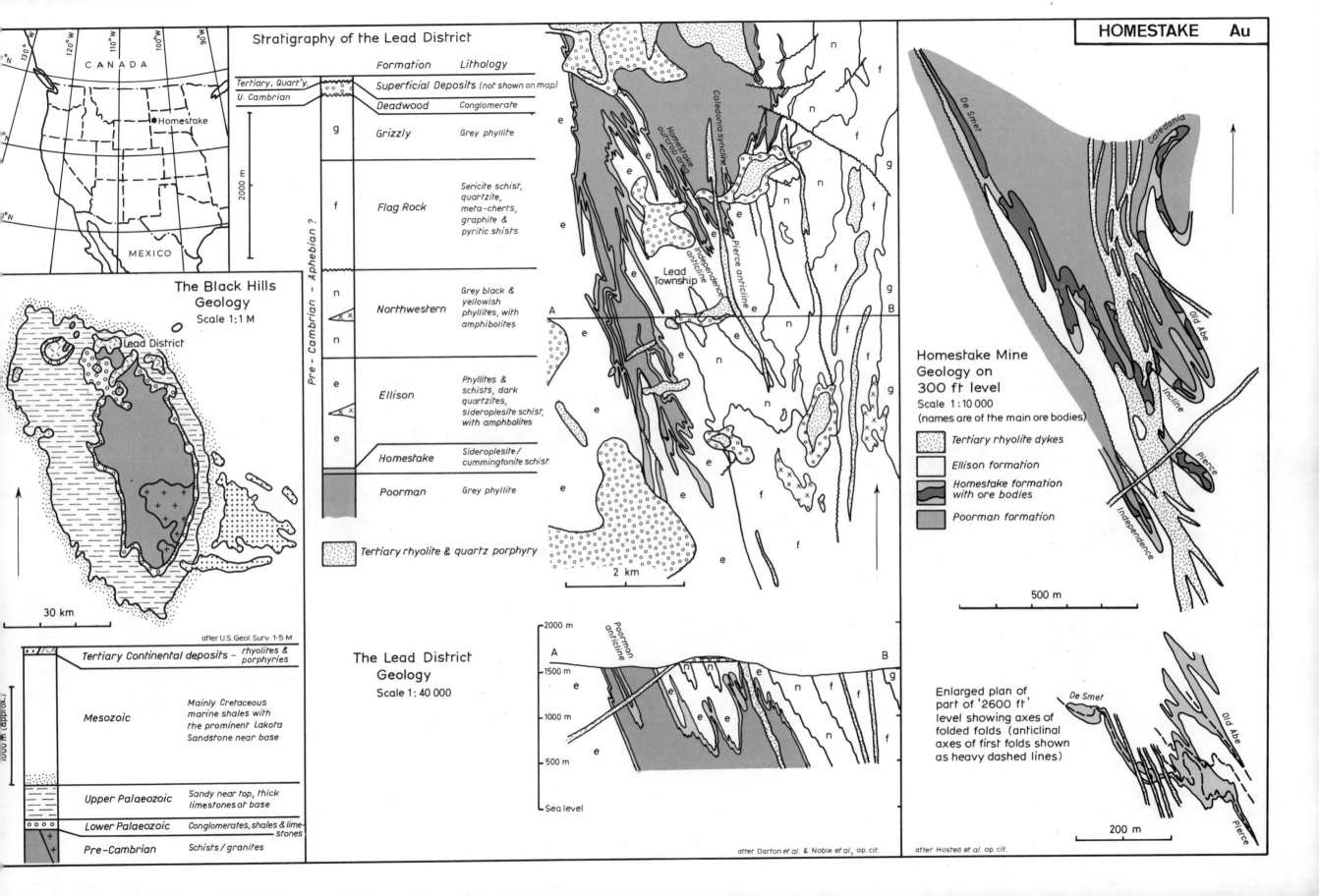

Stratigraphy of the Lead District

	Formation	Lithology
Tertiary, Quart'y.	Superficial Deposits (not shown on map)	
U. Cambrian	Deadwood	Conglomerate
g	Grizzly	Grey phyllite
f	Flag Rock	Sericite schist, quartzite, meta-cherts, graphite & pyritic shists
n	Northwestern	Grey black & yellowish phyllites, with amphibolites
e	Ellison	Phyllites & schists, dark quartzites, sideroplesite schist, with amphbolites
	Homestake	Sideroplesite/cummingtonite schist
	Poorman	Grey phyllite

Pre - Cambrian - Aphebian ?

Tertiary rhyolite & quartz porphyry

The Black Hills Geology
Scale 1:1 M

Lead District

30 km

after U.S. Geol. Surv. 1·5 M

	Tertiary Continental deposits -	rhyolites & porphyries
	Mesozoic	Mainly Cretaceous marine shales with the prominent Lakota Sandstone near base
	Upper Palaeozoic	Sandy near top, thick limestones at base
	Lower Palaeozoic	Conglomerates, shales & limestones
	Pre-Cambrian	Schists / granites

CANADA

Homestake

MEXICO

2000 m

The Lead District Geology
Scale 1: 40 000

Lead Township

Caledonia syncline

Homestake outcrop area

Independence anticline

Pierce anticline

A   B

2 km

Poorman anticline

A   B

2000 m
1500 m
1000 m
500 m
Sea level

after Darton et al. & Noble et al., op. cit.

Homestake Mine Geology on 300 ft level
Scale 1:10 000
(names are of the main ore bodies)

De Smet

Caledonia

Old Abe

Incline

Pierce

Independence

Tertiary rhyolite dykes

Ellison formation

Homestake formation with ore bodies

Poorman formation

500 m

Enlarged plan of part of '2600 ft' level showing axes of folded folds (anticlinal axes of first folds shown as heavy dashed lines)

De Smet

Old Abe

Pierce

200 m

after Hosted et al. op. cit.

The Bunker Hill Silver Deposit – U.S.A.

Bunker Hill has a special place in the history of the United States of America as the great and persistent producer of silver. This mine has significance for economic geologists as one of the few vein deposits to be continuously worked well into the twentieth century. It is the largest, and possibly the richest, of the mines of the legendary Cœur d'Alene district.

Location

The Bunker Hill deposits are centred on latitude 47° 31′ N, longitude 116° 09′ W at about 1000 m altitude. They lie 3 km south-west of the mining town of Kellogg in Shoshone County, Idaho.

Geographical Setting

The Cœur d'Alene district lies on the western flank of the northern Rocky Mountains west of the Bitteroot Divide. It is a youthful, mountainous area with wooded peaks reaching 2000 m in altitude, interspersed with flat-floored valleys descending to 650 m. Most of the area was originally coniferous forest with a few groves of alders and aspens, by fast-flowing mountain streams. A great deal of this was ravaged by a fire in 1910 and today has been replaced by scrub.

The area suffers a mountain continental climate with summer temperatures of 30°C, and winters in which heavy snow usually falls and temperatures descend to − 15°C. To this day the bulk of the population in the area is concentrated in the towns, Kellogg, Wallace, Burk, Mullan, where mines have been developed.

History

Despite encountering a little resistance from the few Indians who inhabited the area, the U.S. Army drove a road over the district in 1854 and, as was so usual in America, this was followed by prospectors. In 1878, A. J. Prichard found gold in what became known as Prichard Creek and a rush developed for alluvial gold which was short lived. But others, among them Phil O'Rourke, N. S. Kellogg, Con Sullivan and Jacob Goetz, searched up the gullies. They are credited with the discovery of the Bunker Hill deposit in 1885. Mining at the then rapid rate of 500 tonnes per day, began at once, and was increased when the Northern Pacific Railway Company built a link to the area. Between 1892 and 1899 disastrous labour troubles hit the area, so much so that, fearing for the silver in their coinage, the Federal Government sent in troops to restore order. One result of this was the setting up of a 'Labour Bureau' which has persisted to the present day. Thus, the Cœur d'Alene district, and with it Bunker Hill, has a special place in the history of American capitalism.

Geology of the Cœur d'Alene District

The district is a strongly tectonized sector of that part of the Rocky Mountains formed of rocks of the Belt Super-group. These shallow-water clastic sediments are 8000 m thick and pose something of a geochronological problem because they are overlain by Cambrian sediments, but are cut by veins containing uranium minerals apparently 1·25 Ga in age.

At the bottom of the Belt Super-Group is the Pritchard Formation, a monotonous clastic sequence the base of which is not seen, and is very difficult to subdivide. Better understood is the Ravalli Group that has been divided into a series of formations that can be correlated, enabling a plausible structural interpretation of the area to be made. The sequence is terminated stratigraphically by the Striped Peak Formation (part of the Missoula Group) which is eroded and unconformably overlain by fossiliferous Cambrian rocks.

The rocks of the district are highly folded and faulted. There are a series of north–south trending folds of early age, much modified by large-scale faulting. The largest fault in the area is the WNW-trending Osburn Fault that can be traced for 800 km from Montana to Washington State, and is interpreted as having a dextral movement of 26 km. Other major faults are recognized in the area, as well as a host of subordinate sub-parallel ones often more closely related to the mineralization than the major faults. Intruded into the Belt rocks are a number of bodies (known as the 'Gem Stocks') of granodiorite and monzonite of mid-Cretaceous age, similar to the large Idaho Batholith that outcrops some distance to the south.

The consistent altitude of the peaks in the area suggests that it is a deeply dissected peneplane. This was probably early Tertiary in age, and the dissection by a network of rivers was greatly influenced by the damming effect of the Colombia River Basalts in the west and the Quaternary glaciation. Most of the major valleys have older alluvial terraces, some metres above their present floodplains.

The Cœur d'Alene Mining District

Hundreds of vein and lode deposits occur in an area 50 km by 25 km and are, for the most part, fracture fillings and replacements associated with the multitude of local fault systems that strike parallel or sub-parallel to the great Osburn Fault. Silver dominates the economic mineralization but, in addition, there is lead, zinc, copper and a little gold. Accompanying minerals include pyrite, arsenopyrite, pyrrhotite, quartz siderite and barite.

Some of the deposits are found in close association with the contact zones of the Gem Stocks, but most occur in belts up to 30 km away from them. The deposits are zoned. Very roughly, there are zones of silver/zinc deposits with pyrrhotite and a silicate gangue near the Gem Stocks; silver/copper deposits with siderite gangue in the east and south-east, and a series of zones of silver/lead deposits in the east and west. The zones do not show a particularly clear relationship, either to the Gem Stocks or to the major faults, but the pattern of distribution is more systematic if the effect of the Osburn Fault is removed. Some bedded veins of gold/quartz ore occur, and both the older and younger alluvium in the valleys contain gold.

The Bunker Hill Deposit

This consists of a number of ore shoots associated with north-west trending faults in a complex area south of the Osburn Fault. The host rocks are ripple-marked sandstones and siltstones at the top of the Revett Formation and the overlying St. Regis Quartzite. The beds dip almost vertically, and are interrupted by a series of faults that dip to the south-west. The mineralization is in the form of fracture zones associated with the smaller faults between the major faults, particularly where lithological differences affect the geometry of the fault planes. There are three main facies of mineralization that are separated both in space and time. The main one is the Bunker Hill facies consisting of infilling and replacement by galena, pyrite, tetrahedrite and some chalcopyrite with a gangue dominated by siderite. The silver is present in both galena and tetrahedrite. The Blue Bird facies occurs along fault zones in the main strike direction and is similar in mineralogy to the Bunker Hill except that sphalerite is more important than galena, and quartz more important than siderite. The third facies is the Jersey, which is associated with cross-cutting veins, of a silver-rich galena and pyrite mineralization with a quartz gangue.

Size and Grade

Since its discovery in 1885, Bunker Hill has gone through many changes in mining technology, as well as changes in ownership and economic climate and, with such a complex arrangement of irregular ore shoots, it is difficult to give an accurate estimation of size. Over 30 Mt of ore have been mined, and the total size may be as high as 40 Mt. The average grade is about 100 g/t silver and between 4 and 5% each of lead and zinc.

Bunker Hill is only one of the deposits of the Cœur d'Alene district that has produced over two billion dollars worth of silver since its discovery.

Geological Interpretations

Most people are agreed that the mineralization was emplaced in several stages, by hydrothermal solutions of medium temperature, into space created by the complex tectonic events which have affected the area. The interaction of faulting with both the variable lithology of the belt rocks and with the early folds, have helped to create the reservoirs in which the solutions accumulated, depositing minerals both in the space created and by replacement. However, there are some problems. One deposit in the district at the Success Mine shows evidence that the mineralization was formed after the emplacement of one of the Gem Stocks which clearly has a mid-Cretaceous age; but lead isotopes from the district show ages of 1·25 Ga. Opinion is divided. On the one hand some authors suggest that the mineralization was due to magmatic hydrothermal solutions emanating from bodies of which the Gem Stocks are outcropping examples, the lead having been derived from upper crustal sources, where it had lain since Pre-Cambrian times. Others hold that the mineralization has accumulated throughout the history of the area, remobilized from low-grade accumulations originally deposited in the belt rocks. The Gem Stocks may only represent evidence of high heat-flow (and therefore enhanced groundwater circulation) at one stage in the history of the area.

Mining

The Bunker Hill Company operates extensive underground workings accessed by long haulage tunnels from their installation at Kellogg. The present production is 450 000 t/a, averaging 115 g/t silver and about 10% lead/zinc combined. The ore is treated at Kellogg by flotation and some smelting and refining is carried out on site.

Further Reading

HOBBS, S. W. and FRYKLUND, V. C. (1968), The Cœur d'Alene District, Idaho. In: *Ore Deposits of the U.S., 1933–1967* (Ridge, J. D., ed.), (Garton–Sales Volume) Chap. 22. [An up-to-date review with good bibliography to this area.]

HOBBS, S. W. *et al.* (1965), *Geology of the Cœur d'Alene District, Shoshone County, Idaho*. U.S. Geol. Surv. Prof. Paper 478, 139pp. [An up-to-date account of the general geology with excellent maps.]

FRYKLUND, V. C. (1964), Ore Deposits of the Cœur d'Alene District, Shoshone County, Idaho. U.S. Geol. Surv. Prof. Paper 445, 103pp. [The companion volume to Hobbs (1965) describing the ore deposits.]

RANSOME, F. L. and CALKINS, F. C. (1908), *The Geology and Ore Deposits of the Cœur d'Alene District, Idaho*. U.S. Geol. Surv. Prof. Paper 62, 203pp. [Although old, this contains a thorough methodical description.]

UMPLEBY, J. B. and JONES, E. L. (1923), Geology and Ore Deposits of the Shoshone County, Idaho. *U.S. Geol. Surv. Bull.*, **732**, 156pp. [Contains much more detail on the Bunker Hill Deposit.]

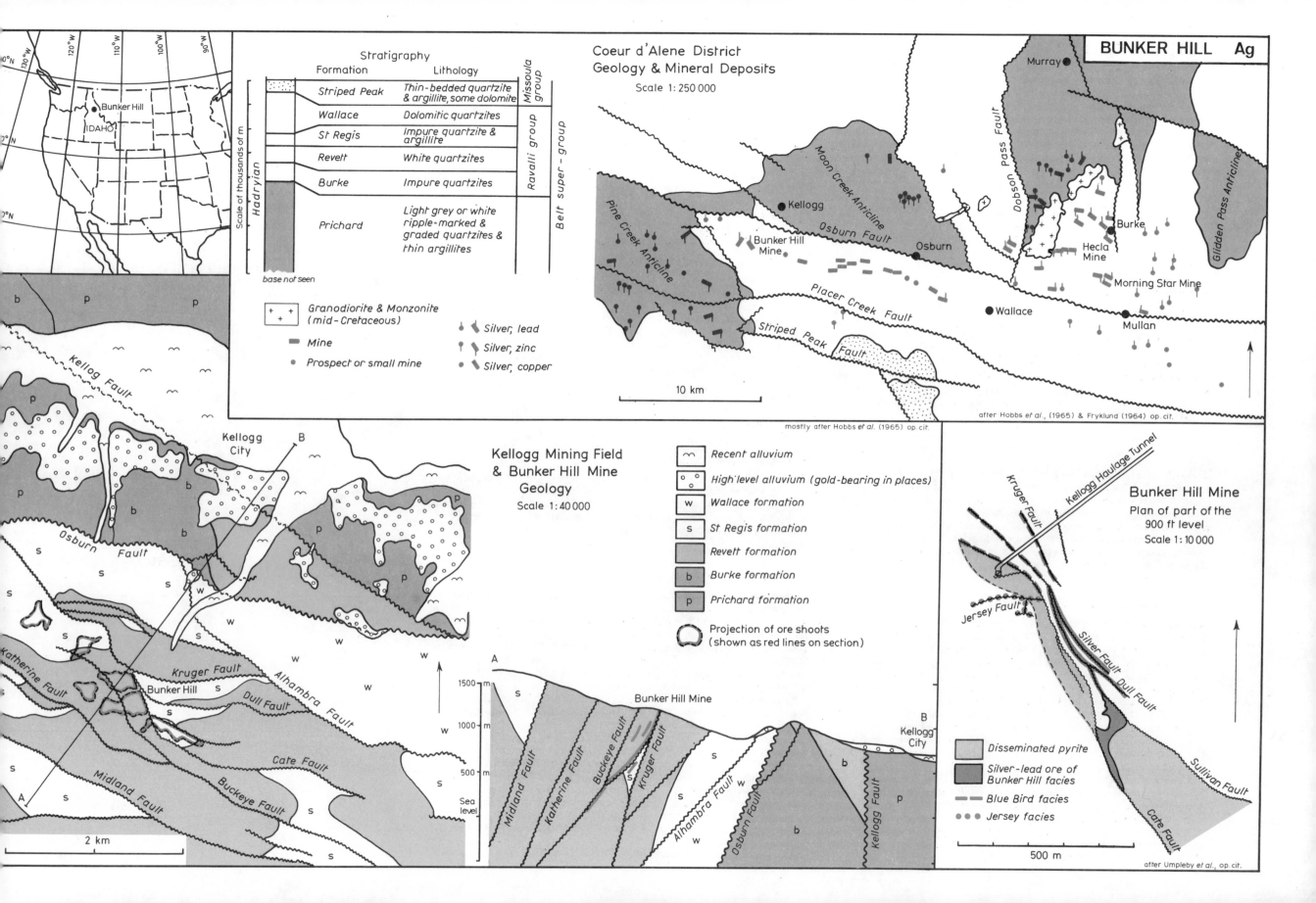

Stratigraphy

Formation	Lithology		
Striped Peak	Thin-bedded quartzite & argillite, some dolomite	Missoula group	
Wallace	Dolomitic quartzites	Ravalli group	Belt super-group
St Regis	Impure quartzite & argillite		
Revett	White quartzites		
Burke	Impure quartzites		
Prichard	Light grey or white ripple-marked & graded quartzites & thin argillites		

Scale of thousands of m — Hadryian

base not seen

+ + Granodiorite & Monzonite
+ (mid-Cretaceous)

━━ Mine

● Prospect or small mine

Silver, lead

Silver, zinc

Silver, copper

Coeur d'Alene District Geology & Mineral Deposits

Scale 1: 250 000

Murray
Dobson Pass Fault
Moon Creek Anticline
Glidden Pass Anticline
Kellogg
Osburn Fault
Burke
Pine Creek Anticline
Bunker Hill Mine
Osburn
Hecla Mine
Placer Creek Fault
Morning Star Mine
Striped Peak Fault
Wallace
Mullan

10 km

after Hobbs et al., (1965) & Fryklund (1964) op.cit.

mostly after Hobbs et al. (1965) op.cit.

Kellog Fault
b   p   p

Kellogg Mining Field & Bunker Hill Mine Geology

Scale 1:40 000

Kellogg City
Osburn Fault
s   s   s
w
p
b
p
p
w
w
Katherine Fault
Kruger Fault
Bunker Hill
Alhambra Fault
Dull Fault
w
Cate Fault
s
Midland Fault
Buckeye Fault
s
s

2 km

∼∼ Recent alluvium

∘∘ High level alluvium (gold-bearing in places)

w Wallace formation

s St Regis formation

Revett formation

b Burke formation

p Prichard formation

⬭ Projection of ore shoots (shown as red lines on section)

A
1500 m
1000 m
500 m
Sea level

s
Midland Fault
Katherine Fault
Buckeye Fault
Kruger Fault
s
Bunker Hill Mine
Alhambra Fault
w
Osburn Fault
s
b
w
B Kellogg City
b
p
Kellogg Fault

Bunker Hill Mine

Plan of part of the 900 ft level

Scale 1: 10 000

Kellogg Haulage Tunnel
Kruger Fault
Jersey Fault
Silver Fault
Dull Fault
Sullivan Fault
Cate Fault

500 m

Disseminated pyrite

Silver-lead ore of Bunker Hill facies

━ ━ Blue Bird facies

• • • Jersey facies

after Umpleby et al., op.cit.

The El Salvador Porphyry Copper Deposit – Chile

Because of its high altitude in the Andes, and the dry climate of the Atacama Desert, El Salvador is so well exposed that it has become one of the most completely studied porphyry copper deposits of South America. It shows most of the features of a classic porphyry copper deposit.

Location

El Salvador lies on the side of the Mountain of Indio Muerto at latitude 26° 15′ S, longitude 69° 34′ W and a mean altitude of 3050 m. It is 115 km by rail east of the port of Chañaral, and 25 km from the older mine of Potrerillos.

Geographical Setting

The deposit is in a range of mountains above the coastal plain of Chile, and separated by a high plateau from the even higher range that forms the frontier with Argentina. The landscape is mostly bare rock, with sparse mountain vegetation in the deep gullies. El Salvador is in the Atacama Desert, though not its driest part, receiving about 100 mm precipitation a year, mostly as snow in the cold winters. Although the summer is bright with long periods of sunshine, the temperature averages only 10°C. Besides the two mining camps of El Salvador and Potrerillos, there is little human activity in the area.

History

Turquoise was worked by the Incas from the outcrop area of the mineralization and the fact became known to the early Spanish conquistadores of Chile. Modern interest in the area followed the development of Potrerillos, and was first reported on in 1922. The area was investigated in detail by geologists from Potrerillos in 1944 and 1945, but it was not until five years later that any drilling was done; and even then the difficulty of access held up the work so that it was not until 1954 that the main ore zone was intersected. El Salvador became a producing mine in 1959.

Geological Setting

Both El Salvador and Potrerillos lie in an area of Upper Cretaceous and Lower Tertiary volcanic rocks, 200 km long and 50 km wide. The basement of the area is not known, but the Upper Cretaceous Cerrillos Formation is over 5 km thick consisting of a lower member of clastic and pyroclastic rocks, andesitic in composition, and an upper member consisting largely of andesite lava flows. The Cerrillos was folded into a broad arch along a northerly axis and then deeply eroded. The eroded surface of the Cerrillos was infilled and partly overlain by two later sequences of volcanic rocks. The Lower Hornitos Formation of Palaeocene age is dominated by a thick (400 m) sequence of rhyolitic ignimbrites, and is unconformably overlain by the Indio Muerto Formation consisting of rhyolite flows and pyroclastic rocks dated as early as Eocene.

There are numerous intrusive rocks in the area. In the vicinity of El Salvador there are rhyolite domes (one of which forms the peak of Indio Muerto) rhyolite plugs, and a complex series of porphyritic granodiorites with which the mineralization is associated. These, together with numerous dykes, are late Eocene in age. There are some areas of locally derived gravels; the remnants of late Tertiary erosion levels. On the lower ground alluvium fills the valley floors.

Geology of the El Salvador Deposit

There are several mineralized zones around Indio Muerto, but the main ore zone is associated with the southern lobe of a granodiorite porphyry complex, west of the peak of Indio Muerto. These intrusions cut the Cerrillos and Hornitos Formations and a series of rhyolite bodies. The Indio Muerto Formation has been eroded away from this area, and is exposed only on the lower ground to the south.

Three phases of granodiorite–porphyry intrusion are recognized; the X, K and L intruded in that order, recognizable by variations in texture and including a number of breccia zones. A wide area round the complex has suffered some degree of propylitization, and the deeper parts of the granodiorite complex and its immediate wallrocks have suffered alteration that includes the development of potash–feldspar, biotite and anhydrite. Above the zone of potash–feldspar alteration is a zone where the rocks have been sericitized and pyritized and, in addition, large parts of the upper zones have been also affected by argillic alteration that masks the earlier sericitization (and even this is modified in places by kaolinitization). Several sets of veinlets cut both the intrusions and the surrounding wallrocks and, to a greater or lesser extent, these contain the sulphide mineralization. This is distinctly zoned with a bornite/chalcopyrite core also containing molybdenite, a chalcopyrite/pyrite transition zone, and an extensive pyritized mantle. Most of the exposed parts of the intrusive complex is shot through with pebble dykes, narrow and wide zones filled with sub-angular or rounded fragments in a fine-grained matrix of similar composition.

The upper part of the mineralized zone has been leached of its sulphide minerals leaving a staining and impregnation of goethite and/or jarosite, the relative distribution of which tends to reflect the sulphide zoning below. At the base of the leached zone is an extensive 'blanket' of chalcocite mineralization, covering and extending beyond, the underlying primary sulphide mineralization.

Size and Grade

The mine was started on reserves of 270 Mt containing 1·6% copper, but this included only the richer ore in the chalcocite blanket because of the relatively high cost of opening a mine in such an inaccessible place. The size of deposit as a whole is probably more like 1000 Mt of about ½% copper. The ore worked contains 0·15 g/t gold, 1·5 g/t silver and 0·02% molybdenum.

Geological Interpretations

El Salvador has been the subject of a special geological study, and the Company geologists concerned have concluded that events began with the intrusion of the X and K porphyries which terminated with a hydrothermal phase to form the potash–feldspar alteration, the majority of the bornite and chalcopyrite veinlets and the halo of propylitization. Fluid inclusion studies suggest that these processes took place at temperatures from 600°C to 380°C. The intrusions are regarded as having been derived by the partial melting of the mantle, largely because the average $^{87}Sr/^{86}Sr$ ratios is 0·704. The L porphyry came later and modified the mineralization and alteration zones. During this phase the upper zones were sericitized and argillitized, and the pyrite mantle formed.

At this stage the complex was probably capped by a volcano, which exploded periodically, with consequent boiling of the magma below, and the formation of channelways, carrying gas, that became the pebble dykes.

It is concluded that as the complex developed, more and more water from the surrounding formations became involved, so that, whereas the earlier phases of potash–feldspar alteration and associated mineralization were formed from magmatic water, the sericitization and argillic alteration were caused by the action of local formation water that was drawn into the system.

Quite soon after these events, the overlying rocks were eroded away leaving the mineralized zone exposed to deep weathering; and as the surface was eroded down, pyrite was oxidized producing an acid groundwater that dissolved out copper, and re-emplaced it further down as chalcocite by replacement of chalcopyrite and bornite. During this process iron from the sulphides was deposited as goethite and jarosite, and the upper zones kaolinized.

Mining

From 1959 to 1971 the Anaconda Company operated a partly underground and partly open-pit mine accessed by a series of adits at various levels up the mountain, producing during that time 73 Mt of ore at a grade of 1·5% copper. Ore was extracted along the lowest haulage adit to a flotation mill. Since 1971 the mine has passed into the hands of the Chilean State which is beginning to re-establish the mine.

Further Reading

GUSTAFSON, L. B. and HUNT, J. P. (1975), The Porphyry Copper Deposit at El Salvador, Chile. *Econ. Geol.*, **70**, 857–912. [A comprehensive description and general interpretation.]

PERRY, V. D. (1960), History of El Salvador Development. *Mining Eng.*, **12**, 339–82. [Traces the history of exploration up to the opening of the mine.]

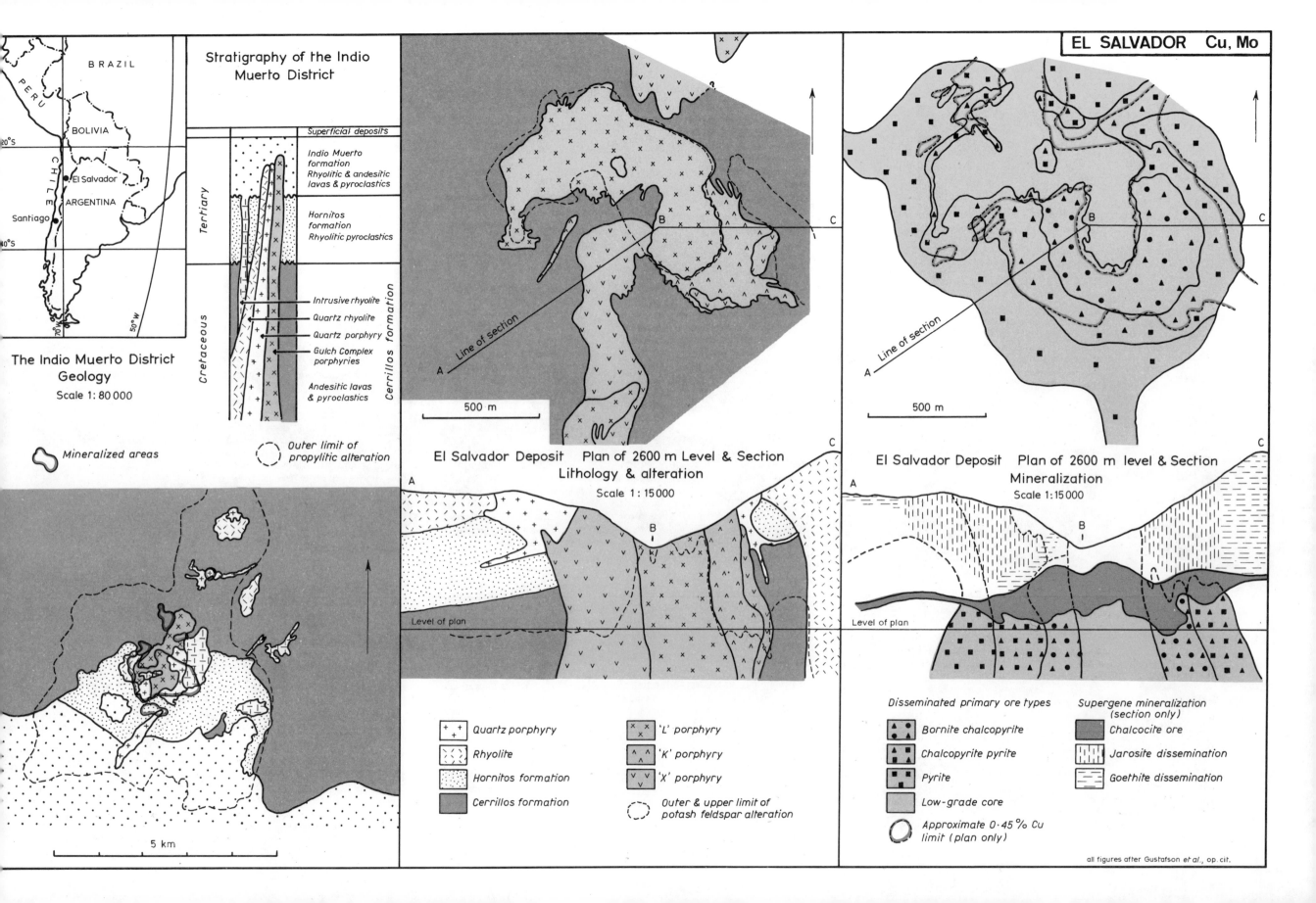

The Indio Muerto District
Geology

Scale 1 : 80 000

Mineralized areas

Outer limit of propylitic alteration

Stratigraphy of the Indio Muerto District

Superficial deposits

Indio Muerto formation
Rhyolitic & andesitic lavas & pyroclastics

Hornitos formation
Rhyolitic pyroclastics

Intrusive rhyolite
Quartz rhyolite
Quartz porphyry
Gulch Complex porphyries

Andesitic lavas & pyroclastics

EL SALVADOR   Cu, Mo

El Salvador Deposit   Plan of 2600 m Level & Section
Lithology & alteration

Scale 1 : 15 000

El Salvador Deposit   Plan of 2600 m level & Section
Mineralization

Scale 1:15 000

Quartz porphyry

Rhyolite

Hornitos formation

Cerrillos formation

'L' porphyry

'K' porphyry

'X' porphyry

Outer & upper limit of potash feldspar alteration

Disseminated primary ore types

Bornite chalcopyrite

Chalcopyrite pyrite

Pyrite

Low-grade core

Approximate 0·45 % Cu limit (plan only)

Supergene mineralization (section only)

Chalcocite ore

Jarosite dissemination

Goethite dissemination

all figures after Gustafson et al., op. cit.

The Chuquicamata Copper Deposit – Chile

Chuquicamata is the largest copper deposit in the world and would demand our attention for no other reason, but it has many interesting geological features as well. It is a 'porphyry-copper' type deposit with a rather atypical pattern of alteration, and the example of the type with best development of 'oxide-ore'. The drying out of the climate in the Atacama desert region has left a large body of soluble copper mineralization and has preserved a considerable amount of soluble copper in adjoining alluvium; an accumulation known as the Exotica deposit.

Location

Chuquicamata is situated at latitude 22° 17′ S, longitude 68° 55′ W and at 2840 m altitude in the province of Antofagasta, 150 km east of the port of Tocopilla.

Geographical Setting

Chuquicamata is in the northern part of the Atacama Desert and in the foothills of the high Andes. The land around is mature topography with hills formed by the outcrops of plutonic rocks and alluvial-filled valleys between. The climate is very dry, the annual rainfall averages less than 100 mm, all of which falls in one or two rainstorms. The winter is windy with cold days and freezing nights, and the summer is warm with maximum temperatures (midday) of up to 35°C. There is practically no vegetation except for a few desert bushes and there is no human activity at Chuquicamata apart from the mine. Twenty kilometres south of the mine flows the Loa river, fed by meltwaters from the high Andes in spring, on which is situated the modest town of Calama. This town is on the main road and railway from Antofagasta over the mountains to Bolivia, and provides the mine with communication to the coast.

History

Copper objects and ornaments made from secondary copper minerals are found in local Indian graves, so it is assumed that Chuquicamata was known before the arrival of the Conquistadores. Modern interest in the deposit dates from the opening up of 1880s. Several attempts were made to mine minerals from veins by British and Chilean companies, but the veins were narrow and the operation not a success. However, the news of bulk open-pit mining by steam shovel spread quickly from Bingham Canyon. In 1915 open-pit mining began, developed later when the company was acquired by the Anaconda Company (of Butte fame). The mine was expanded on several subsequent occasions, but a major deepening of the open-pit after the Second World War necessitated building drainage tunnels and a new plant to treat the sulphide ore, and it was during investigations for these projects that the so-called Exotica ore body was found by accident.

Geological Setting

The Andes Mountain Range is the type 'Cordillera', a sequence of plutonic and volcanic rocks of the calc-alkaline suit, associated with relatively immature sediments, welded along the edge of a major continental block. Palaeozoic meta-sediments and plutonic rocks are known from a number of the horsts produced by block-faulting; but for the most part, the region is made up of plutonic, volcanic and sedimentary rocks that become progressively younger inland. The coastal belt is largely of Jurassic age, further inland the predominating rocks are Cretaceous. By the time one reaches Chuquicamata, Tertiary rocks begin to predominate and these are succeeded in turn by Quaternary volcanic rocks and the active volcanic range of the high Andes.

At Chuquicamata, Cretaceous volcanic rocks and sediments are invaded by early Tertiary granodiorites and both covered in part by late Tertiary and Quaternary clastic sediments. As a result of the dry climate many of the inland or semi-inland drainage basins contain player-lakes, and large areas of the surface sediments are cemented by salt and gypsum. Between Chuquicamata and the sea are found some of the famous sodium nitrate deposits.

Porphyry copper deposits occur along the whole of the Andes but Chuquicamata is to some extent isolated.

Geology of the Chuquicamata Deposit

Most of the copper mineralization is contained in an altered and fractured zone at the southern end of the Chuquicamata Granodiorite. At the southern end of the deposit, the intrusive is hidden under a thick pile of poorly consolidated and poorly sorted gravels, sands and silts.

The Chuquicamata Granodiorite is porphyritic with large orthoclase phenocrysts up to 50 mm in length in a groundmass of plagioclase, hornblend, quartz and biotite with an average grain size of about a centimetre. Two adjoining intrusives, the Elena and the Fortuna, are similar in composition, and the three are probably phases of a multiple pluton.

A large fault separates the Fortuna from the Chuquicamata Granodiorite, but the contact with the Elena is gradational. The intrusive rocks cut a complex of sediments and metamorphic rocks that include some of Upper Cretaceous age, so the granodiorites are presumed to be early Tertiary.

The Chuquicamata intrusive is extensively fractured and altered. Several systems of fractures are present and are much more numerous than the ones indicated diagrammatically on the adjoining map. Many of the fractures are sinuous, having the form of shear-zones at the extremities, and open extension fractures in the centre. Many of these extension fractures are filled with copper mineralization. There are two predominating types of alteration; silicification and sericitization. Much of the granodiorite has been subjected to relatively mild alteration in which the feldspars have been affected by change to sericite, but the original igneous texture remains readily visible. The true zones of sericitic alteration are ones in which the rock is extensively shattered, all feldspar broken down into sericite and traversed by a fine network of quartz veinlets. In the siliceous (or silicified) zones the original nature of the rock has been completely obliterated by strong crushing, and the rock consists of aggregates of quartz, cross-cutting quartz veinlets, and some interstitial sericite. These two types are, to some extent, gradational one to another and the siliceous zones correspond to those of intense fracturing, the sericitic zones being generally marginal to them. The other type of alteration is intense silicification which has generated the so-called 'flooded rock'. In these zones the original texture of the rock is apparent and it is not greatly crushed or fractured; the feldspars are not much affected by sericitization, but the whole rock is invaded by a new generation of quartz, making it very hard and compact.

The mineralization is of two types; 'sulphide' and 'oxide'. In the sulphide zone, there are quartz molybdenite veinlets in the sericitic alteration areas. Pyrite occurs throughout the ore zone along with the two principal ore minerals, enargite and chalcopyrite. Most of the sulphides are in veins or veinlets rather than truly disseminated.

Chalcocite occurs in the upper parts of the sulphide zone representing a rather diffuse and not well-developed enrichment blanket.

The 'oxide' ore is Chuquicamata's most spectacular feature. It occurs as one distinct mass above the sulphides and mixed with the upper parts of the sulphide zone. The dominant mineral is brochantite but, near the surface, 5–10% of the copper is present as antlerite, chalcanthite or atacamite. Unlike other porphyry copper 'oxide ores' there is a general paucity of carbonates and oxides. The leached capping that overlies the ore zone (and separates the main brochantite body from the sulphide zone) contains a great deal of jarosite and a number of normally soluble minerals, mainly sulphates of iron.

Some way to the south of the mine is a zone of Quaternary gravels cemented by copper minerals, particularly chrysocolla. This is separately mined and is known as the Exotica ore body.

Size and Grade

Down the depths of exploration, the deposit contains of the order of 4 Gt of ore containing over 1% Cu. For the first half century of mining the grade worked was 1·6%.

Geological Interpretation

Chuquicamata is a porphyry copper, but perhaps not as 'typical' an example as El Salvador (q.v.). It seems possible however that it may have had the usual development alteration and mineralization zones which were obliterated at a late stage by the intense development of sericite and silica. In so far as the mineralization is in veins and very dominated by fractures which form a series of systems, it resembles Butte. It may be that, like Butte, the space for the introduction of the mineralizing solutions was created by tectonic fracturing rather than the hydrothermal breccia formation of the typical porphyry copper.

The development of the brochantite ore is unusual, and is thought to result from the local climatic and geomorphological changes during the Quaternary. It is now thought that the brochantite ore zone was originally a very well-developed chalcocite enrichment blanket. A sudden rejuvenation of the topography and lowering of the groundwater level caused the chalcocite to become oxidized, but the copper was not removed in solution because there was too little pyrite remaining to generate the required acid groundwater (or that any acid produced was neutralized by sericite and feldspars). At the same time, the climate became much dryer and the lack of groundwater allowed the oxidized copper to remain as sulphates and chlorides. On the west side of the ore body where the underlying sulphide ore is richer in pyrite, it is supposed that the original chalcocite blanket contained some unreplaced pyrite that provided enough acid solution to remove the copper. Part of this went to form the present chalcocite ore (which is thicker in this area) and part escaped along the fissures into the groundwater of the valley to the south, where it became precipitated as a coating around pebbles.

Mining

Chuquicamata is mined by a division of the Chilean State Copper Corporation (CODELCO) and produces 440 000 t of copper per year, representing about half the production of Chile, and one-twelfth that of the western world. The ore is mined from a large open-pit and sent to one of two plants. The 'oxide' ore is crushed, ground and placed in large leaching tanks where dilute acid solution removes the soluble copper minerals; the copper is then recovered electrolitically. The sulphide ore is sent to a flotation plant, and the resulting sulphide concentrates are smelted on site and refined along with the oxide-ore product.

Further Reading

TAYLOR, A. V. (1935), *Ore Deposits at Chuquicamata, Chile*, pp.473–84 in XVI Int. Geol. Congr. Copper Resources of the World. Washington 1935, 2 vols. 855pp. [The most comprehensive work on the deposit, well-illustrated and informative.]

LOPEZ, V. M. (1939), The Primary Mineralization at Chuquicamata, Chile, S.A. *Econ. Geol.*, **34**, 674–711. [Contains the result of more extensive study of the primary ore and the alteration than Taylor.]

JARRELL, O. W. (1944), Oxidation at Chuquicamata, Chile. *Econ. Geol.* **39**, 251–86. [A thorough work on the unusual oxide ore of the deposit with the results of some experimental work on the subject.]

PERRY, V. D. (1952), Geology of the Chuquicamata Orebody. *Mining Eng.*, **4**, 1166–8. [A brief but useful and very readable survey, more up-to-date than Taylor or Lopez. One of a series of papers in the same volume that covers all aspects of the Chuquicamata operation.]

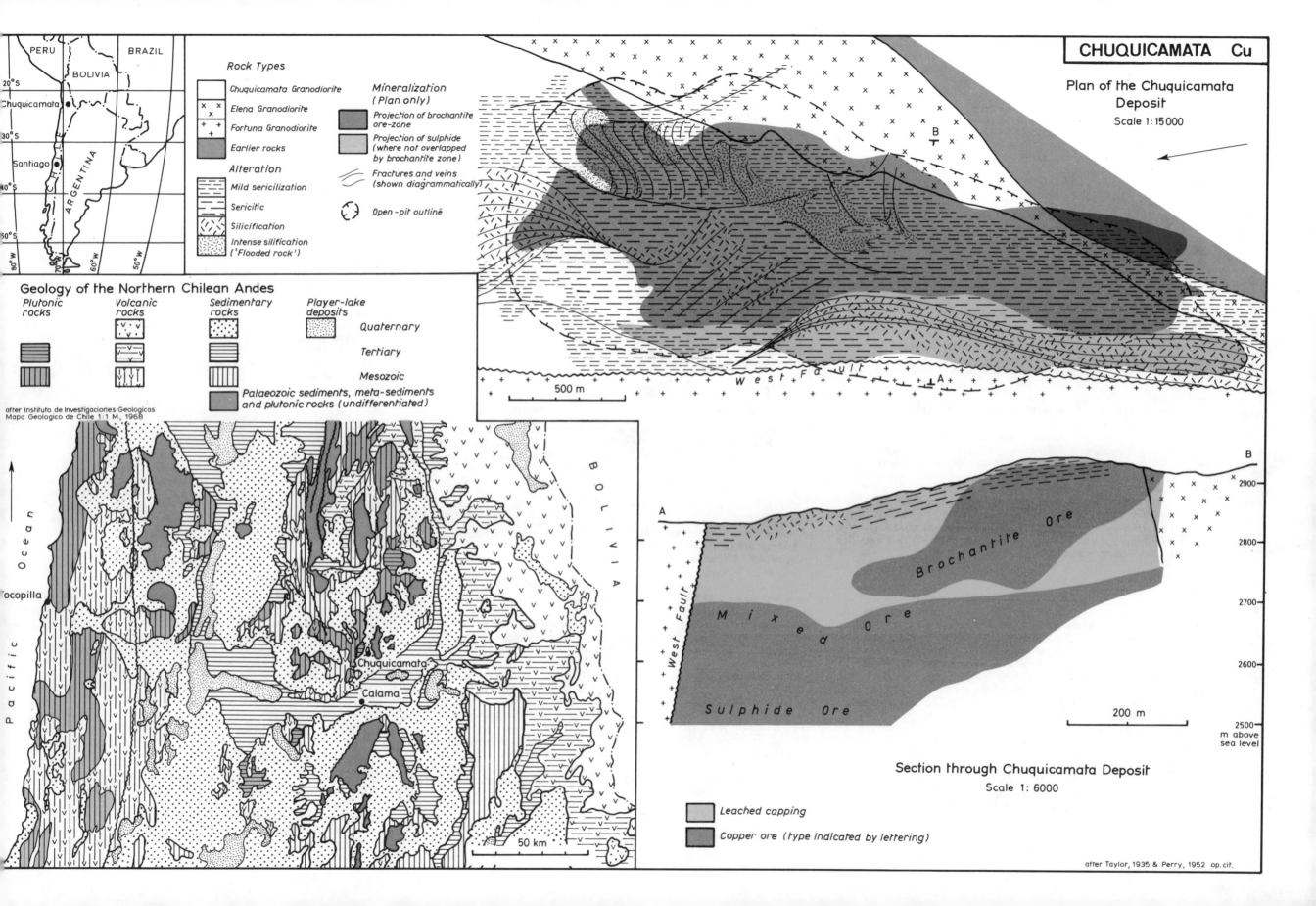

CHUQUICAMATA Cu

Plan of the Chuquicamata Deposit
Scale 1:15 000

Rock Types
- Chuquicamata Granodiorite
- Elena Granodiorite
- Fortuna Granodiorite
- Earlier rocks

Mineralization
(Plan only)
- Projection of brochantite ore-zone
- Projection of sulphide (where not overlapped by brochantite zone)
- Fractures and veins (shown diagrammatically)
- Open-pit outline

Alteration
- Mild sericilization
- Sericitic
- Silicification
- Intense silicification ('Flooded rock')

Geology of the Northern Chilean Andes

Plutonic rocks   Volcanic rocks   Sedimentary rocks   Player-lake deposits
- Quaternary
- Tertiary
- Mesozoic
- Palaeozoic sediments, meta-sediments and plutonic rocks (undifferentiated)

after Instituto de Investigaciones Geologicas
Mapa Geologico de Chile 1:1 M, 1968

Section through Chuquicamata Deposit
Scale 1:6000

Brochantite Ore

Mixed Ore

Sulphide Ore

West Fault

2900
2800
2700
2600
2500 m above sea level

200 m

- Leached capping
- Copper ore (type indicated by lettering)

after Taylor, 1935 & Perry, 1952 op.cit.

PERU BRAZIL BOLIVIA
Chuquicamata
Santiago
CHILE ARGENTINA
20°S
30°S
40°S
50°S
80°W 70°W 60°W 50°W

Pacific Ocean
Tocopilla
Chuquicamata
Calama
BOLIVIA
50 km

500 m
West Fault

The Bingham Canyon Copper Deposit – U.S.A.

Bingham was the first of the so-called 'porphyry copper' deposits to be worked. It was at Bingham that the technology and tradition of large open-pit low-grade copper mining began which still dominates the production of the metal. Among 'porphyry coppers' Bingham is normal in that the copper sulphides occur disseminated in altered granodiorite porphyry, but it is unusual in the proportion of mineralization that occurs in the sedimentary host rocks of the intrusive.

Location

The Bingham Canyon deposit occurs at latitude 40° 13′ N, longitude 112° 09′ W at 2100 m altitude, in Salt Lake County, Utah, 35 km south-west of Salt Lake City.

Geographical Setting

The copper deposit is situated almost at the head of Bingham Canyon, a steep-sided valley, one of many that have been formed down the sides of the Oquirrh Mountains. These mountains, which reach an altitude of 3000 m rise above an alluvial plain which forms the northern part of Utah, at the centre of which is the Great Salt Lake.

The climate is warm and semi-arid. It can freeze in winter, but the summers are hot, averaging 23°C with midday maxima in the high thirties. Rainfall averages under half a metre, falling almost entirely in spring and autumn. The sole permanent river in the area being the Jordan which drains the Utah into the Great Salt Lake. The vegetation of the Oquirrh mountains is confined to the scrub oak and aromatic bushes, but it was originally thickly forested with Red Pine.

Bingham is largely a mining area, and there is little else in the mountains, but fertile plains are not far away. The original town of Bingham Canyon has been overrun by mining activities, and a new town called Copperton built lower down the valley. Bingham is conveniently situated near Salt Lake City and communication routes.

History

George Ogilvie, is credited with the first discovery of mineralization in 1863. Alluvial gold was discovered in the Canyon in 1864, and was all but exhausted by groups of Californian miners by 1871. In the mean time, Ogilvie's original discovery had developed, with some difficulty, into the Old Jordan Silver–Lead Mine. A turning-point came in 1870 when the railway finally connected Salt Lake City to the rest of the country. Production of lead–silver ore began and in 1897, exploratory drives in the Highland Boy Mine intersected zones of high-grade copper ore.

A decade before this discovery Col. Enos Wall became interested in the large area of copper staining over the outcrop of the Bingham Canyon Stock. Although the Highland Boy discovery helped, it took another decade of drilling and driving to convince investors that Wall's project was worthwhile. Wall's idea was to dispense with the cost of shaft sinking, pumping, ventilation, etc. and mine the whole rock-mass at low grade from the surface using the potential of the newly invented power shovel.

In 1906 the steam shovels of the Utah Copper Company began digging the leached capping from the west side of the Canyon, so beginning the first large open-pit porphyry copper mine, which set the pattern for copper mining in the twentieth century. By the First World War, Bingham had become the world's largest copper mine.

Geological Setting

Bingham occurs in what is known as the Basin and Range tectonic province, not far from the edge of the stable block of the Colorado Plateau. The Oquirrh Mountains form a horst complex of folded Palaeozoic rocks bounded by northerly trending faults.

The principal rocks of the area are sediments of Pennsylvanian age (referred to as the Oquirrh Group) composed of quartzites, with a number of important limestone members, which were folded during the Mesozoic. The deposit itself is contained in a subsidiary horst, bounded by the northwest trending Bear and Occidental Faults, which bring up a section of the north-easterly trending Copperton Anticline.

The mineralization is in, or associated with, a stock of porphyritic granodiorite. In fact, there are two stocks exposed at Bingham and a number of dykes, sills, and small bodies. Almost all the mineralization is associated with the Bingham Canyon stock which is extensively altered, while the southerly Last Chance stock (which is hardly altered at all) seems to be barren. The age of the intrusives is middle Eocene.

During the later part of the Cainozoic the region was eroded, and substantial thicknesses of continental sediments deposited eventually forming the inland drainage basin, at the centre of which is the Great Salt Lake.

Three kinds of mineralization have been mined in the Oquirrh Mountains. Alluvial gold was worked in the canyons and valleys, sulphide ore bodies containing silver-rich galena and sphalerite occur in the Pennsylvanian limestones, and the third type is the porphyry copper mineralization of the Bingham Canyon stock and its surroundings.

Geology of the Bingham Canyon Porphyry Copper

The Bingham Canyon deposit is a roughly triangular zone of disseminated and veinlet copper sulphide mineralization, about $1\frac{1}{2}$ by $2\frac{1}{4}$ km in plan, and is known to continue below the present-day pit bottom, already over 300 m below the original surface. The majority of the ore is in the altered parts of the Bingham granodiorite stock, but a substantial proportion is in its host rocks.

The original rock of the stock was probably a medium-grained porphyritic granodiorite containing about 30% quartz, orthoclase and plagioclase, in the ratio of about 1:4. However, alteration has given the mineralized rock the bulk chemistry of a granite with approximately the inverse of the original feldspar ratio. Alteration is zoned; in the central and most intensely mineralized part of the stock there is a strong development of hydrothermal orthoclase associated with biotite. Sericite occurs throughout the stock and extends beyond the limits of the potash–feldspar zone. Clay mineral alteration occurs in irregular patches. Both the stock and the surrounding sediments are strongly fractured, forming breccia pipes in places of very variable size from a few centimetres to 250 m in diameter.

The primary mineralization consists of disseminated sulphides and narrow veinlets amounting to between 1 and 4% of the rock. In the centre of the ore-zone chalcopyrite, bornite and molybdenite are most abundant; pyrite becomes more abundant towards the outside where chalcopyrite is the dominant valuable mineral. A more varied sulphide mineralogy exists in the peripheral zone, with enargite, galena, sphalerite and tetrahedrite. The original copper ore worked at Bingham was the so-called chalcocite blanket. This occurred beneath a limonite–hematite–jarosite leached capping of between 8 and 100 m thickness, and was composed of disseminated chalcocite of the order of two or three times the average grade of the underlying primary ore. Associated with this, were some quantities of carbonate, oxide and native copper.

Size and Grade

It is known that the pit has produced about 1·3 Gt of ore, containing about 0·8% Cu, and is still producing and the total reserves even at a somewhat lower average grade, must exceed 2 Gt. The copper ore also yields about 0·05% MoS_2 recoverable amounts of gold and silver, and the surrounding mines in the limestones yield lead and zinc.

Geological Interpretations

There are so many porphyry copper deposits in the world which have been studied that enough evidence has now accumulated to be reasonably certain about their mode of formation. Bingham conforms in most respects to models that have been proposed.

It is generally concluded that the mineralization (including the alteration of the host rocks and the peripheral lead–zinc ores) was emplaced by fluids generated, at a late stage, during the cooling of the granodiorite stock. In the case of Bingham, there is a striking parallelism between the outline of the copper zone, its internal structure and the structure of the surrounding sediments. The fracturing and brecciation of the intrusive and surrounding quartzites, which created the space into which the sulphide and alteration minerals were emplaced, was probably caused in part by the boiling of the volatile phase of the granodiorite magma, although it must have been strongly influenced by the pre-intrusive structure of the area.

The 'chalcocite blanket' was formed by the well understood process of secondary enrichment. Oxidizing pyrite produced an acid solution which dissolved copper from the primary sulphides, and the copper in solution replaced the primary copper sulphides by chalcocite at a lower level where the oxidizing effect of the atmosphere is less strong. At Bingham this seems to have taken place in two stages, having been restarted after a late uplift of the Oquirrh Mountains. This probably accounts for the high grade of the ore (1·6–2%) that was exposed below the leached capping in the early days.

Mining

The whole Bingham field is operated as the Utah Division of the Kennecott Copper Corporation. The open-pit now 3 km long, $2\frac{1}{2}$ km wide and over 300 km deep, is one of the largest man-made excavations in the world, and is capable of producing 35 Mt/a of ore. About two and a half times this quantity of waste rock has to be removed at the same time. The ore is taken through a haulage tunnel to a milling and smelting complex. Chalcopyrite and molybdenite are recovered by flotation from ground ore, and separated by a similar process. Precious metals are recovered during the electrolytic refining of the smelted copper.

Further Reading

PETERS, W. C. et al. (1966), Geology of the Bingham Canyon Porphyry Copper Deposit, Utah. In: *Geology of the Porphyry Copper Deposits, Southwestern North America.* (Titley, S. R. and Hicks, C. L., eds.), pp.165–75, University of Arizona Press, Tucson, Arizona. [The best modern description in a very useful volume that includes descriptions of the major deposits in the region.]

BEESON, J. J. (1917), The Disseminated Copper Ores of Bingham Canyon, Utah. *Am. Inst. Min. Metall. Petroleum Eng. Trans.,* **54**, 356–401. [Probably the most comprehensive work on the deposit, interesting because written just a few years after production began in earnest, but a little old from the point of view of geological interpretation.]

BOUTWELL, J. M. et al. (1905), Economic Geology of the Bingham Mining District, Utah. *U.S. Geol. Surv. Prof. Paper,* **38**, 411pp. [The original major work on the area, written before the opening up of the copper deposit, but contains good descriptions of the peripheral lead–silver deposits.]

BUTLER, B. S. et al. (1920), The Ore Deposits of Utah. *U.S. Geol. Surv. Prof. Paper,* **111**, 672pp. [Contains a description of the region and pp.338–62 is a shorter and updated version of Boutwell et al., 1905.]

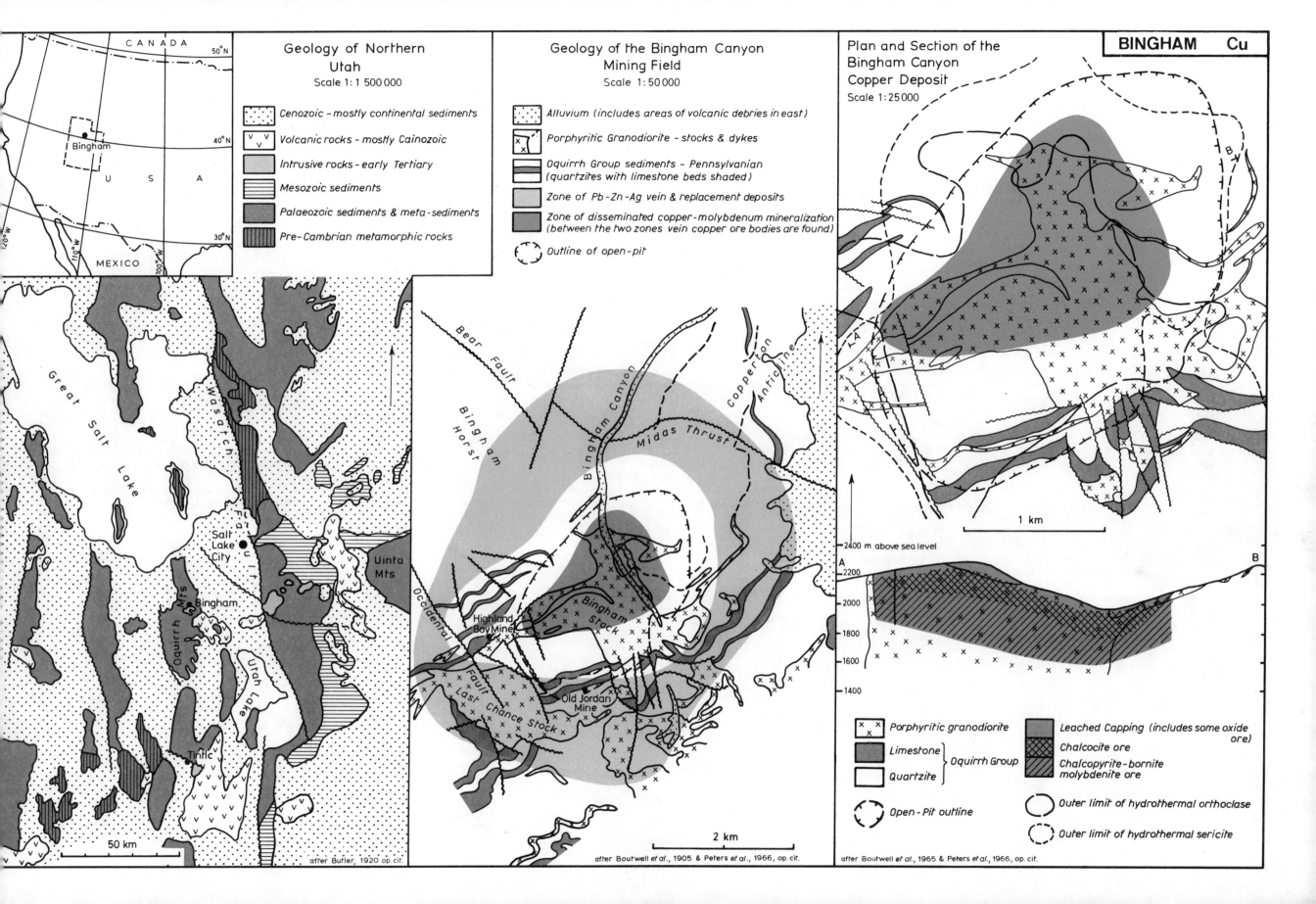

BINGHAM Cu

Geology of Northern Utah
Scale 1 : 1 500 000

Cenozoic – mostly continental sediments
Volcanic rocks – mostly Cainozoic
Intrusive rocks – early Tertiary
Mesozoic sediments
Palaeozoic sediments & meta-sediments
Pre-Cambrian metamorphic rocks

CANADA
50°N
40°N
30°N
Bingham
U S A
MEXICO

Great Salt Lake
Wasatch
Salt Lake City
Uinta Mts
Oquirrh Mts
Bingham
Utah Lake
Tintic

50 km

after Butler, 1920 op. cit.

Geology of the Bingham Canyon Mining Field
Scale 1 : 50 000

Alluvium (includes areas of volcanic debries in east)
Porphyritic Granodiorite – stocks & dykes
Oquirrh Group sediments – Pennsylvanian (quartzites with limestone beds shaded)
Zone of Pb-Zn-Ag vein & replacement deposits
Zone of disseminated copper-molybdenum mineralization (between the two zones vein copper ore bodies are found)
Outline of open-pit

Bear Fault
Bingham Horst
Bingham Canyon
Copperton Anticline
Midas Thrust
Occidental Fault
Last Chance Stock
Highland Boy Mine
Bingham Stock
Old Jordan Mine

2 km

after Boutwell et al., 1905 & Peters et al., 1966, op. cit.

Plan and Section of the Bingham Canyon Copper Deposit
Scale 1 : 25 000

1 km

2400 m above sea level
A
2200
2000
1800
1600
1400
B

Porphyritic granodiorite
Limestone } Oquirrh Group
Quartzite
Open-Pit outline
Leached Capping (includes some oxide ore)
Chalcocite ore
Chalcopyrite-bornite molybdenite ore
Outer limit of hydrothermal orthoclase
Outer limit of hydrothermal sericite

after Boutwell et al., 1965 & Peters et al., 1966, op. cit.

The Climax Molybdenum Deposit – U.S.A.

Just before the First World War it was discovered that the addition of the rare metal molybdenum could form an alloy steel with superior properties for toolmaking. For many years the world's supply came largely from this deposit. It is one of the most important deposits in the famous 'Colorado Mineral Belt', and in many respects resembles a 'porphyry copper' deposit, many of which also contain molybdenite. Climax was the home mine of a company of the same name, which was one of the partners of the merger that formed the international mining and metallurgical corporation, AMAX.

Location

Climax is situated at 3500 m altitude in the Tenmile Mountains, at latitude 39° 20′ N, longitude 106° 08′ W, in Lake County, Colorado, 100 km southwest of the state capital of Denver.

Geographical Setting

Climax lies on the continental divide in the centre of a series of ranges that form the southern prolongation of the Rocky Mountains. The land rises to over 4000 m but descends to 3000 m in the adjoining valleys. The terrain is rocky, almost barren of vegetation and shows much evidence of recent glaciation. The climate is cold and continental; mean temperatures rarely exceeding freezing point, and almost all of the 500 mm of precipitation during the year falling as snow. The ground is permanently frozen in some of the valleys. Climax is just a mining camp, but 16 km to the south the much older mining town of Leadville is more developed. The headwaters of rivers that flow to the Mississippi and the Colorado occur near the mine, and a main highway passes by, over Freemont Pass on its way from the Great Plains, westwards to Salt Lake City.

History

Prospectors, who blazed the trails over the Rocky Mountains in the nineteenth century, found many mineralized districts in the ranges west of Denver, among them such famous ones as Leadville, Cripple Creek, Idaho Springs; and with all this activity it is no surprise that the bare rock outcrops around Climax were known from at least as long ago as 1879 when Charles Senter brought back specimens of a strange grey metallic mineral from a leached capping on Bartlett Mountain. The mineral was identified as molybdenite by the then professor of Colorado School of Mines, but it was not until 1912 that it was realized that a tool steel, superior to the carbon manganese steels, could be made using a small quantity of molybdenum. A small attempt to mine Climax was made between 1915 and 1919, the mineral being concentrated at Leadville by the then newly discovered flotation process. In 1924, a major mining venture began to supply an expanding market.

Geological Setting

A large part of the Park Range and the adjoining Front Range is composed of middle Proterozoic biotite-schists and gneisses, (known as the Idaho Springs Group) which contain a series of foliated granites known as the Silver Plume Granites. The structure of these rocks is complex; a north-east trend of folds is detectable, but is modified by later folding events that followed the deposition in the area of Palaeozoic sediments. Two groups of these sediments are present at Climax: the largely clastic Sawatch Formation of Cambrian age and the Pennsylvanian to Permian, Minturn Formation. No Mesozoic strata survive at Climax. The important Mosquito Fault cuts the mineralized area at Climax. It is traceable for some distance away from the mine itself and seems to have had a long history of movement, having controlled sedimentation during the Palaeozoic (perhaps even earlier), and was strongly reactivated during late Tertiary times.

The mountainlands of Colorado were invaded by magmatic rocks during the Tertiary. These were, for the most part, acid rocks; stocks of granodiorite, hyperbyssal intrusions and volcanic centres with large thicknesses of rhyolites and pyroclastic rocks.

The Colorado Mineral Belt is a zone of mineralized districts that is between 25 and 80 km wide and stretches from the town of Boulder southwestwards for 300 km to the San Juan Mountains, where it disappears under the cover-rocks of the Colorado Plateau. Many of these districts are associated with Tertiary magmatic rocks but some are older, and it has been suggested that the belt is controlled by some deep feature of the crust that has a Pre-Cambrian ancestry. The prolongation of the belt, south-west of the Colorado Plateau, intersects the richest part of the Arizona porphyry–copper province.

It is clear that faulting affected the area during the late Tertiary and Quaternary, and the uplifted blocks, such as the Tenmile Mountains, were deeply eroded, particularly by glaciation.

Geology of the Climax Deposit

The mineralized zone of Climax is roughly circular, about 1·8 km in diameter, and in the form of a capping over a series of stocks and dyke complexes that intrude the Idaho Springs metamorphic rocks. It is truncated on the west side by the Mosquito Fault which is thought to have a downthrow of nearly 3000 m, and a dextral movement of 750 m. Both the Minturn Formation rocks of the downfaulted western block and the Pre-Cambrian of the eastern block are invaded by dykes and sills of porphyritic micro-granodiorites (known locally by the American petrographic term of quartz monzonite porphyry).

Associated with the mineralized zone are four groups of intrusive rocks, all of which are medium- to fine-grained granitic rocks composed of quartz, orthoclase, albite and biotite. They vary to some extent in texture, and are thought to be of Oligocene age. The earliest is the Southwest Stock, succeeded by the Central Stock, and a co-axial one known as the 'Aplitic Phase', with which are associated a swarm of dykes known as the Intra-Mineral (because they seem to have been intruded between two main phases of mineralization). Finally, there is a coarse-grained porphyritic granite phase, again co-axial, with the Central Stock and associated with a further dyke-swarm.

Alteration is common, intense and pervasive in places. Many parts of the stocks are altered by the introduction of hydrothermal potash-feldspar. Several masses of high silica rock occur capping the stocks, thought to be intense zones of silicification that obliterate the contacts between the various intrusive pases. Sericitization occurs locally associated with the latest stage of hydrothermal activity.

Four phases of mineralization are recognized, each associated with a phase of intrusion, the first three forming thick, cap-like ore bodies. The earliest (and uppermost) Ceresco body has almost been eroded away, but large parts of the two lower (and younger) bodies are preserved, known as the Upper and Lower. The mineralization is much the same in each, consisting of a zone of irregular veinlets, a few millimetres to a centimere in width, with rather diffuse boundaries and containing quartz, a little feldspar and fine-grained molybdenite. A small proportion of the ore is present in larger veins (or disseminated in dykes and pegmatized zones of the granites) or as clots in the high-silica rock, or along joint surfaces. Fluorite and topaz are found in some of the veinlets. In most parts of the ore bodies (but particularly in the Upper) later quartz-sericite veinlets contain appreciable amounts of pyrite, fluorite, topaz, small quantities of wolframite, and even smaller quantities of cassiterite. The ore bodies have a central, lower molybdenum-rich zone, and an outer capping of more pyrite-rich tungsten-bearing ore. These two zones overlap in the Lower ore body, but are to some extent separate in the Upper. If the Ceresco ore body ever had a tungsten zone, it has been eroded away. (It should be pointed out that the form of the ore bodies is to some extent a function of the grade being mined, and the remarks here refer to the bodies that yield the present mine average grade of 0·4% MoS_2.)

The fourth phase of mineralization, associated with the latest granitic phase, consists of a series of veinlets and veins containing quartz, sericite, pyrite and a host of minor minerals that are of no importance economically, and is generally referred to as the 'barren stage'.

The ore zone was uplifted along the Mosquito Fault in the late Tertiary and eroded, and it was at this time that as much as 120 m of the upper parts of the zone was partially oxidized, yielding a complex range of secondary molybdenum minerals closely associated with iron oxide (formed from the pyrite). Some of the oxide material is ore. Much of the original oxide zone, and a good deal of the primary ore, was removed by glaciation in the Quaternary, leaving only the outer roots of the Ceresco ore body, but most of the other two.

Size and Grade

Because Climax has been mined over a period during which molybdenum became of increasing importance metallurgically, and during which mining technology developed, the grade of ore worked has changed from 0·9% MoS_2 in the 1920s, to less than 0·4% MoS_2 today. The deposit contains about 1·4 Mt of Mo, of which about half has been removed, the total mass of ore above a cut-off of 0·2% MoS_2 being of the order of 550 Mt.

Geological Interpretations

Climax has been likened to a porphyry copper without copper, and indeed many examples of that type of deposit contain recoverable quantities of molybdenite. However, there are some important differences. The parent rock is more alkaline than is general for porphyry coppers, the total sulphide content much lower, and the geometry of the alteration zones rather different. None the less, Climax seems to have been formed by the intrusion of a multiple stock of wet granitic magma carrying a range of exotic elements, which were deposited during a series of late hydrothermal stages along fractures formed by the stresses associated with the intrusions. As compared with the 'porphyry copper' deposits, it is possible that meteoric water in the host rocks of the stocks was not an important factor in developing a hydrothermal phase. There are other examples of similar deposits, and one can perhaps speak of a Climax type of deposit.

Mining

Climax is exploited by the mining company of the same name, a subsidiary of AMAX Corporation. The mine is accessed through a series of haulage adits and internal shafts, and the ore is mined largely by block caving methods. The ore is hauled to a mill on the west side of the Mosquito Fault, where it is ground fine, and the molybdenite recovered by flotation. By-products such as wolframite, cassiterite, some pyrite and zircon are also recovered, and the tailings from the molybdenite flotation are treated to recover molybdic oxide from the oxidized ore. The mill currently handles 40 000 t/d, producing concentrates containing about 30 000 t of Mo per year. Despite the cold climate, open-pit mining has recently been started to augment underground production.

Further Reading

WALLACE, S. R. et al. (1968), Multiple Intrusion and Mineralization at Climax, Colorado. In: *Ore Deposits of the United States, 1933–1967.* (Ridge, J. D. ed.), pp.606–40. American Inst. Min. Met. Pet. Eng., New York. [Probably the best modern description, with good bibliography.]

CLARK, K. F. (1972), Stockwork Molybdenum Deposits in the Western Cordillera of North America. *Econ. Geol.,* **67**, 731–58. [Comprehensive discussion of the type of deposit with a very full bibliography.]

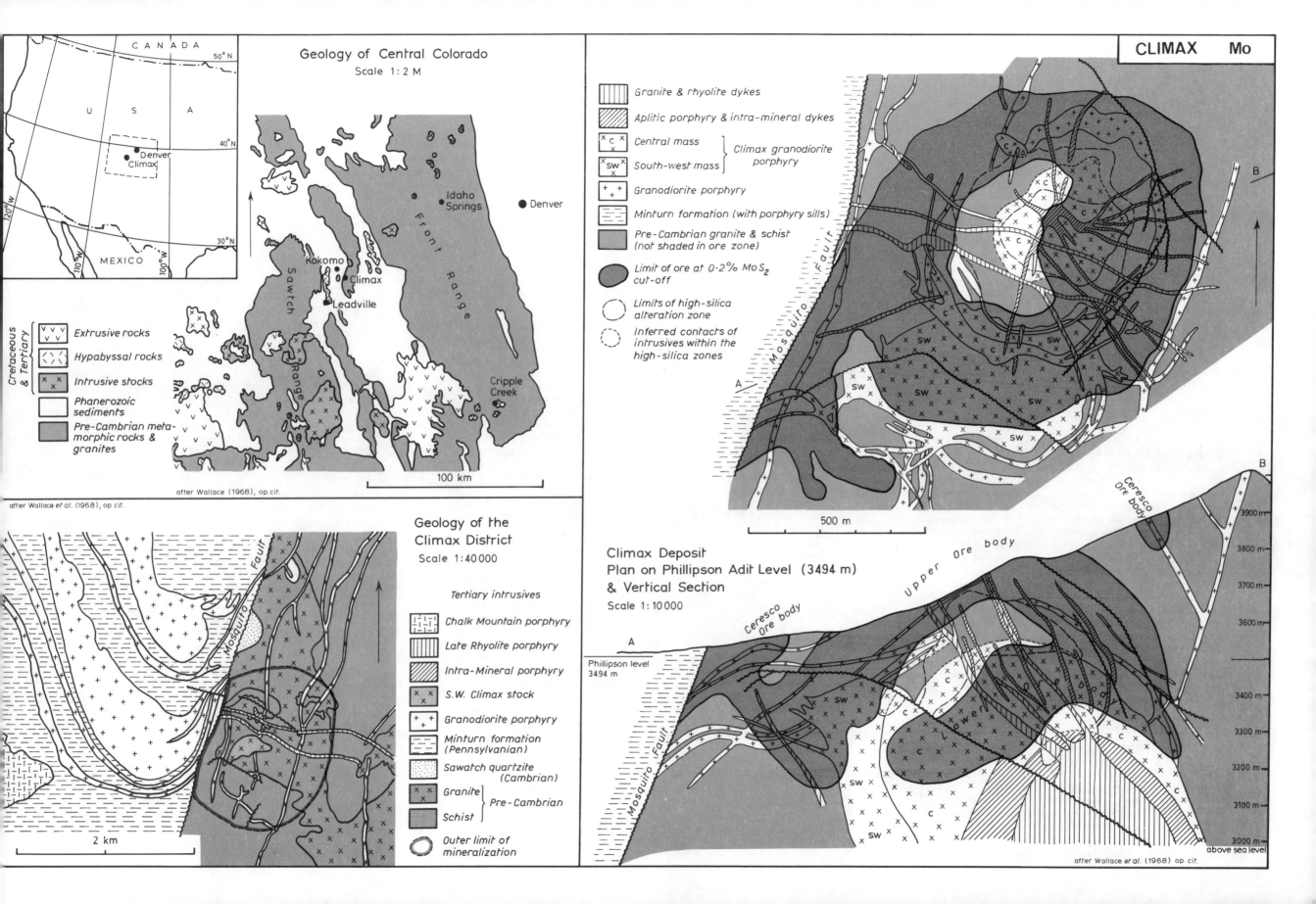

Geology of Central Colorado
Scale 1 : 2 M

after Wallace et al. (1968), op.cit.

Crefaceous & Tertiary
- v v / Extrusive rocks
- < > / Hypabyssal rocks
- x x / Intrusive stocks
- Phanerozoic sediments
- Pre-Cambrian meta-morphic rocks & granites

after Wallace (1968), op.cit.

100 km

Geology of the Climax District
Scale 1 : 40 000

Tertiary intrusives
- Chalk Mountain porphyry
- Late Rhyolite porphyry
- Intra-Mineral porphyry
- S.W. Climax stock
- Granodiorite porphyry
- Minturn formation (Pennsylvanian)
- Sawatch quartzite (Cambrian)
- Granite } Pre-Cambrian
- Schist }
- Outer limit of mineralization

2 km

CLIMAX   Mo

- Granite & rhyolite dykes
- Aplitic porphyry & intra-mineral dykes
- x C x / Central mass } Climax granodiorite porphyry
- x SW x / South-west mass }
- + + / Granodiorite porphyry
- Minturn formation (with porphyry sills)
- Pre-Cambrian granite & schist (not shaded in ore zone)
- Limit of ore at 0·2% MoS₂ cut-off
- Limits of high-silica alteration zone
- Inferred contacts of intrusives within the high-silica zones

500 m

Climax Deposit
Plan on Phillipson Adit Level (3494 m)
& Vertical Section
Scale 1 : 10 000

Phillipson level 3494 m

3900 m
3800 m
3700 m
3600 m
3400 m
3300 m
3200 m
3100 m
3000 m
above sea level

after Wallace et al. (1968) op.cit.

The Butte District – U.S.A.

Once described as the 'richest hill on earth', this district is something of a legend among the mineral deposits of the American West. It has some resemblance to a porphyry copper deposit, but much of the ore is in the form of large individually exploitable veins. It is perhaps the world's greatest classic hydrothermal vein deposit.

Location

The district is centred on the town of Butte, located at latitude 46° 01′ N, longitude 112° 32′ W and 1790 m altitude in Silver Bow County, Montana, 100 km south-west of the state capital of Helena.

Geographical Setting

Butte is in the centre of the Rocky Mountains. The name comes from Big Butte, a 1923 m high hill that overlooks the whole area. Although high, the area is not very rugged, the ground consisting of a mature topography with rounded boulder-strewn hills and shallow valleys. The area is close to the 'continental divide' and, although originally pine forest with a little mixed woodland and grassland, mining and smelting have turned it into a desolate landscape that is just beginning to recover. The climate is continental and influenced by the altitude; warm summers, very cold winters but relatively little snow. Annual precipitation is about 300–400 mm. Butte township itself is no more than a large mining camp, and most of the population and surface activity is centred in Anaconda, 32 km to the west. Here there is an airport, and main trans-continental roads and railways pass through *en route* from the Dakotas to Seattle.

History

In 1864, a passing traveller noticed gold and the following year gold diggers rushed to the area: but the gold was all but exhausted in two years. However, during those years, silver ore was found in the weathered outcrops of large quartz veins, and Butte suddenly became an important mining district. In 1882, an exploratory cross-cut from a shaft that had been named 'the Anaconda' intersected a very large vein of rich chalcocite ore and, coinciding as it did with Edison's development of the electric lamp, Butte became almost at once the most important copper deposit in the world.

At the turn of the century Butte was the site of a series of famous power struggles between rival claim owners. The mining law of Montana gave the surface claim holders the right to follow a vein wherever it went, and a series of cases came before the courts, aimed at establishing title to the faulted extensions of veins. The chief beneficiaries of these troubles were the lawyers, the Anaconda Mining Company, and a handful of geologists who were retained to give evidence about the faulting of veins. Modern mining geology can be said to have been born at Butte, and it is one of the best studied deposits of all.

There was important change in the 1950s to bulk mining of veined ground by block caving and, in 1955, large open-pit workings began (although vein mining still continues on a large scale).

Geological Setting

Butte is located in the south-west corner of the Boulder Batholith, a complex of intrusions ranging in composition from diorite to granodiorite formed during the Cretaceous. To the south are exposed in a series of large inlyers, metamorphic rocks of Pre-Cambrian age (probably Archean). To the north is the immensely thick, shallow water clastic sequence of the Belt Super-Group; to the east is a great series of Palaeozoic and Mesozoic sediments that were strongly folded at about the time of the intrusion of the Boulder Batholith. In size the Boulder is overshadowed by the much larger Idaho Batholith some distance to the west, but the former has proved much more important for

mineralization. Shortly after the emplacement of the intrusion, volcanic rocks were extruded, and large parts of the area (including part of the Butte district) are covered by rhyolites and dacites, ranging in age from late Cretaceous to Palaeogene.

The structure of the area has a complex history. There is a considerable amount of infolding of the Phanerozoic sediments with the Pre-Cambrian, and the former is thrust over the latter in several places. The great Osburn wrench fault, with which the Cœur d'Alene district is associated, crosses the Belt Super-Group and, if it were projected beneath the Phanerozoic cover, would pass through the centre of the Boulder Batholith. Butte thus lies at something of a tectonic 'cross-roads', and this could account for some of its unique features.

Geology of the Butte District

The district itself is a roughly circular area 7 km across, and is more or less terminated on the eastern side by the Continental Fault. The western part is covered by rhyolites, and the east and south are buried beneath up to 100 m of alluvium. The essential host rock of the mineralization is a hornblende–biotite–granodiorite (see footnote) of even, medium- to coarse-grain size and a radiometric age of 78 Ma. It is intruded by small bodies of an aplitic variety of the same rock, and contains pegmatitic zones. It is cut by quartz-porphyry dykes that are probably feeders to, or associated with, the overlying rhyolites.

There are two stages of mineralization, known in the literature as the pre-main and main stages. The pre-main stage is largely confined to a central area (not exposed in the higher levels) and consists of small short, random veinlets, averaging 25–250 mm in width, and from 1–10 m long. They are simple quartz veins containing molybdenite and chalcopyrite with an envelope of potash feldspar, biotite and sericite alteration. The biotite from these alteration envelopes has a radiometric age of 63 Ma, and the veinlets of this stage are cut by barren quartz veins without alteration envelopes.

The main stage of mineralization consists of several systems of veins of which the two most important are the east-west trending Anaconda, and the north-west trending (and later) Blue veins. The Anaconda veins are the largest and most productive. They dip steeply north from surface, but often turn to a southerly dip in depth. They have a fairly consistent strike just north of east, except in some regions to the eastern end of the district where they suddenly change to a mass of small veins with a south-easterly strike. These are the so-called 'horse-tail' vein areas that are the locus of modern mass-mining methods. The Anaconda vein ore-shoots are very extensive, and average between 6–10 m in width, and locally ore pods up to 30 m width have been found. The Blue veins which offset the Anacondas (usually with a sinistral wrench movement) are narrower and the ore shoots less persistent. Both vein systems are cut by a number of faults, particularly those of the flat-dipping Rarus Fault zone.

The mineral content of the veins is strongly zoned, and three principal zones are recognized. The Central Zone contains at depth the pre-main stage molybdenite veinlets and (in the Anaconda veins) particularly rich shoots of chalcocite–enargite ore. This gradually gives way to shoots containing chalcopyrite (with local areas of chalocite and bornite) associated with minor amounts of sphalerite termed the Intermediate Zone. The Peripheral Zone is dominantly of sphalerite-rhodochrosite mineralization with small (but valuable) quantities of silver minerals. Most ore-shoots contain some pyrite. All the veins are surrounded by envelopes of alteration. Generally the granodiorite is sericitized next to the vein, followed by argillic and propylitic zones. Sericitization is intense and pervasive in the Central Zone (where the walls of the veins are often silicified). The widths of the alteration envelopes becomes less and less towards the peripheral zone. Sericite from this stage has a radiometric age of 58 Ma.

The mineralization has been affected by weathering down to about 60–100 m; deeper in some cases along water-bearing fissures. Thick gossanous quartz masses are found at surface associated with which is native silver. There is usually a leached zone followed by very rich 'sooty' chalcocite in

the Central and Intermediate, and manganese oxide minerals in the Peripheral zone. In the smaller valleys draining the area high-silver alluvial gold was found, presumably derived from the erosion of the weathered vein outcrops.

Size and Grade

In the years from 1880 to 1964, the district produced 300 Mt of ore which yielded 7·3 Mt copper, 2·2 Mt zinc, 1·7 Mt manganese, 0·3 Mt lead, 20 000 000 kg silver, 78 000 kg gold and important amounts of cadmium, bismuth, selenium, tellurium and sulphuric acid. Reserves remaining are very considerable, of the order of 10 Mt of high-grade copper and silver ore in the vein mining areas, and 500 Mt of low-grade copper ore in the open-pit areas.

Geological Interpretations

Hardly anyone disputes that the Butte mineralization is ultimately derived from the Boulder Batholith. It is clear that the mineralization of the outer zones was gradually overtaken by the rich copper mineralization, but that the copper–molybdenum mineralization in the centre was the earliest. There is a 20 Ma gap between the emplacement of the host granodiorite, and the alteration accompanying the main stage. It seems that the Boulder Batholith had a long history. The pre-main stage mineralization was probably a late-stage product of the host intrusion. The form of the main-stage mineralization strongly suggests, however, that the host rock was consolidated and fractured prior to the emplacement of the mineralization, and perhaps associated with one of the later phases of intrusion in the Batholith. The evidence suggests mineralization by heated waters passing outwards from the centre through a fracture system early in the period of fracturing. The emplacement of this intrusive complex seems to have taken place during the late Cretaceous tectonic events, at the intersection of the Archean Belt Super-Group hinge-zone; the Osburn wrench fault, and the 'Front' of the structures that affect the Phanerozoic sediments.

Mining

In the past, almost all vein-mining methods have been used at Butte. Today most ore is produced by underground cut-and-fill stoping, or open-pit mining. The Anaconda Company who now effectively control the district, operates the large Berkeley pit at a rate of 15 Mt/a and several underground mines. There are a number of mills operating, producing copper and zinc sulphide flotation concentrates for shipment to the company's smelters at Anaconda town. Electrolytic refining takes place at Great Falls, 200 km to the north-east, where there is an hydroelectric power station.

Further Reading

MAYER, C. *et al.* (1968), Ore Deposits at Butte, Montana. In: *Ore Deposits of the U.S.A. 1933–67* (Ridge, J., ed.), pp.1373–1416. New York Am. Inst. Min. Met. & Pet. Eng. Inc. 2 vols., 1880pp. [Good and well-illustrated modern summary with fairly comprehensive bibliography.]

SALES, R. H. (1914), Ore Deposits at Butte, Montana. *Am. Inst. Min. Eng. Trans.*, **46**, 4–106. [The main descriptive text of the early vein-mining days by the man who devoted most of his life to the district.]

WEED, W. H. (1912), Geology and Ore Deposits of Butte District, Montana. *U.S. Geol. Surv. Prof. Paper*, **74**, 262pp. [The original standard work on the area, a little out-of-date but still interesting.]

SALES, R. H. (1964), *Underground Warfare at Butte*. Caxton Printers Ltd., Caldwell, Idaho, 77pp. [Should not be missed if you can find a copy; lucid tale of the struggle for control of claims; better than most 'westerns'.]

Footnote. The American literature uses the term 'quartz monzonite' for these rocks, but it is reported to contain 24% quartz, 20% orthoclase, 47% andesine with biotite and hornblende, which corresponds to a granodiorite in present-day international terminology.

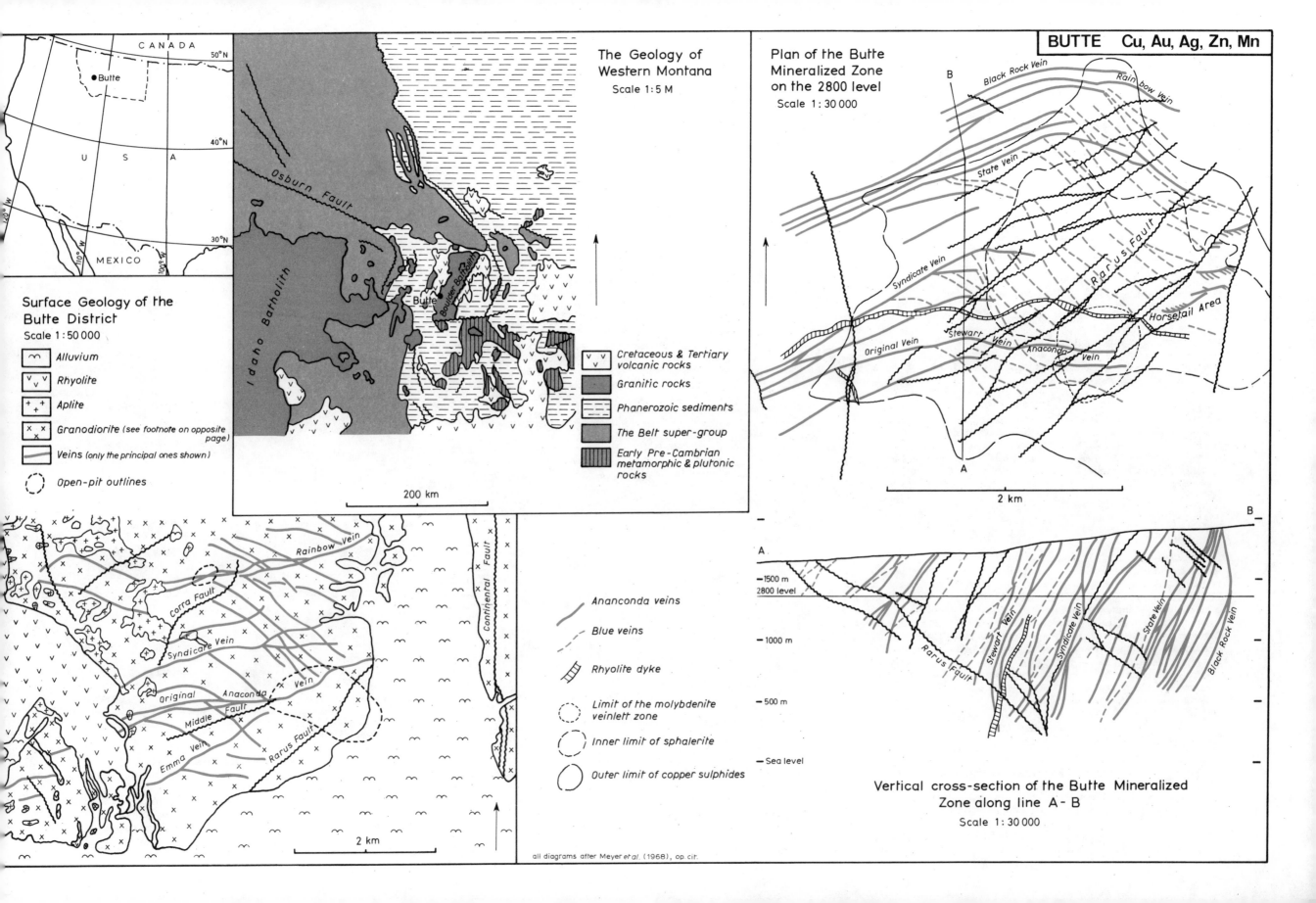

BUTTE Cu, Au, Ag, Zn, Mn

The Geology of
Western Montana
Scale 1:5 M

Plan of the Butte
Mineralized Zone
on the 2800 level
Scale 1:30 000

Surface Geology of the
Butte District
Scale 1:50 000

⌒	Alluvium
v v v	Rhyolite
+ +	Aplite
x x	Granodiorite (see footnote on opposite page)
	Veins (only the principal ones shown)
⊙	Open-pit outlines

v v	Cretaceous & Tertiary volcanic rocks
	Granitic rocks
	Phanerozoic sediments
	The Belt super-group
‖‖‖	Early Pre-Cambrian metamorphic & plutonic rocks

200 km

CANADA

Butte

U S A

MEXICO

Osburn Fault

Idaho Batholith

Boulder Batholith

Butte

Black Rock Vein
Rainbow Vein
State Vein
Syndicate Vein
Rarus Fault
Original Vein
Stewart Vein
Anaconda Vein
Horsetail Area

2 km

Rainbow Vein
Corra Fault
Syndicate Vein
Continental Fault
Original Vein
Anaconda Vein
Middle Fault
Emma Vein
Rarus Fault

	Anaconda veins
	Blue veins
‖‖	Rhyolite dyke
⊙	Limit of the molybdenite veinlett zone
◡	Inner limit of sphalerite
◡	Outer limit of copper sulphides

2 km

Vertical cross-section of the Butte Mineralized
Zone along line A-B
Scale 1:30 000

B

A

1500 m
2800 level

Rarus Fault
Stewart Vein
Syndicate Vein
State Vein
Black Rock Vein

1000 m

500 m

Sea level

B

all diagrams after Meyer *et al.* (1968), op. cit.

The Santa Eulaila Deposit – Mexico

This is perhaps the best example of the 'chimenea y manto' type of deposit (literally 'chimney and cloak') sometimes called 'pipe and flat' in English. Such deposits are found all along the central part of the American Cordillera. They are among the most extraordinary deposits in the world, having the morphology of a limestone cave system, but an infilling that often resembles a contact metasomatic deposit.

Location

The mining township of Santa Eulalia (which is 2 km south-west of the main ore bodies) is situated at latitude 28° 36′ N, longitude 105° 53′ W and 2000 m altitude, 22 km from the state capital of Chihuahua.

Geographical Setting

Chihuahua State is 'basin and range' country; flat dusty plains separating long 'sierras' or ranges of rugged hills. The highest points on the ranges exceed 2000 m in altitude and the surrounding plains are some 600 m lower. The plains are drained by seasonal rivers that join the Rio Conchos, a tributary of the Rio Grande. The climate is arid sub-tropical, with a rainfall of 350 mm/a, winter average temperatures of 8°C and summers averaging 21°C (with much higher mid-day maxima). Most of the rain falls during summer storms.

Santa Eulalia is one of North America's oldest mining towns, and has very good communications, being connected to the main Mexico City–Juanez railway (and to the Mid-West of the U.S.A.) as well as being almost on one of the country's main highways. Santa Eulalia is principally a mining town of a few thousand inhabitants, but only 22 km away is the much larger and historic city of Chihuahua, with one of the largest and most richly endowed cathedrals in Mexico.

History

Santa Eulalia is said to have been the Aztec city of Tenochtetlan that was sacked by Cortes and his Conquistadores in 1521. It is hardly likely that the Aztecs did not know of the deposits in their midst, but there are no records of discovery until 1591 when the San Antonio ore bodies were first worked. However, the attempts were not successful and serious mining did not begin until 1702. Eighteenth-century mining seems to have been better because between 1738 and 1750 it is recorded that Chihuahua Cathedral was built with money raised by a royalty on silver produced. Mining has continued on an ever-increasing scale through to the present day. Modern methods of mining and treatment were introduced by American companies early in this century, and the mines passed into Mexican hands in 1965 as part of a general move towards national control.

Geological Setting

The eastern part of the Mexican Cordillera that passes through Chihuahua State is essentially a horst-and-graben country. In the horsts, corresponding to the sierras, are exposed a variety of late Palaeozoic and Mesozoic sedimentary rocks, of which Cretaceous carbonate sequences are by far the most important. These rocks include monotonous un-fossiliferous, as well as skeletal limestones, bituminous limestone marls and evaporite sequences, broadly similar to those found in the Gulf Basin to the north-east. However, during the Tertiary the area was affected by vulcanism associated with the development of the Cordillera. Great thicknesses of lavas and pyroclastics were poured out over the Mesozoic sediments. The sources of these lavas were plutons, some of which are known at depth in the area, although few are exposed. The vulcanism was accompanied by strong uplift and block-faulting which led, later on in the Tertiary, to erosion and the formation of great thicknesses of continental sediments derived from the volcanic rocks. The Mexican Cordillera is richly endowed with mineral deposits. In the more deeply-eroded western part, where the underlying plutons have been exposed, porphyry–copper type deposits are found. In the eastern part, most of the deposits are in sediments (particularly the Mesozoic carbonates) among which the 'chimenea y manto' deposits are widespread.

Geology of the Deposit

The deposit consists of two groups of ore bodies separated by a barren zone about 2 km in width. (They could be regarded as two separate deposits.) At surface the ground is largely covered by sediments with a high proportion of volcanic material and lava flows. Among these are rhyolite flows, agglomerates, tuffs, tuffites, conglomerates and sands, the latter containing a mixture of volcanic material and limestone fragments. Beneath these (which are known in the area as the 'capping group') are carbonate sediments that form a gentle dome structure. The stratigraphic control is poor, but by comparison with other areas, it is thought that the middle and upper parts of the Cretaceous are represented. The Limestone Group can be divided into an upper, skeletal unit with a prominent fossiliferous horizon at its base, followed by a dark non-skeletal unit, and a basal unit of black bituminous skeletal limestones and marls. Known only from boreholes, there is a unit of anydrite and marl below the limestone group, which is penetrated by a massive granodiorite that forms the core of the dome structure.

Sills of dolerite and micro-diorite are found at two particular levels in the limestones in the western part of the field, and original sill-like and dyke-like bodies of micro-granite are found throughout the area. One very prominent sill with many off-shoots exists below the ore bodies in the western part of the field, and a large dyke runs alongside the main ore bodies in the eastern part.

The ore bodies are very irregular but can be divided into two types; those that run along the bedding of the limestone (the mantos) and those that cross the bedding (the chimeneas). The composition of these bodies is of three types, known locally as normal sulphide, silicate and oxide ore. The sulphide ore contains 45–60% iron sulphide, dominantly pyrrhotite (but with pyrite in the higher levels of the deposit) accompanied by 9–29% silver-rich galena, 10–40% iron-rich sphalerite (48% Zn) and small amounts of arsenopyrite, chalcopyrite, fluorite and silicates.

The so-called silicate ores contain the same assemblage of minerals but, in addition, a substantial proportion of iron, manganese and calcium silicates (knebelite, fayalite and ilvaite) with some chlorite. The combined galena and sphalerite content of these ores is as low as 5%. A large quantity of the ores at Santa Eulalia are oxide. These are mixtures of cerussite, smithsonite, hydrated iron and manganese oxides, gypsum and chalcedonic silica. They contain a high proportion of lead, much less zinc, and are very rich in silver. This material is very variable, both in mineral composition and texture, and there are many bodies of ore that are mixtures of oxide and sulphide. In some cases caves have been found, at the bottom of which are rubbles of limonite, cerussite and smithsonite with lumps of un-oxidated sulphide ore, and silicates.

In the eastern group of ore bodies, cassiterite is found which is present both in the primary and oxidized ore. There the primary ore contains a more characteristic assemblage of calc-silicates; garnet, hedenbergite and epidote. In this part of the field the oxide ore is relatively rich in vanadium, both descloizite and vanadinite being present.

Alteration associated with the ore bodies is very slight, but careful stratigraphic correlation has shown that there is a thinning of the limestone in the immediate vicinity of the ore, presumably due to solution.

Size and Grade

The Santa Eulalia ore bodies have yielded over 35 Mt of ore, but are approaching exhaustion. Sulphide ore averages 12% Pb, 11% Zn and 230 g/t Ag and oxide ore averages 13% Pb, 2% Zn and about 350 g/t Ag. The ores of the eastern part of the field contain between 1·5 and 4% tin and some vanadium.

Geological Interpretations

It is very hard to resist the impression that the Santa Eulalia ore bodies occupy a cave system (indeed, if one did not know that the plans and sections were of a mine, one would inevitably conclude that they were made by a spaeleologist). On the other hand, cave fillings consisting of sulphides and a curious assemblage of silicates is, to say the least, unusual. A variety of theories have been put forward to explain these deposits, all of them depending ultimately on a source of mineralizing fluids from the underlying granodiorite, or the dykes and sills that are found in the limestone. Some authors have suggested that the evaporites were, in part, digested and remobilized by the intrusives, yielding a more mobile mineralizing agent. Others have drawn attention to the existence of hydrogen sulphide in the darker members of the Limestone Group, that could have precipitated metals from chloride-rich solutions. An association of silver-rich galena and iron-rich sphalerite in intimate contact with pyrrhotite, is usually taken to indicate high temperatures of formation, and the existence of cassiterite and calc-silicate minerals in the eastern ore bodies would also indicate this. It is clear that the sills and dykes and, to some extent, joints and fractures related to the doming of the limestones, has controlled the position of the ore bodies (the same is true of many ordinary cave systems). On the one hand these deposits may be thought of as an unusual variety of hydrothermal replacement deposit, or, as an incipient contact metasomatic one in which the mineralizing fluids found their way into a pre-existing solution cave system. The oxide ore is simpler to understand as being formed by surface oxidation above the water-table. It is noticeable that oxidation is most effective where the Capping Group has been eroded away, and one may suppose that it took place as the cover was progressively removed in very late Tertiary or Recent times.

Mining

Originally there were many independent mines at Santa Eulalia, but they are now operated more or less as one. The complex nature of the ore bodies has led to the use of a wide range of different mining underground methods. Ore is hoisted up several different shafts. Much of the ore, particularly the oxide ore, is smelted directly, but over the years methods of beneficiation were introduced, and selective flotation is used to treat the mixed sulphide ores.

Further Reading

HEWITT, W. P. (1968), Geology and Mineralization of the Main Mineral Zone of the Santa Eulalia District, Chihuahua, Mexico. *Am. Inst. Min. Eng. Trans.*, **241**, 228–60.

HEWITT, W. P. (1943), Geology and Mineralization of the San Antonio Mine, Santa Eulalia District, Chihuahua, Mexico. *Biol. Soc. Am. Bull.*, **54**, 173–204. [These two papers contain a fairly comprehensive documentation of the deposit but Hewitt's genetic conclusions have not gone unchallenged – see the critical essay by Ridge (1972) (see Further Reading, p. 11)]

GONZALEZ-REYNA, J. (1956), *Memoria Geologico – Minera del Estado de Chihuahua*. XXth Int. Geol. Cong. Contributions, 28pp. [Pp. 198–211 of this admirable volume contain a concise description of the deposit, but unfortunately it is not illustrated.]

PRESCOTT, B. (1916), The Main Mineral Zone of the Santa Eulalia District, Mexico. *Am. Inst. Min. Eng. Trans.*, **51**, 57–99. [Although old, this paper contains a fuller description of the oxide ore than in Hewitt. It was also the paper that established Santa Eulalia as the type example of 'chimenea y manto' deposits.]

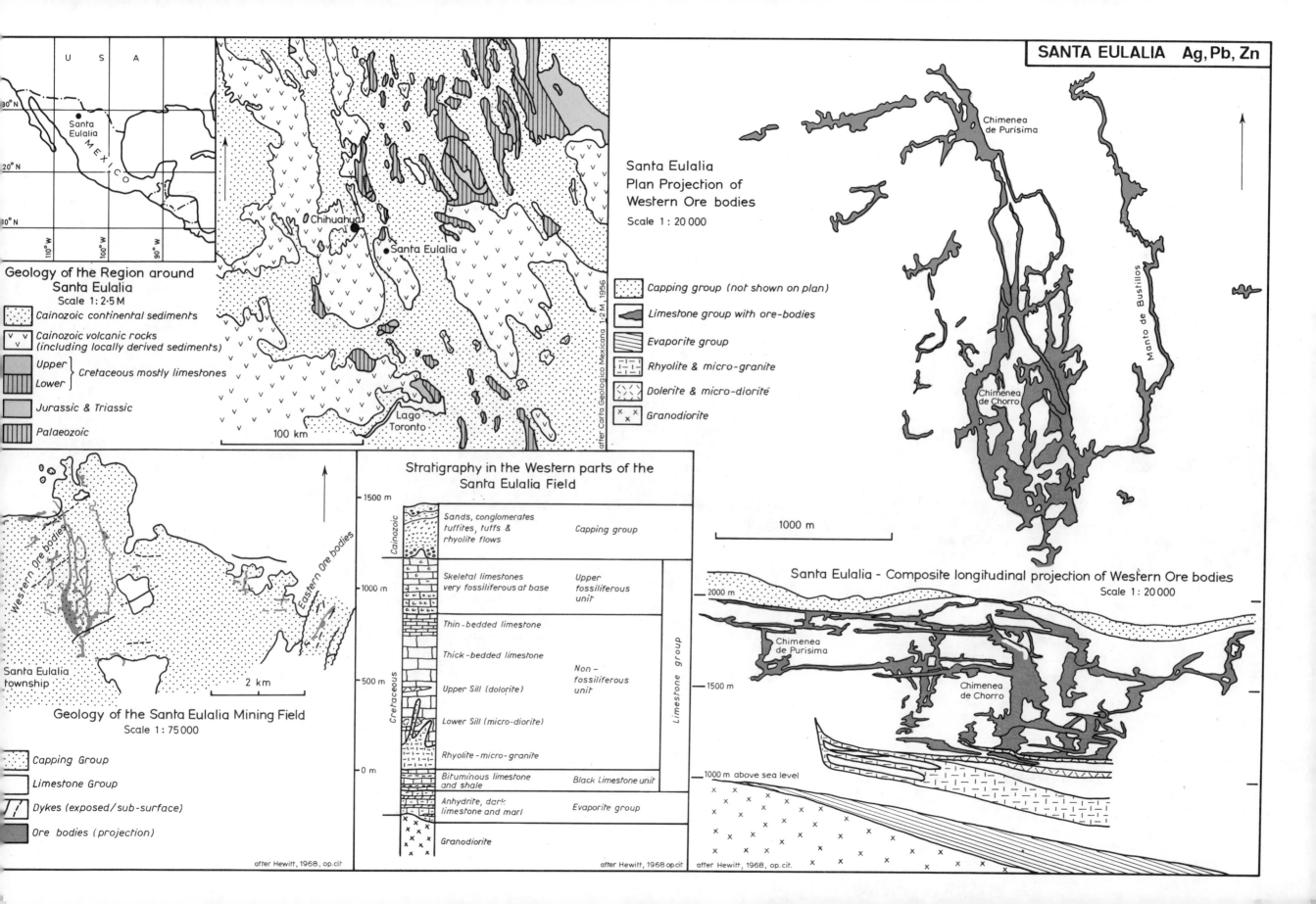

SANTA EULALIA Ag, Pb, Zn

Geology of the Region around Santa Eulalia
Scale 1: 2·5 M

- Cainozoic continental sediments
- Cainozoic volcanic rocks (including locally derived sediments)
- Upper } Cretaceous mostly limestones
- Lower
- Jurassic & Triassic
- Palaeozoic

100 km

after Carta Geológico Mexicana 1:2 M, 1956

U S A

MEXICO

Santa Eulalia

Chihuahua

Santa Eulalia

Lago Toronto

Santa Eulalia Plan Projection of Western Ore bodies
Scale 1: 20 000

Chimenea de Purisima

Manto de Bustillos

Chimenea de Chorro

- Capping group (not shown on plan)
- Limestone group with ore-bodies
- Evaporite group
- Rhyolite & micro-granite
- Dolerite & micro-diorite
- Granodiorite

1000 m

Geology of the Santa Eulalia Mining Field
Scale 1: 75 000

Western Ore bodies

Eastern Ore bodies

Santa Eulalia township

2 km

- Capping Group
- Limestone Group
- Dykes (exposed/sub-surface)
- Ore bodies (projection)

after Hewitt, 1968, op.cit.

Stratigraphy in the Western parts of the Santa Eulalia Field

Cainozoic	Sands, conglomerates tuffites, tuffs & rhyolite flows	Capping group
Cretaceous	Skeletal limestones very fossiliferous at base	Upper fossiliferous unit
	Thin-bedded limestone	
	Thick-bedded limestone	Non-fossiliferous unit
	Upper Sill (dolerite)	
	Lower Sill (micro-diorite)	
	Rhyolite-micro-granite	
	Bituminous limestone and shale	Black Limestone unit
	Anhydrite, dark limestone and marl	Evaporite group
	Granodiorite	

1500 m
1000 m
500 m
0 m

Limestone group

after Hewitt, 1968 op.cit.

Santa Eulalia - Composite longitudinal projection of Western Ore bodies
Scale 1: 20 000

Chimenea de Purisima

Chimenea de Chorro

2000 m

1500 m

1000 m above sea level

after Hewitt, 1968, op.cit.

The South–West England District – U.K.

The south-western peninsula of England has had a long history of mineral production; tin and copper were worked by the Ancients; it was one of the world's leading copper districts in the early part of the Industrial Revolution, later becoming important for tin and tungsten, and today it is one of the most important kaolinite producers. These are typical hydrothermal deposits associated with granite and show all the characteristic features of zoning, alteration and polyphase mineralization.

Location

The district is 150 km long and 40 km wide, centred on latitude 50° 30′ and longitude 4° 30′ W, comprising the county of Cornwall and part of Devon.

Geological Setting

The south-western peninsula of England is a land of open moors and steep-sided coastal valleys. The highest point is 621 m above sea level on Dartmoor, and a large part of the area is rolling plateau. Along the south coast there are several prominent drowned valleys. The climate is humid, influenced by the south-westerly weather and the gulf stream. It rarely freezes or snows in winter, the average temperature being 5°C; and in summer the average temperatures are about 17°C. The average annual rainfall is just over a metre.

The lower land is fertile, and good pastures exist, but on the moorlands the soils are acid and waterlogged. Fishing and other forms of maritime activity were the traditional occupations in the area, but this has now been replaced by tourism as the major industry.

History

The earliest written record of tin mining in south-west England is by the historian Diodorus Siculus during the first century B.C. Tin, copper and other mineral substances have continued to be exploited throughout the present millennium. By the beginning of the eighteenth century gunpowder, and the development of the water-wheel, enabled underground mining of veins to develop. In 1745, William Cookworthy discovered a good clay deposit on Tregonning Hill, and in 1775 Josiah Wedgwood started transporting china clay from Cornwall to the Midland potteries.

The invention of the steam engine enabled mining to continue to much greater depths, and the district became one of the largest tin- and copper-mining fields in the world, but was outclassed at the end of the nineteenth century by the new discoveries of copper in the U.S.A. and tin in South-East Asia. However, some mines still continue. China clay working developed and, since the Second World War, has become a major industry.

Geological Setting

The south-western part of England is part of a large belt of rocks known as the Hercynian, that runs from the Atlantic seaboard of England eastwards through the centre of Europe.

In the district itself the dominating rocks are slates, quartzites and volcanic rocks of Devonian age. These were folded along an east–west axis and are well cleaved. The sequence is overlain in the north-east by Carboniferous turbidites and by Permo-Triassic red beds. Towards the end of the period of tectonism that affected the area during the Permian, a series of plutons of granite were intruded into the Devonian and Carboniferous sediments. There is geophysical and geological evidence that these interconnect in depth to form a batholith. The mean radiometric age of the intrusions is 290 Ma. Each granite outcrop area is surrounded by a zone of contact metamorphism.

In the Tertiary, the district was eroded to form a sequence of erosion surfaces, the most persistent of which (believed to be of Pliocene age) is at about 120 m above the present sea level. One small sedimentary basin formed in the east known as the Bovey Basin, filled with a sequence of lancustrine and esturine sediments.

The area remained a region of tundra during the Quaternary, at the end of which some modest areas of alluvial and aeolian sands were formed, and adjustments in the sea level drowned the valleys on the southern coast.

Geology of the Mineral Deposits

The mineral deposits of this district consist of veins, alteration zones and others derived from them. The great variety of veins that occur can be grouped into five types. The first are relatively small pods, veins of vein-swarms that contain feldspar, quartz, wolframite, arsenopyrite, other sulphides and sometimes cassiterite and fluorite. The second type are vein-swarms, individual veins being narrow and bordered by greisen (a quartz-white mica rock). Such veins contain chlorite, tourmaline, with some cassiterite and wolframite, the surrounding rock being strongly sericitized and usually kaolinized. The third type (from which most of the tin has been mined) are persistent, narrow and filled with quartz, cassiterite, some wolframite and sulphides, principally arseno-pyrite, iron-rich sphalerite and stannite. They are bordered by a zone of intense chloritization and tourmalinization which is, in turn, bordered by sericitized wallrock and a zone of reddening. These veins often pass gradually into the fourth type associated with less intense alteration containing chalcopyrite, pyrite, sphalerite, some fluorite and hematite. The fifth type of vein has a carbonate and quartz gangue with galena, sphalerite, pyrite, a host of minor sulphides (some silver, others cobalt-bearing) and in a few cases pitchblend.

These vein types are arranged zonally (in the order described from inside to outside) around the main mineralized centres. The feldspathic and the greisen-bordered veins occur almost entirely in fractures in granite, but the tin and copper veins occupy faults that traverse both granite and the surrounding metasediments. In a few places, the veins are cut by micro-granite dykes or small off-shoots of the granite, but most of the veins post-date the intrusions. In most places the veins cut, are cut by, or stop at cross-faults that strike approximately at right angles to the veins and in several parts of the district there were complex sequences of movements along these two sets of faults that continued throughout the period of mineralization.

China clay deposits are found in several of the granites, particularly St. Austell. They are mostly funnel-shaped and associated with vein-swarms of the second type. The rock consists of a soft mixture of altered feldspar, quartz and mica with a fair proportion of kaolinite and sericite. The boundaries are very irregular in most instances, and the amount and purity of the kaolinite is best at the tops of the deposits. There are also zones of what is known commercially as china-stone; a variety of sericitized granite containing fluorite, at one time mined for the manufacture of glazes and enamels.

Alluvial tin occurs in the valleys draining the main mineralized fields and in unconsolidated sediments out to sea. Beds of 'ball clay' are found in Oligocene sediments of the Bovey Basin.

Size and Grade

It is estimated that the district has produced about 2·5 Mt of tin metal and about the same quantity of copper from ores containing the order of 1% tin or 2% copper. The mines at present working have reserves totalling about 7 Mt containing about 1% tin. Over 50 Mt of china clay has been produced, and the reserves are many times more. The kaolinite makes up about 15% of the rock from which it is extracted.

Geological Interpretations

There is general agreement that the mineralization in the district was emplaced by hydrothermal solutions and that, in some way, these were associated with the formation of the granite plutons. There is evidence that the feldspathic and greisen-bordered veins, were formed at high temperatures, just below the granite solidus, and that the others were formed at progressively lower temperatures. Much of the mineralization is in fractures that formed after the folding of the Palaeozoic rocks, but essentially as part of the same tectonic event. The role played by the intrusion of the plutons on the formation of the fracture systems is a matter of dispute. The spacial relationship of the mineralized fields to the highest parts of the plutons suggests that the solutions came from residual magmas in the granite, but stable-isotope evidence suggests that much of the water associated with the mineralization was meteoric. Furthermore, the association of the mineralized fields with Devonian meta-volcanics and argillaceous sediments is at least as notable as their association with the tops of plutons. There are two possibilities; derivation of the solutions from the residual magma of the granites formed, as a result of the absorption by them of water from the rocks into which they were intruded; or by leaching and redeposition of trace elements in the sediments and meta-volcanics by water caused to circulate by the heat of the granites.

Both field relationships and isotope evidence suggest that the kaolinite deposits were formed by weathering, but on the other hand, in depth the association with vein-swarms is universal. It seems probable that hydrothermal alteration produced zones of sericitized (and perhaps partly kaolinized) granite, and that during the Tertiary the amount of kaolinite was enhanced by weathering. Where altered (probably sericitized) granite was eroded from Dartmoor into the Bovey Basin, ball-clays seem to have formed in the stagnant, organic-rich environment of an estuary.

Mining

The veins of the south-west England district were mined by traditional underground methods and the ore hoisted up shafts that usually ran down the veins. Most of the cassiterite (and wolframite) was recovered by gravity concentration methods. Similar methods were used for copper minerals, but even as late as the beginning of this century, hand picking was a regular method of concentration. More recently flotation has been used to treat some of the mixed sulphide/oxide ores.

China clay is almost all mined by hydraulic methods, forming a slurry that is classified, using a combination of rake or spiral classifiers, settling tanks and cyclones. These processes separate the different grades of clay from the waste material, and the clay is then filtered and pressed into blocks. Many byproducts are made from the waste material, including insulating mineral wool and cement building blocks.

Further Reading

DINES, H. G. (1956), *The Metalliferous Mining Region of Southwest England* (2nd edn., 1969), Memoirs of the Geol. Survl of G. Britain, London, H.M. Stationery Office, 2 vols. 795pp. [The standard work on the area, but a bit verbose for all but the specialist.]

HOSKING, K. F. G. and SHRIMPTON, G. J. (ed.) (1964), *Present Views of some Aspects of the Geology of Cornwall and Devon.* 150th Anniversary Edition of the Royal Geological Soc. of Cornwall, Penzance. Royal Geol. Soc. Cornwall, 330pp. [A very good review volume which includes a good paper on the mineralization by the first editor.]

BRISTOW, C. M. (1968), Kaolinite Deposits of the United Kingdom of Great Britain and N. Ireland. *XXIII Int. Geol. Cong.* – Prague, Vol. 15, pp.275–88. [Despite its title this paper is largely about the deposits of Cornwall and Devon and the best on the subject. The remaining papers in the same volume were all presented at a symposium on Kaolinite Deposits of the World.]

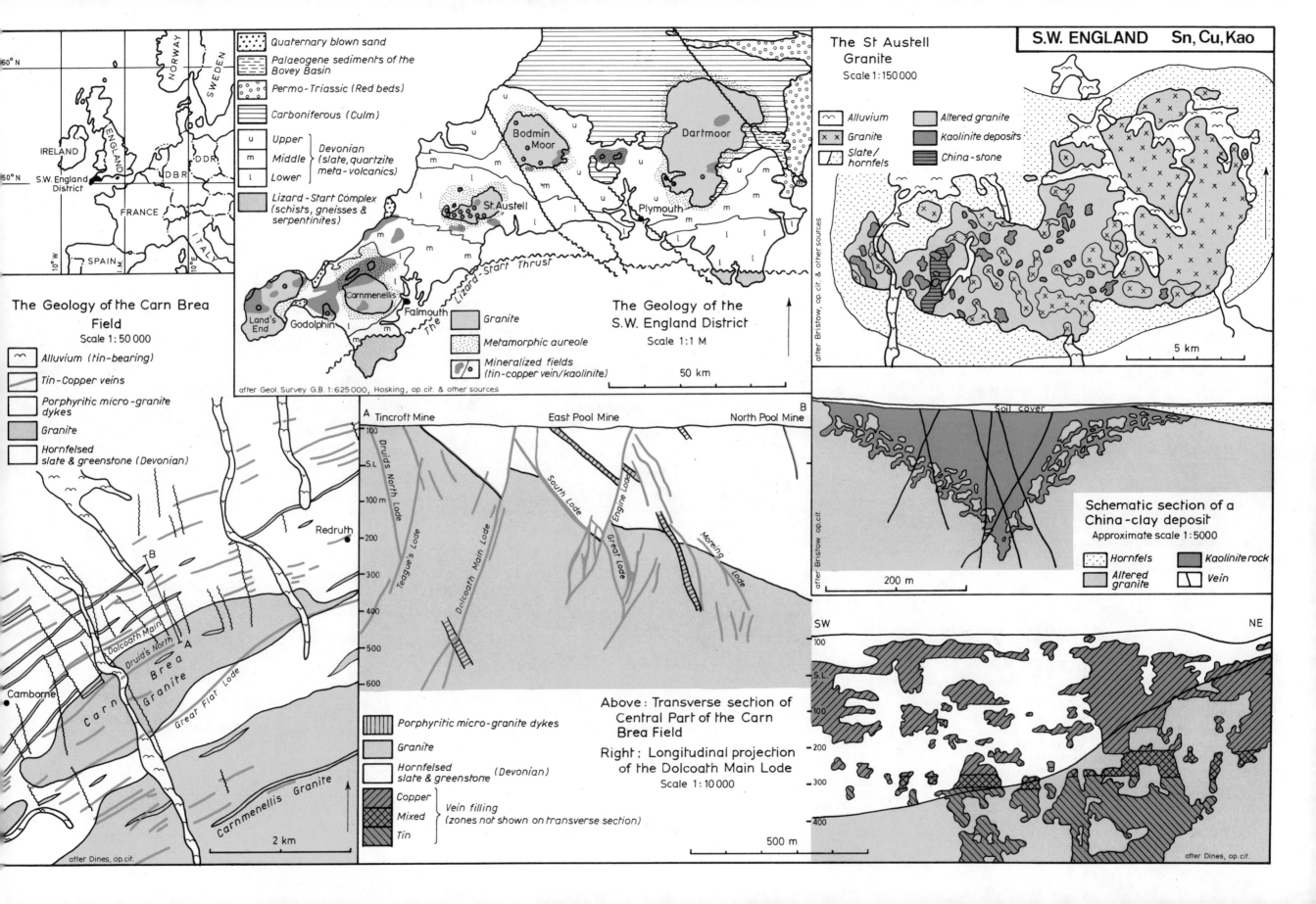

The Pine Creek Tungsten Deposit – U.S.A.

Contact metasomatic deposits are a fairly clearly defined type in which the ore occurs as a replacement of the contact metamorphic zone between an intrusive and its wallrock. Several economic minerals are found in such deposits, among which scheelite is one of the most important, particularly where the wallrock is limestone. Pine Creek is a good example of the type and is also the most important tungsten deposit in the U.S.A.

Location

Pine Creek Mine is situated at latitude 37° 24′ N, longitude 118° 42′ W and at over 3000 m altitude, 30 km west of the town of Bishop in Inyo County, California.

Geographical Setting

The mine lies on the side of Morgan's Creek, a former glacial valley, below Mt. Morgan (4190 m) close to the eastern edge of the highest part of the Sierra Nevada. Pine Creek itself descends to the broad Round Valley and eventually the major drainage channel of the area, Owens River, which separates the Sierra Nevada from the gentler White Mountains to the east. It is an area of high relief and varied topography, from bare rugged mountains to dusty plains. The climate is semi-arid, being in the rain-shadow of the Sierra, and much influenced by altitude. Winter averages are just above freezing on the plains, and just below in the mountains. Rainfall in the spring and autumn varies between 150 mm and 450 mm and the mountains around the mine receive on average 4·5 m of snow. On the plains, the vegetation is sparse grassland and scrub, and in the mountains bare rock predominates, but with some good forest in the valleys. Few people live in the area; the main town of Bishop (pop. 3000) is at the centre of the lower land where there is some ranching, but there is almost nothing in the mountains apart from the mining camps. Pine Creek lies between the Sequoia–Kings Canyon, and the famous Yosemite National Parks. The Owens River valley is an important water supply area for Los Angeles, 370 km to the south. Bishop has an airport, and is connected to other parts of the country by the main U.S. Highway 6.

History

Many travellers passed through this country during the great migrations at the end of the Civil War, and there were numerous reports of small alluvial gold finds in the area. The first major find, however, came in 1881 when the Poleta Mine was found in the White Mountains. This, and one or two other gold properties, were rich enough to maintain interest in the area, and systematic prospecting, and small-scale alluvial working continued. In 1913, three prospectors located a gold showing in a range of hills in Round Valley and had a great deal of difficulty cleaning concentrates because of a white mineral of high density. The local government assayer identified the offending mineral as scheelite. Stimulated by the sudden surge in demand for tungsten during the First World War, a number of deposits were found in the Bishop district, including the largest, Pine Creek, by Sproule and Vaughan in 1916. Because of difficult access, Pine Creek was not extensively developed before the slump that followed the war. There was sporadic working during the twenties and early thirties, which at least showed the full extent of the deposit, so that by the time demand increased at the outbreak of the Second World War, a full-scale mine could be opened, which has continued to operate to this day.

Geological Setting

Pine Creek is on the eastern margin of the Sierra Nevada Batholith. Prior to its intrusion the area underwent a series of geological changes which have left a complex series of belts of deformed, and at times metamorphosed, sediments that range in age from the late Pre-Cambrian to early Mesozoic. In the Bishop district these are represented by two groups of rocks. The first are the late Pre-Cambrian to lower Cambrian sediments in the White Mountains, and the second is a group of metamorphosed sediments and volcanics, now found in the Sierra Nevada itself. The latter are too deformed and metamorphosed for proper stratigraphic grouping or dating, but are believed to be Palaeozoic. Three main lithological groups are present: meta-volcanic rocks (for the most part originally andesites and andesitic pyroclastics); a series of quartzose meta-sediments now largely converted to biotite and felspathic hornfelses; and the host rock of the tungsten ore, marble. The marbles vary between fine-grained black to coarse-grained white types.

The Sierra Nevada Batholith is a large and very complex series of intersecting and cross-cutting plutons. In the Bishop district three rock types are found: granites, granodiorites and more basic rocks that vary from quartz-diorite to hornblende gabbro. The metamorphic rocks appear as 'septa' between individual plutons, and as 'roof pendants' representing the remnants of the original roof of the batholith. Along the contacts the marbles are usually converted to a pale-coloured calc-silicate rock containing diopside, wollastonite and a little garnet. These zones are 50–100 m wide.

The age of the batholith is about 100 Ma, that is to say mid-Cretaceous, and by the mid-Tertiary the area had become a major volcanic area. This is represented in the district by an extensive area of rhyolitic tuffs, pummice and ash-fall deposits. A number of basaltic cinder-cones and flows occur, for the most part of Pleistocene age. The uplift and unroofing of the batholith in the late Tertiary produced much debris, and large parts of the area are covered by great alluvium-filled valleys, of which Owens Valley is one.

The Sierra Nevada was heavily glaciated in the Quaternary, in spite of its low latitude, and small glaciers hang above Pine Creek to this day. Many of the mountain valleys such as Morgan's Creek are extensively covered by moraine and scree.

Apart from the tungsten deposits, the Bishop area contains a number of small quartz-vein gold deposits, and down on the volcanic ground are some modest deposits of pummice and perlite.

Geology of the Pine Creek Deposit

The deposit consists of a series of irregular ore bodies immediately on the contact between a belt of marble and one of the plutons of the batholith. The marble is grey or white, and in the neighbourhood of the mine, almost wholly converted to a calcite–wollastonite–diopside hornfels. At immediate contact, are zones of a rock of more complex mineralogy, known in the American terminology as 'tactite'. (Elsewhere it might be termed 'skarn' but the American term is used here.) The Pine Creek tactite is a fine-grained hard rock composed of hedenbergite, grossularite-andradite garnet, epidote, idocrase with some calcite and wollastonite. The zones of tactite are somewhat irregular and show a preference for certain beds in the marble.

The ore minerals are scheelite and molybdenite, but small traces of other minerals are found, including powellite, chalcopyrite, pyrrhotite and several other sulphides. A little gold is present, probably associated with the sulphides since it reports in the copper-smelter products.

There are five ore bodies, all very irregular, and all in effect zones of the tactite rich enough to be mined either for tungsten or molybdenum, or both. Those illustrated are based on assay limits of 0·4% WO_3 and 0·4% MoS_2. Scheelite is more widespread than molybdenite, which is confined to certain parts where quartz-veining is most common, and in the dense quartz-vein zones. The scheelite occurs as small interstitial grains down to 100 mm in diameter as flat grain boundary fillings, or occasionally as well-formed crystals over 10 mm across. The scheelite is relatively free of molybdenum and its molybdate polymorph, powellite, occurs as a separate phase.

At outcrop, the ore bodies are oxidized and leached, and evidence on the surface of tungsten mineralization is confined to the occurrence of various tungsten minerals that are difficult to identify; scheelite only outcrops in more rapidly eroding gulies. Molybdenite, however, fares better in the oxidized zone, and the first discoverer of the deposit mistook it for ore of lead–zinc or silver, presumably because of the superficial similarity of molybdenite to sulphides of those metals, particularly when fine-grained. Mining on the upper levels is complicated by a number of solution cavities in the marble that tend to follow the contact with the tactite.

Size and Grade

Ore above the cut-offs of 0·4% WO_3 and 0·4% MoS_2 averages 0·7% WO_3. (Were it economic to do so, practically all the tactite could be mined at a grade of 0·25% WO_3.) Published plans indicate the presence of well over 10 Mt of ore at the stated grade, but the total tonnage is probably well in excess of this. Deposits of this type are generally small, and Pine Creek is larger than the average.

Geological Interpretations

Pine Creek is an example of a contact-metasomatic (sometimes called a 'pyrometasomatic') deposit. The carbonate host rock was first heated and deformed as granitic magma pushed its way up to its present position, and this resulted in the transformation of the marble into a calc-silicate hornfels. It is clear that the formation of the tactite took place at a relatively late stage by the diffusion of material from the residual magma into the wallrocks. The scheelite is perhaps best regarded as a normal mineral formed at this stage. Local fracturing and the formation of quartz veins seems to have followed, and much of the molybdenite and sulphide mineralization can be regarded as hydrothermal.

Mining

Pine Creek is mined by a subsidiary of Union Carbide Nuclear Corporation which operates an underground mine and mill capable of producing 600 000 t/a. Ore is mined from open long-hole stopes, and passed down to the Zero Adit Level (287 m) where it is hauled by trains to a crusher at the adit portal. From there crushed ore is sent by aerial ropeway down to the mill in Pine Creek itself, 466 m below. At the mill the ore is ground, and passed through two stages of flotation to concentrate first the sulphides, and then the scheelite. The sulphide concentrates are further treated, to separate molybdenite and chalcopyrite. The scheelite concentrates and tailings pass over shaking tables to recover coarse-grained mineral which is free of powellite and can be sold. The finer-grained fraction of the scheelite flotation concentrate is treated chemically to separate tungsten from molybdenum; the former, being sold as 'artificial' scheelite, the latter as molybdic oxide.

Further Reading

BATEMAN, P. C. (1956), *Economic Geology of the Bishop Tungsten District, California.* California Div. of Mines Special Reports 47. 87pp. [A good general account of the deposits, well-illustrated.]

BATEMAN, P. C. (1965), Geology and Tungsten Mineralization of the Bishop District, California. *U.S. Geol. Surv. Prof. Paper*, **470**, 208pp. [A longer more detailed work than the 1956 paper.]

GREY, R. F. *et al.* (1968), Bishop Tungsten District, California. In: *Ore Deposits of the U.S.A., 1933–1967* (Ridge, J. D., ed.), pp.131–54. Am. Inst. Min. Met. Pet. Eng., New York. 2 vols, 1880pp. [A concise account by a group of geologists from the operating company as opposed to Bateman of the Geological Survey.]

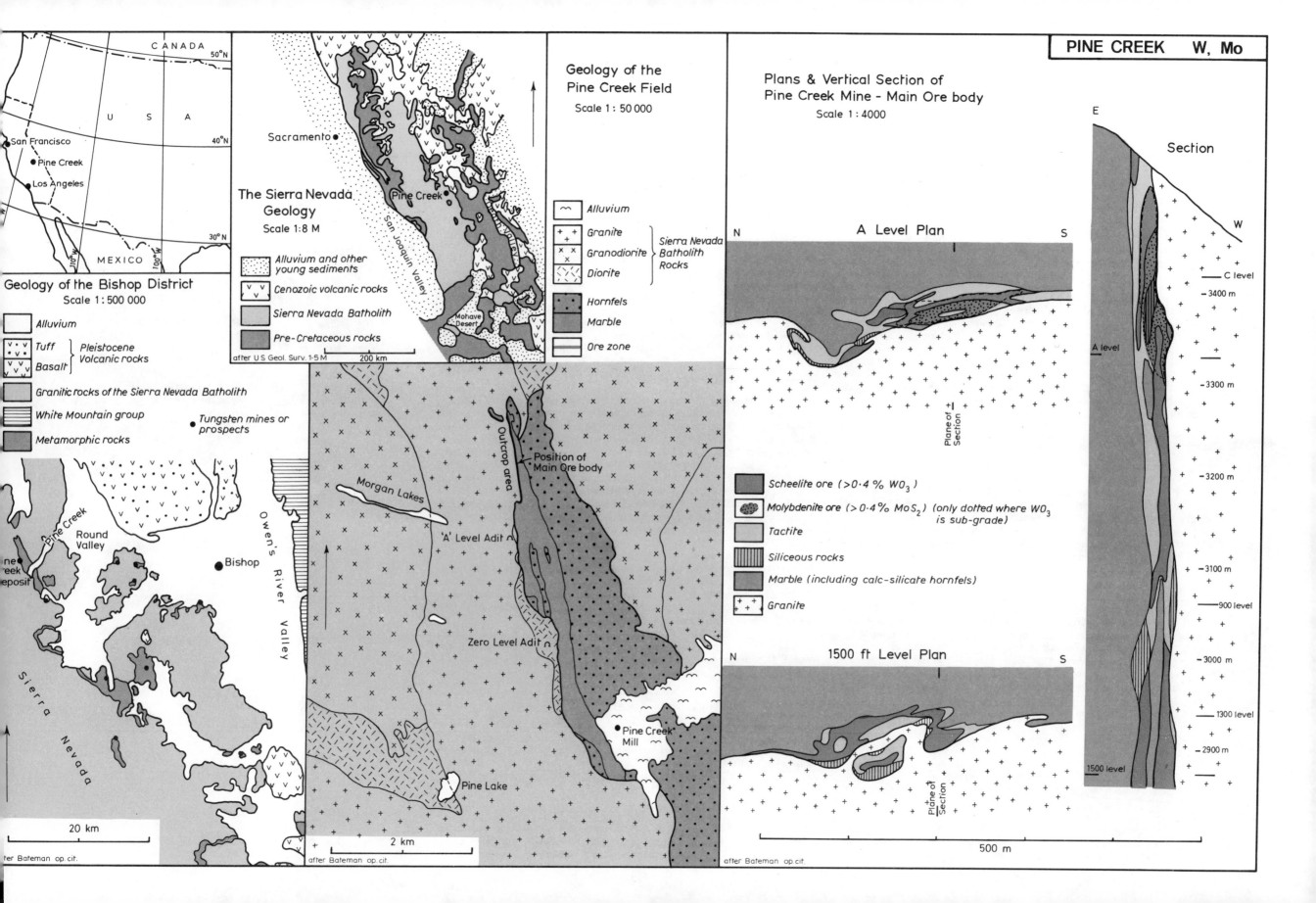

Geology of the Pine Creek Field

Scale 1 : 50 000

The Sierra Nevada Geology

Scale 1:8 M

Alluvium and other young sediments

v v Cenozoic volcanic rocks

Sierra Nevada Batholith

Pre-Cretaceous rocks

after U.S Geol. Surv. 1·5 M   200 km

CANADA   50°N

U S A

San Francisco

Pine Creek

40°N

Los Angeles

30°N

MEXICO

Sacramento

Pine Creek

San Joaquin Valley

Mohave Desert

~ Alluvium

+ Granite

× Granodiorite   Sierra Nevada Batholith Rocks

/ Diorite

Hornfels

Marble

Ore zone

Geology of the Bishop District

Scale 1 : 500 000

Alluvium

v v Tuff   Pleistocene Volcanic rocks

· Basalt

Granitic rocks of the Sierra Nevada Batholith

White Mountain group

Metamorphic rocks

Tungsten mines or prospects

Pine Creek

Round Valley

ne reek eposit

Bishop

Sierra Nevada

Owen's River Valley

Morgan Lakes

Outcrop area

Position of Main Ore body

'A' Level Adit

Zero Level Adit

Pine Creek Mill

Pine Lake

20 km

2 km

after Bateman op.cit.

after Bateman op.cit.

Plans & Vertical Section of Pine Creek Mine – Main Ore body

Scale 1 : 4000

A Level Plan

N   S

Plane of Section

Scheelite ore (>0·4 % WO_3)

Molybdenite ore (>0·4 % MoS_2) (only dotted where WO_3 is sub-grade)

Tactite

Siliceous rocks

Marble (including calc-silicate hornfels)

+ + Granite

1500 ft Level Plan

N   S

Plane of Section

500 m

after Bateman op.cit.

Section

E

W

— C level

—3400 m

A level

—3300 m

—3200 m

—3100 m

900 level

—3000 m

1300 level

—2900 m

1500 level

The Bikita Pegmatite – Rhodesia (Zimbabwe)

Complex pegmatites are thought to be derived from granite magma and are notable for their richness in exotic constituents such as lithium, beryllium, tantalum and niobium. This example, one of the largest in the world, is situated in one of the oldest greenstone belts of the Rhodesian Shield, a typical environment of this kind of deposit.

Location

Bikita Mine is situated at latitude 19° 57' S, longitude 31° 26' E and at 1180 m altitude, 69 km east of Fort Victoria.

Geographical Setting

The mine lies on the so-called 'middle veld', a somewhat flattish country but with prominent ranges of hills where harder rocks outcrop. The land ranges in altitude between 1000–1500 m draining gently to the south and the Limpopo Valley. The climate is sub-tropical but with a relatively large range of temperature from day to night; 33°C to 14°C in summer (January), 26°C to 7°C in winter. Rainfall averages 660 mm per year mostly falling in summer storms, and although originally savannah bushland with some tall trees, much of the area is now cleared for cattle ranching, maize or tobacco growing, even citrus groves where irrigated. Fort Victoria is a modest town, and a rail-head with connections to Beria. The Bikita mining field itself supports a small mining township in an area of scattered farms.

History

Henry Koestlich is recorded as having discovered the Nigel Tin Mine in 1909, and throughout the first half of the century the area was sporadically worked for eluvial cassiterite, and became known as the Bikita Tinfield. During this period prospectors and government geologists discovered the wealth of other minerals in the pegmatite outcrops, culminating in trial shipments of beryl and lithium minerals in 1949 and 1950. Systematic mining dates from 1953, since when Bikita has become a lithium and beryl producer of world importance.

Geological Setting

The Victoria Schist Belt is 80 km long and between 5 km and 12 km wide oriented in a NE–SW direction. It takes the form of a synclinorium of supercrustal rocks that have been divided by unconformities into three super-groups. The lowest, or Sebakwian Super-Group, consists of schists and amphibolites, but is not represented in the eastern part, which is entirely composed of rocks named the Bulawayan and Shamvaian Super-Groups. By far the greater part of the belt in the vicinity of Bikita is composed of the Bulawayan, divided into a lower, Sedimentary Group; pelitic schists, banded ironstones and carbonate rocks; and an upper Volcanic Group, pillow lavas, chlorite schists, hornblende schist and epidiorites. The Volcanic Group, in particular, contains many serpentinite lenses and pegmatites of spectacular size and economic importance. Infolded into the Bulawayan are pelitic schists, carbonate rocks and occasional acid volcanic rocks of the Shamvaian. The entire belt has been deformed into tight folds, cleaved and metamorphosed to the greenschists (and locally to the amphibolite) facies.

The Schist Belt is surrounded by granitic rocks, including gneisses and migmatites, and plutons of massive granite. Where observable, the granitic rocks show sharp intrusive contacts against the schists, but it is tempting to suggest (as many have) that the granites are the remobilized product of the pre-schist basement. The pegmatites cut the rocks of the schist belt, and contain minerals that have an average radiometric age of 2·65 Ga, thus placing the supercrustal rocks well into the Archean.

Large dykes and sills of dolerite cut all other rocks, some are perhaps Phanerozoic in age, but apart from locally derived eluvial cover and river alluvium, the later history of the area has left no record.

Geology of the Bikita Mining Field

Fifteen pegmatite bodies have been described in an area 4·5 km by 2 km. Their host rocks are the chlorite schists and epidiorites of the Bulawayan Volcanic Group, folded round a dome of pelitic schists and quartzites, of the Sedimentary Group, all exhibiting a persistent NE–SW and steep dipping cleavage. The pegmatites are lenticular in form with shallow dips, traversing the host rocks in a way apparently unrelated to the regional structure. They tend to be lobe shaped up-dip, and thin out to a wedge down-dip; their complexity of outcrop on the map being due in large measure to the effect of topography. Some intersect or touch one another, and the larger ones have offshoots. Those that have been explored in detail show a zonal arrangement of mineral assemblages, dominated by albite, microcline and quartz but with substantial amounts of lithium minerals; lepidolite, zinnwaldite, ambligonite, spodumene, petalite, eucriptite and bikitaite, with a wealth of accessory minerals including beryl, tantalite, pollucite, microlite and cassiterite.

The minerals more resistant to weathering, beryl, tantalite and cassiterite, are found in eluvial deposits near the pegmatite outcrops.

The Bikita Main Pegmatite

This, the largest and most economically important of the group, is 1600 m long, 30–60 m thick, and is known down-dip for a distance of over 300 m. The outcrop strikes almost north–south, dips to the east at 15° in the south and up to 45° in the north. It has sharp contacts (though with no visible contact effects) with the adjoining greenstones, and internally is strongly zoned with some spectacular features. On the footwall border sugary albite contains spots, or clumps, of lepidolite crystals that have earned it the local name of 'spotted dog', and is followed by mica-rich bands with large greenish beryls and tantalite, passing into a strongly banded albite/lepidolite zone. Higher up in the Lower Intermediate Zone a remarkable rock occurs, of which 50% may consist of rounded lenses of lepidolite 150–400 mm across, locally known as the 'cobble zone'. Towards the north the two intermediate zones change to petalite with lenses of spodumene and eucriptite. The Core is only present in the southern part, but is over 300 m in length consisting of dense felspathic lepidolite, or almost pure lepidolite, sometimes with a mantle of pure quartz. In contrast to the Lower, the Upper Border Zone is largely composed of giant microcline crystals. Cassiterite is found in small quantities in the wall zones and the mantle of the Core.

Though the Bikita Main Pegmatite is a good example of a zoned complex pegmatite, it differs in two ways from other examples elsewhere in the world. It contains no open vugs, and no evidence of crosscutting channelways and replacement textures.

Size and Grade

In 1960 the mines on the Main Pegmatite were declared to have reserves of almost 6 Mt averaging 2·9% Li_2O, but only 2·6 Mt at an average grade of 4·13% Li_2O was in the form from which lepidolite, petalite, spodumene or eucriptite could be recovered by hand picking. Parts of the Lower Wall Zone contain up to 0·5% BeO corresponding to about 5% beryl.

Geological Interpretations

All the authors that have described this area conclude that the pegmatites were intruded into the greenstones and schists, and that they are derived from residual magmas of the granites that surround the schist belt. Evidence suggests that the magmas were intruded in a single step, containing a wealth of exotic constituents from which the various assemblages were formed by differential crystallization outwards from the walls.

Mining

The Main Pegmatite is mined by the Bikita Minerals (Pvt.) Ltd. Although three levels of underground drives and cross-cuts have been made for exploration and sampling, the bulk of the production comes from open-pits along the length of the outcrop, with an average stripping ratio of 1:1. The different mineral zones are selectively mined and sent to crushing, screening, and hand-picking plants. The mine produces a variety of products, as far as possible monomineralic, and at various qualities and grain sizes, according to demand. The products are shipped by road and rail to other countries for chemical refining.

Further Reading

WILSON, J. F. (1964), *The Geology of the Country around Fort Victoria.*

MARTIN, H. J. (1964), *The Bikita Tinfield.* Published together as *Southern Rhodesian Geol. Surv. Bull.,* **58**, 47pp. [The standard work on the area.]

SYMONS, R. (1961), Operation at Bikita Minerals (Private) Ltd. Southern Rhodesia. *Trans. Inst. Min. Metall.,* **71**, 120–72. Reprinted in Pegmatites in S. Rhodesia – a symposium. S. Rhodesia Section of Inst. of Min. & Met. 1963, pp.58–74. [A comprehensive description by the manager of the day.]

TYNDALE-BISCOE, R. (1951), The Geology of the Bikita Tin-Field, S. Rhodesia. *Trans. Geol. Soc. South Africa,* **54**, 11–26. [A timely paper that stimulated commercial interest in the area.]

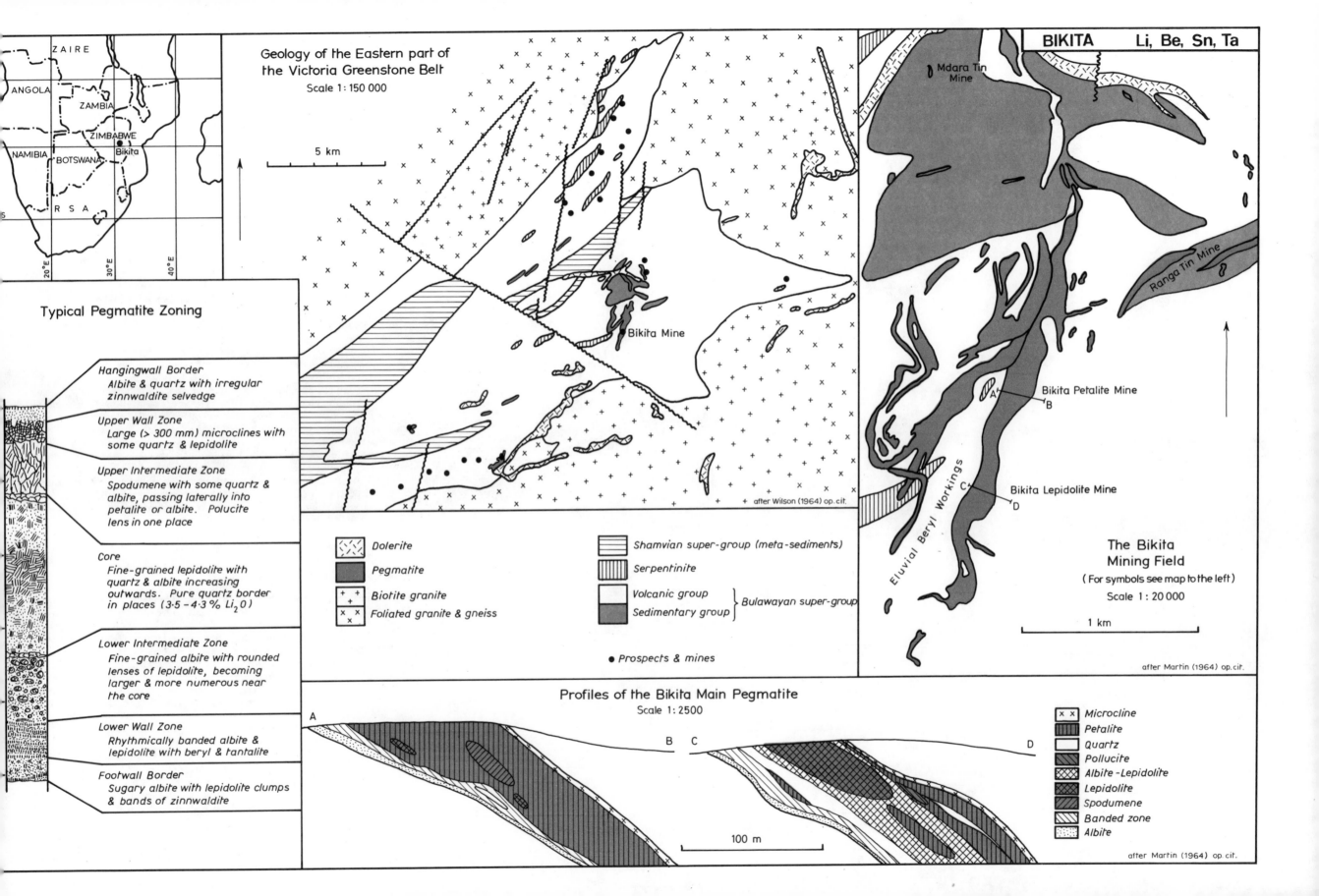

BIKITA   Li, Be, Sn, Ta

Geology of the Eastern part of the Victoria Greenstone Belt
Scale 1:150 000

5 km

Bikita Mine

+ after Wilson (1964) op.cit.

Typical Pegmatite Zoning

Hangingwall Border
Albite & quartz with irregular zinnwaldite selvedge

Upper Wall Zone
Large (> 300 mm) microclines with some quartz & lepidolite

Upper Intermediate Zone
Spodumene with some quartz & albite, passing laterally into petalite or albite. Polucite lens in one place

Core
Fine-grained lepidolite with quartz & albite increasing outwards. Pure quartz border in places (3·5 – 4·3 % Li_2O)

Lower Intermediate Zone
Fine-grained albite with rounded lenses of lepidolite, becoming larger & more numerous near the core

Lower Wall Zone
Rhythmically banded albite & lepidolite with beryl & tantalite

Footwall Border
Sugary albite with lepidolite clumps & bands of zinnwaldite

Mdara Tin Mine

Ranga Tin Mine

Bikita Petalite Mine

Eluvial Beryl Workings

Bikita Lepidolite Mine

The Bikita Mining Field
(For symbols see map to the left)
Scale 1: 20 000

1 km

after Martin (1964) op.cit.

Dolerite

Pegmatite

Biotite granite

Foliated granite & gneiss

Shamvian super-group (meta-sediments)

Serpentinite

Volcanic group
Sedimentary group
} Bulawayan super-group

● Prospects & mines

Profiles of the Bikita Main Pegmatite
Scale 1:2500

A

B   C

D

100 m

after Martin (1964) op.cit.

Microcline

Petalite

Quartz

Pollucite

Albite-Lepidolite

Lepidolite

Spodumene

Banded zone

Albite

ZAIRE

ANGOLA

ZAMBIA

ZIMBABWE

Bikita

NAMIBIA

BOTSWANA

R S A

20° E   30° E   40° E

Section Four: Mineral Deposits in Basic and Ultrabasic Magmatic Rocks

Introduction

It is convenient to distinguish between the depositional environments of mineral deposits in basic and ultrabasic magmatic rocks and those in other igneous and volcanic rocks, because the rocks themselves have rather different characteristics, and perhaps a somewhat separate origin. Also these environments contain deposits of several minerals that are unique to them, such as nickel minerals, chromite, platinum and asbestos.

Rocks of this group have certain things in common. They are rich in magnesium and iron, and consist of various mixtures of olivine, pyroxine and plagioclase. They seem to originate as fluid magmas and as they cool, various oxides and silicates crystallize in a fairly well-established order. Since the composition of the crystals formed is different from that of the magma, the magma at any stage during crystallization is undergoing a process of change which in turn affects the composition of the minerals that next crystallize. Furthermore, the oxides and ferromagnesian silicates are, in general, more dense than the magma and tend to sink, so removing them from the local chemical system. Thus, these rocks generally show differentiation leading to layering and zoning of a far more pronounced nature than is displayed by the more acid magmatic rocks, or basic ones that cool more rapidly. Contamination by wallrocks and hybridization with other magmas may be important, but not as a general rule.

As in all magmatic rocks, volatile elements and compounds play an important part in altering the course of crystallization, most dramatically in the case of carbon dioxide in the carbonatites. Sulphur and, to some extent, arsenic, antimony, selenium and tellurium appear to be present in small traces in these magmas. During the course of crystallization there comes a critical stage where sulphur becomes so concentrated in the magma that it unmixes as a separate sulphide melt, and at this stage an important partitioning takes place between the elements in the two melts which is vital to the origin of mineral deposits. Furthermore, the sulphide melt formed remains liquid long after the most silicate minerals have become solid.

As far as associated mineral deposits are concerned, basic and ultrabasic rocks can be classed into four groups; two of them, the large layered basic intrusives and the anorthosite complexes, are rare in the crust, although often of considerable size. The other two, the peridotite and serpentinite bodies, and the carbonatites and kimbertites, are small but numerous in certain belts or areas of the crust.

The Layered Basic Intrusions

The largest and best known of these is the Bushveld Complex in South Africa. This multiple sheet-like intrusion underlies 125 000 km² in the north Transvaal and contains over 80% of the world's known resources of platinum. The complex is markedly differentiated from bottom to top and is strongly layered in places, and the layering is so consistent that a 'stratigraphy' is traceable for hundreds of kilometres round its outcropping edge. Some layers are rich in oxide minerals, particularly chromite near the base, and ilmenite and titaniferous magnetite near the top. In addition to layers, the latter is found as irregular lenses and pipe-like bodies cutting the layering, and some of these are important because they contain workable vanadiferous magnetite. At certain critical positions in the sequence of rocks, sulphide minerals appear. The 'Merensky Reef' is the most important, a single layer of coarse-grained or pegmatoid norite which contains chromite, pyrrhotite, pentlandite and, most important, platinum metals, largely in the form of the arsenide speryllite. More irregular bodies of massive or disseminated sulphide sometimes cross-cutting the layering, and pipe-like bodies very rich in platinum associated with iron-rich olivine occur in the lower zones.

Just north of the Bushveld Complex is the 'Great Dyke', not so large but in many ways more curious. It is not a dyke at all but a series of very elongated elliptical intrusions that in cross-section taper down to a 'root'.

The series of intrusions are in a straight line giving the outcrop the appearance of a long dyke that traverses the Rhodesian Shield for over 500 km. The individual intrusions are broadly similar, consisting of a strongly layered dunite/pyroxenite sequence at the base, passing quite sharply into gabbro. The most important feature of this remarkable body is the persistent layers of chromite. The lower parts of the complexes also contain nickel, copper and platinum metals, but although in enormous quantity, they are too low in grade for economic recovery at the present time.

There has been a great deal of speculation about the origin of these two unique intrusions. Their composition is consistent with an origin from the large-scale melting of the upper mantle. The Great Dyke is commonly supposed to be formed along a great and deep fracture in the crust. There are however many other deep fracture systems in the crust, but none occupied by an intrusion of this sort. Clearly something strange took place in Southern Africa on two occasions in the Pre-Cambrian.

The Sudbury Complex in Ontario, Canada and the somewhat similar Noril'sk Complex in northern Siberia, though smaller than the two southern African giants, are in many ways just as singular. Sudbury is a differentiated (but not layered) gabbroic intrusion that is associated with some extremely fractured wallrock zones at its base, and a sequence of very high-energy pyroclastic rocks at its top. It contains, usually in irregularities of the floor or in fractured rocks beneath, a series of large lenses of massive or disseminated sulphide ore, containing pyrrhotite, pentalandite, chalcopyrite, magnetite and a range of minor minerals that give the ore important quantities of platinum metals, gold, selenium and other elements. Many other modest-sized gabbroic intrusions are known in the world, and although they commonly contain segregations of copper–nickel–iron sulphides, they come nowhere near the scale on which these occur at Sudbury and Noril'sk. On the whole, other gabbro intrusions seem to have been emplaced more passively. As is usual with geological objects that have no equals, the lack of the facility to make comparisons leads to a great deal of uncertainty and speculation about their origin. Both in the case of Sudbury and Noril'sk, it has been suggested that they are rather special kinds of eruptive complexes formed by melting of the upper mantle, initiated by the impact of a large meteorite, and thus more a 'lunar' phenomenon that a 'terrestrial' one.

The essential feature of these deposits from the metallogenic point of view, is that a basic magma containing enough sulphur can form a sulphide-melt rich in certain chalcophile metals and that, because this melt remains as such after the solidus of the silicate has been passed, the localization of the sulphide deposits depends very much on the structural history of the intrusion, and its surrounding rocks at the later stages of cooling.

Ultrabasic Rocks

An almost exactly similar assemblage of minerals and elements is found in certain bodies of ultrabasic rocks, dunite, peridotite, and serpentinite. Many of the best examples of such deposits are found in the Archean greenstone belts, but they are not necessarily confined to them. The host rocks of such mineralization may be intrusive or lava flows, and the mineralization usually occurs in irregularities of the floor; although as one would expect, deformation may modify the distribution of mineralization.

It is the exception rather than the rule that ultrabasic magmas contain sufficient sulphur for the segregation of a sulphide melt. In most rocks of this group the elements like nickel and copper are contained as traces in the lattices of other minerals, particularly olivine. Such rocks form the primary material for the formation of the lateritic nickel deposits.

There are a number of different types of ultrabasic body. The two most important are those found as small lens-like intrusions or lava flow-sequences in association with the pillow lavas, that usually dominate the stratigraphically lower parts of the Archean greenstone belts. A second type is generally found in zones of great tectonic complexity in some of the younger mobile belts. The latter group is interesting because they are now interpreted by some geologists as fragments of oceanic crust that become tectonically emplaced along destructive plate edges. These rocks form part of the association of rocks known as ophiolites, along with other rocks of possible oceanic origin such as pillow lavas, radiolarites and pelagic black shales. Such rocks are found in belts along with highly deformed sediments, not uncommonly with metamorphic rocks of 'blue schist' facies. Some bodies of rock of this type occur in the so-called 'melange' zones which are found at intervals along the Alpine–Himalayan belt, and consist of areas of completely different types of rock, mixed together and separated by faults in a way that almost defies tectonic and stratigraphic interpretation.

Ultrabasic rocks also contain accumulations of chromite. Broadly these are of two types: one consisting of layers almost exactly the same as those found in the Great Dyke and Bushveldt Complex, but in general more discontinuous; and a second type known generally as the 'podiform' deposits that are irregular lenses, pipe-like or pod-like bodies in the rock. These usually contain a core of massive chromite, surrounded by a zone of disseminated ore within an envelope of serpentinite.

A group of deposits of magnesium-rich minerals is also found in ultrabasic intrusions, indeed some dunite and serpentinite are worked en masse for the manufacture of magnesium silicate refractories. Talc deposits are found, usually occurring in the marginal zones of serpentinite bodies in deformed rocks, and it is common to find a zone of rock surrounding such bodies composed of talc, magnesiste, brucite with a little quartz. Magnesite is also found as veins or irregular masses in serpentinite. Some of these seem to be associated with the alteration process that serpentinized the original ultrabasic rock, others seem to originate from, or are enriched by, weathering. Finally, one of the most important materials found in these rocks is chrysotile asbestos. This occurs in veinlets through masses of partially serpentinized dunite or peridotite, usually with a close spatial relationship between the asbestos and the serpentinization. The localization of zones of asbestos veining can usually be related to fracturing of the rocks by tensile failure during deformation.

Ultrabasic rocks seem to be primarily a feature of oceanic and island arc regions, and presumably originate from melting of parts of the upper mantle depleted in the lighter elements, or by differentiation of magma. Chromite and nickel–copper–iron sulphides are both products of the crystallization of the magma, in much the same way as those found in the large layered intrusions. The other minerals; asbestos, talc and magnesite, are products of hydrous activity that takes place at the closing stages of cooling, or subsequently.

There are some differences in the type and frequency of mineral deposit found in the two types of ultrabasic complex. Almost all nickel–copper–iron sulphide deposits occur in the Archean type. Chromite is found in both types but predominantly in the ophiolite type. The majority of talc, magnesite and a substantial part of the asbestos deposits occur in the ophiolite type.

The Anorthosite Complexes

There are relatively few of these in the world, and they are almost confined to the later Pre-Cambrian mobile belts. Characteristically they are large bodies of anorthosite with subordinate amounts of banded norite or gabbro. They usually have a structure that suggests strong deformation, and are often associated with charnockites or metamorphic rocks of the granulite facies. Although they may be a form of layered basic intrusive, it has yet to be explained why such large masses of feldspar-rich rocks occur (the Egarsund Complex in Norway probably contains at least 1000 km³ of almost pure feldspar rock). Nor is there any explanation so far of the narrow time range of their occurrence.

The importance of these rock units is that they contain large deposits of good quality ilmenite, particularly at Egarsund in Norway, Alard Lake in Quebec, in the Adirondacks of north-eastern United States. Several similar complexes in other parts of the world contain large quantities of titaniferous iron oxides, usually too rich in titanium for use as an iron ore, but too lean to be mined for titanium.

Peralkaline Complexes, Carbonatites and Kimberlites

These are among the most exotic magmatic rocks. There are some hundreds of peralkaline complexes, mostly cutting shield areas and often in areas or along zones associated with rift systems. Although variable, these are generally circular or elliptical pipe-shaped bodies that contain alkaline pyroxenite and a variety of rock types that are differentiates or modifications of it. They are usually surrounded by a zone of intense potash metasomatism, veining or replacement of the wallrocks with alkali feldspar. This rock is known as fenite. Most of the peralkaline complexes contain an exotic mineralogy. They frequently contain apatite in unusual quantities, minerals like sphene, ilmenite, pyrochlore, uranium and thorium minerals, fluorite and sulphides. Some peralkaline complexes have a core of carbonatite, a carbonate rock that may be largely calcite but can be dolomite, iron-bearing or even strontium-rich carbonate. Carbonatites are rarely simple, and usually show a range from ones that seem to be magmatic, to others that may be formed from hydrothermal fluids.

There is abundant evidence that the peralkaline and carbonatite complexes are feeders to volcanoes, and there are a group of volcanoes in the East African Rift that extrude sodium-calcium carbonate lavas at the present day. It has been suggested that they form by differentiation of a melt that began deep in the mantle below the layers where it is depleted in alkaline and volatile elements. Clearly these complexes are associated with major fracture systems in the crust, but the origin of these and the restricted distribution of the complexes to certain parts of the crust has not been explained.

In these complexes are found economic deposits of apatite, pyrochlore, some radioactive minerals, fluorite, minor amounts of iron–titanium ores and in one case at Palabora in South Africa, copper sulphides.

Kimberlites are often regarded as being related to the peralkaline complexes. The rock itself, kimberlite, is essentially a potash-rich serpentinite volcanic breccia with an immense assemblage of minor minerals such as phlogopite, chrome-dipside, certain types of garnet, titanium minerals and carbonates. Their importance is that sometimes the rock contains diamond (albeit in very small quantities). Kimberlite occurs in the form of pipes and dykes and, like carbonatites, these are found in clusters or zones apparently controlled by deep crustal fractures. (Carbonatites and kimberlites, however, rarely occur in the same area.) Most of the pipes are filled with explosion breccias, but a few are in the form of craters full of pyroclastic rocks and crater lake sediments. The rock is usually full of fragments of the wallrocks and occasional rounded xenoliths of eclogite. It has been suggested that kimberlites begin life as a magma of similar composition to that of the carbonatite complexes, but contain less volatile material and come straight to the surface rather than passing through a long series of differentiations. The occurrence of diamond has been explained by postulating that the magma has its origin deep enough in the mantle for carbon to be stable as diamond rather than as graphite, but some geologists still adhere to the idea that the diamond is produced as a result of the enormous pressure developed during violent eruptions.

The Platinum Deposits of the Merensky Reef – South Africa

The world's largest resource of the platinum-group metals is the Merensky Reef, which has been traced for 250 km around the lower part of the famous Bushveldt igneous complex. It is quite unique in its persistence and extent, and is one of the best examples of a simple magmatic deposit.

Location

The Bushveldt Complex outcrops over a large area, roughly centred on latitude 25° S and longitude 28° E, and the main area of platinum production from the Merensky Reef is centred on the town of Rustenburg (25° 40' S, 27° 15' E) in the Transvaal State of the Republic of South Africa.

Geographical Setting

The Bushveldt Complex is on the high 'veldt' which averages about 1200 m in altitude. It includes some spectacular landforms, particularly the escarpments formed by the hard quartzites that surround much of the complex. Even some of the igneous rocks of the Complex give rise to great cliff-like forms, but much of the Complex is rolling hilly country interspersed with quite large areas of flat plains. The climate is sub-tropical, modified by high altitude, warm or hot, wet, summers, and dry cool winters. The 600 mm rainfall is enough for extensive agriculture, the land having originally been long grass savanna. The Complex is in a well-developed part of the country, the capital, Pretoria, being on the southern edge.

History

Geological mapping and prospecting of the Transvaal began in the 1870s, and among the discoveries were seams of chromite both in the eastern and western parts of the complex. In 1924, A. F. Lombard found platinum in panned concentrates from a river bed on his farm at Maandagshoek near Lydenburg, and the famous geologist Hans Merensky became interested in the find. The alluvial material was traced to an outcrop of dunite that subsequently turned out to be a small, but rich, pipe. Merensky went on to organize a systematic search for platinum, and, within a few years, it was realized that there was a continuous horizon of layered basic rock that was platinum-bearing through its outcrop length; which became known as the Merensky Reef.

Geological Setting

The Bushveldt Igneous Complex is a composite body intruded into a group of Proterozoic sediments, that lie on a crystalline basement of Archean granitoid rocks and schist belts. For the most part the intrusions cut the Pretoria Group, a clastic sequence that includes the Magaliesburg Quartzite, which forms such spectacular landscapes round the edge of the igneous complex.

The Complex is in two parts: a basic and an acid. The basic part is remarkable for its differentiation and layering, so persistent that one can almost speak of a 'stratigraphy'. The Complex seems to have the form of a series of sheet-like intrusions, all more or less similar, resulting in a lobate pattern.

Below the base of the basic sequences are Pretoria Group Sediments intruded by many sills and dykes of gabbroic rocks. The continuous part of the Complex begins with a layered sequence, the Basal Zone, consisting of pyroxenite and peridotite, with some intercalations of norite. At the top of this are a series of layered rocks that commonly contain bands of chromitite which culminate in the so-called Main Chromite Seam, taken as the boundary between the Basal Zone and the Critical Zone. This is a more regularly banded sequence in which each band is pyroxene-rich at the base, and feldspar-rich at the top, but with considerable textural variation. The

Merensky Reef is a coarse-grained band that is generally taken as the top of this zone. The Main Zone which follows is again banded, and includes bands of more even-textured norite, and almost pure anothosite that serve as marker horizons in the sequence. The Upper Zone which follows is somewhat different. It is marked by the reappearance of olivine (albeit much richer in iron than is found in the peridotites of the Basal Zone) an increase in the alkalie content of the feldspars, and the appearance of bands and bodies of vanadiferous magnetite.

Above this are granophyres and granitic rocks (the acid part of the Complex) which do not occur as a regular layered sequence. Large parts of the inner Complex consist of reddish coloured granite.

Throughout the sequence the regular layering is interrupted by coarse-grained or pegmatoid facies in the form of pipes or irregular bodies. Dunite pipes occur in the critical zone, bronzite pipes associated with sulphides in the basal, and magnetite-bearing pegmatoid bodies in the main and upper zones. The granites contain tourmaline and cassiterite-bearing pegmatoid pipes.

The granites of the Complex have a radiometric age of 1·96 Ga, and are thus older than both the sequences of sediments that overlie it; the clastic sediments of the Waterburg Group (which are late Pre-Cambrian) and the Karoo (which is of Palaeozoic age).

A wealth of mineral deposits occur in and around the Complex. Chromite and vanadiferous magnetite occur in the basic part, in addition to the Merensky Reef with its platinum metals, gold, nickel and copper. There is further nickel and copper associated with bronzitites of the Basal Zone, and the pegmatoid pipes of the red granite contain tin. Andalusite has been worked from alluvial deposits derived from the contact metamorphic zone around the Complex.

The Merensky Reef

This remarkable rock-unit is a coarse-grained felspathic pyroxenite of great uniformity and continuity. It appears to be transgressive over the underlying layered anorthosites and pyroxenites, and its base is marked by a thin layer of chromite crystals. Above the chromite is coarse pyroxene that becomes more felspar-rich at the top, where a further, but less persistent chromite layer marks the beginning of the overlying layer.

Near the base of the 'reef' are concentrated a great variety of metallic and sulphide minerals including pyrrhotite, pentlandite, chalcopyrite, ferroplatinum alloys, sperylite and many more. Together these minerals give the 'reef' a content in certain areas of 6 g/t precious metal and 0·3% nickel and copper combined. In the bullion extracted the metal proportions are: Pt 60%, Pd 27%, Ru 5%, Rh 2·7%, Ir 0·7%, Os 0·6% and Au 4%, and the ratio of Ni/Cu is 1:8.

The 'reef' follows the structure of the layered sequence dipping inwards to the centre of the Complex at angles varying between 5° and 30°. The 'reef' varies in thickness but the amount of metal remains fairly constant, and thus only where it is thin can it be worked economically; where it is thick the same metal is distributed through more rock.

In the best known and most mined section round Rustenburg the 'reef' is interrupted by various features. The so-called 'koppies' are areas where the 'reef' is missing over a 'basement high'; and the 'pots' or 'potholes' are circular or elliptical areas which the reef transgresses downwards into the underlying anorthosites, often with high concentrates of sulphides at the base.

At outcrop the 'reef' is weathered to a brownish soil, and in the top few metres from the outcrop the Pt content is a little higher than normal, and much of the sulphide oxidized.

Size and Grade

The Merensky Reef has been traced at outcrop for over 250 km, and in some places proved down-dip to a vertical depth of 2000 m; its thickness is variable but rarely exceeds 0·3 m. Out of this immense volume of rock only a

part can be worked under present economic conditions. One estimate puts the total potential reserve at 41×10^6 kg of precious metal. The usual grade at which it is worked is about 6 g/t over a mining width of 0·7 m.

Geological Interpretations

Clearly the formation of the Merensky Reef is bound up with that of the Complex as a whole. Present opinion is that the Complex was intruded from a magma that was undergoing some differentiation, but was intruded in discontinuous phases. There is an overall trend in the mineralogy and chemistry of the basic rocks that is normal, and the layering of individual rock units is thought to be due to settling out of crystals according to density modified by connection currents flowing in the magma. The Merensky Reef seems to be at the base of a unit that represented new injection of magma, but the reasons for its metal content remain obscure.

To account for this peculiarity some people have suggested hydrothermal processes, but the evidence for this is scanty. Some people are more prepared to think that volatile constituents of the magma may have played a part in forming the various cross-cutting pegmatoid bodies. It is almost certain that volatiles played an important part in the case of the tin-bearing pipes in the red granite.

Mining

There are several mines along the Merensky Reef on the western lobe of the Complex operated by Rustenburg Platinum Mines, Impala Platinum and Western Platinum, which between them have a productive capacity of over 52 000 kg per annum. The 'reef' is usually mined in large longwall or herringbone stopes over a width of 0·7 m or more, taking 0·3 m of the footwall anorthosite, the 'reef' and a few centimetres of the hanging wall. The rock is crushed and ground, and a crude concentrate of sulphides and other minerals made by flotation and gravity concentration methods. This is smelted to a matte, converted to crude copper/nickel alloy which is electronically refined, the residue from which is treated chemically to recover the various platinum metals and gold.

Further Reading

HALL, A. L. (1932), The Bushveldt Igneous Complex of the Central Transvaal. *S. African Geol. Surv. Mem.*, **28**, 510pp. [Although rather old, this was the first and one of the best compilations of factual data about the Complex.]

HAUGHTON, S. H., ed. (1964), *The Geology of Some Ore Deposits in Southern Africa.* Vol.II, Part II. Geol. Soc. S. Africa, Johannesburg. [Vol.II, Part II, contains fine papers about the deposits in the Complex, particularly one on the Merensky Reef by Cousins, pp.225–38.]

WILSON, H. D. B., ed. (1969), Magmatic Ore Deposits. *Econ. Geol. Mono.*, **4**. [A symposium volume containing a number of relevant papers on the Bushveldt and related deposits.]

VISSER, D. J. L. and GRUENEWALDT, G. von, eds (1969), *The Bushveldt Complex.* [Another symposium volume and perhaps the most comprehensive account of modern research.]

MOLYNEUX, T. G. (1972), *A Survey of the Mineral Deposits in and related to the Bushveldt Complex, South Africa.* 24th Int. Geol. Cong. Rept., Sec. 4, pp.225–32. [A nice summary for those hard pressed for time to read.]

BEATH, C. B. (1961), The Exploitation of the Platiniferous Ores of the Bushveldt Igneous Complex with particular reference to the Rustenburg Platinum Mines. *7th Commonwealth Min. Met. Cong. S. Africa. Trans.*, **1**, 216–43. [Perhaps the best account of the mining and metallurgy of the Merensky Reef.]

NEWMAN, S. C. (1963), Platinum. *Trans. Inst. Min. Met.*, **82**, A52–A65. [Interesting account of the economics of the platinum industry with some useful information on the Bushveldt deposits.]

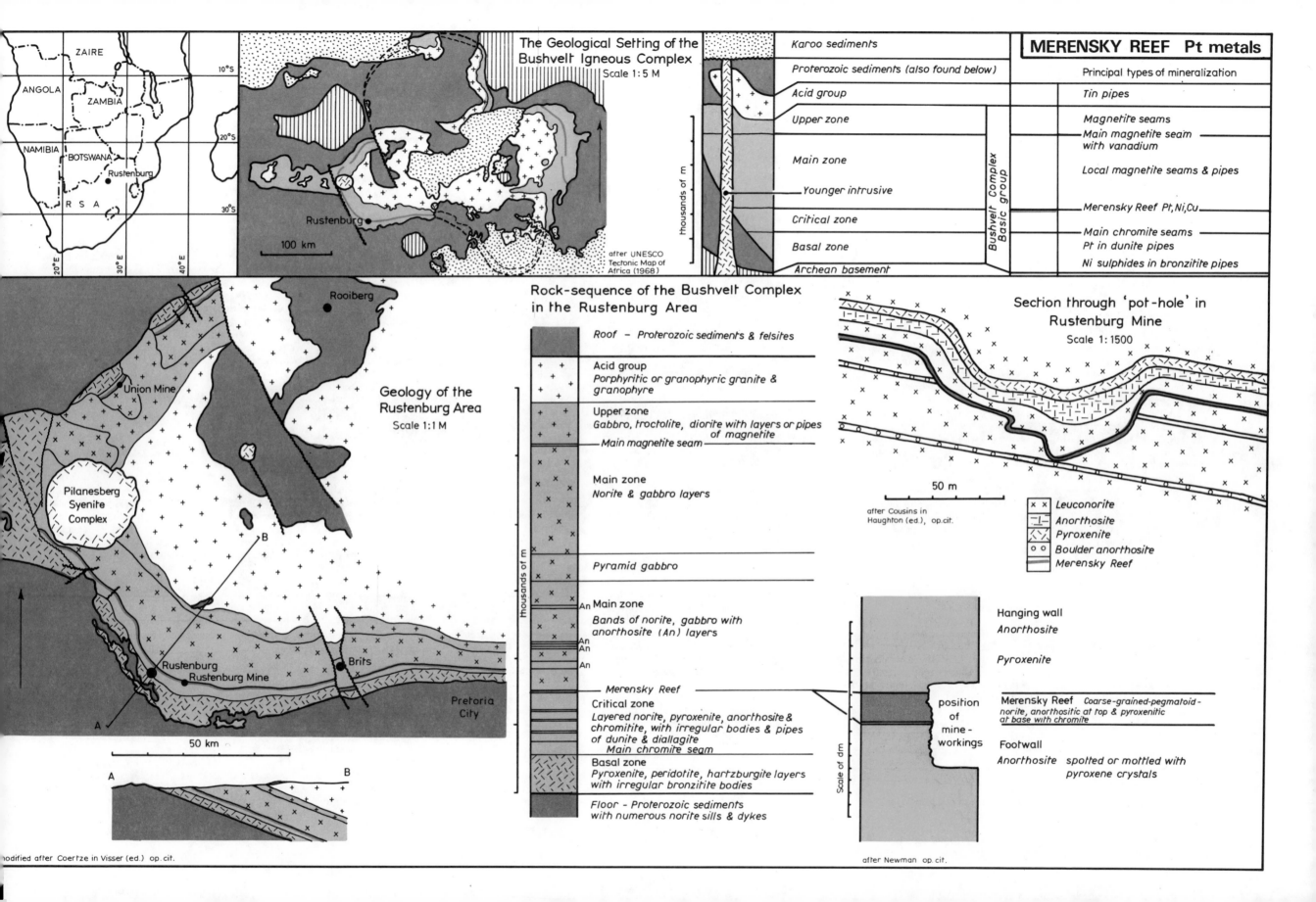

The Geological Setting of the Bushveld Igneous Complex
Scale 1:5 M

after UNESCO
Tectonic Map of
Africa (1968)

Karoo sediments

Proterozoic sediments (also found below)

Acid group

Upper zone

Main zone

Younger intrusive

Critical zone

Basal zone

Archean basement

Bushvelt Complex Basic group

100 km

MERENSKY REEF Pt metals

Principal types of mineralization

Tin pipes

Magnetite seams
Main magnetite seam
with vanadium

Local magnetite seams & pipes

Merensky Reef Pt, Ni, Cu

Main chromite seams
Pt in dunite pipes
Ni sulphides in bronzitite pipes

Geology of the Rustenburg Area
Scale 1:1 M

Rooiberg

Union Mine

Pilanesberg
Syenite
Complex

Rustenburg
Rustenburg Mine

Brits

Pretoria
City

B

A

50 km

A   B

modified after Coertze in Visser (ed.) op.cit.

Rock-sequence of the Bushvelt Complex in the Rustenburg Area

Roof — Proterozoic sediments & felsites

Acid group
Porphyritic or granophyric granite & granophyre

Upper zone
Gabbro, troctolite, diorite with layers or pipes of magnetite
Main magnetite seam

Main zone
Norite & gabbro layers

Pyramid gabbro

An Main zone
Bands of norite, gabbro with anorthosite (An) layers
An
An
An

Merensky Reef

Critical zone
Layered norite, pyroxenite, anorthosite & chromitite, with irregular bodies & pipes of dunite & diallagite
Main chromite seam

Basal zone
Pyroxenite, peridotite, hartzburgite layers with irregular bronzitite bodies

Floor — Proterozoic sediments
with numerous norite sills & dykes

thousands of m

Section through 'pot-hole' in Rustenburg Mine
Scale 1:1500

after Cousins in
Haughton (ed.), op.cit.

50 m

	Leuconorite
	Anorthosite
	Pyroxenite
	Boulder anorthosite
	Merensky Reef

Scale of dm

position
of
mine-
workings

Hanging wall
Anorthosite

Pyroxenite

Merensky Reef *Coarse-grained-pegmatoid-norite, anorthositic at top & pyroxenitic at base with chromite*

Footwall
Anorthosite *spotted or mottled with pyroxene crystals*

after Newman op.cit.

The Chromite Deposits of the Great Dyke – Rhodesia (Zimbabwe)

The Great Dyke is one of the most remarkable geological features known. Over five hundred kiolmetres long and rarely more than ten wide, it traverses the Archean Shield of Zimbabwe (Rhodesia) from north to south. Within layered ultrabasic rocks are several seams of chromite. These are strikingly persistent and together constitute the world's largest resource of that mineral.

Location

The Great Dyke traverses Zimbabwe (Rhodesia) in a direction slightly east of north astride longitude 30° 30′ E between 16° 30′ and 21° S. The capital, Salisbury, is 55 km to the east.

Geographical Setting

The landscape along the Great Dyke is little more than hilly in the south, though it averages 900 m in altitude, but as one goes north the country becomes more and more mountainous, and averages 1 200 m in altitude. Several rivers cut the Dyke, some with deep and spectacular gorges, most of the rivers draining eventually to the Zambezi, except in the south where streams draining the eastern side form part of the Limpopo basin. Generally the centre of the Dyke is hilly, and there is usually a flat swampy zone along its margin and granite hills rise away from the contacts.

Much of the higher parts of the Dyke are wooded, particularly in the north; otherwise the land is long-grass covered. Little vegetation grows on the marginal zones of the Dyke (probably because of toxic soils) and most of the cultivation in the area is on the shield rocks adjoining rather than on the Dyke itself. The climate is sub-tropical high altitude with hot, wet summers and cooler, dry winters; rainfall being 500 mm in the south and over 750 mm in the north. Much of the Dyke is very permeable, and it abounds with springs and streams, and water poses a great problem for mining. The marginal zones of the Dyke are well populated with several towns just off the contact, served by good roads and railways that run between Salisbury and the south, and with a connection to the port of Beira in neighbouring Mozambique.

History

The Great Dyke, is shown on a geological map by Carl Mauch dating from his explorations in 1865 and 1872. The chromite ore outcrops were certainly known to European settlers who began to establish themselves in the 1890s. It was in 1904 that one, J. Popham, a local farmer, pegged a claim on the Selukwe chromite deposit (not actually in the Dyke) which aroused the interest of the local farmers in the mineral. The First World War stimulated the market, and the first claims were staked on the chromite seams of the Dyke. Most of the working remained a sideline of farming for years until the 1950s when deeper, more organized mines were opened, the first flotation and magnetic separation plant being commissioned in 1957.

Geological Setting

The Rhodesian Shield is an area of gneisses and granitic rocks, with several belts of folded schists that were stabilized well over 3 Ga ago. The Dyke cuts two major schist belts, the Selukwe and Hartley. Both of these contain a basal group consiting of dunites, serpentinites and talc schists, passing up into amphibolites, hornblend-schists, epidiorites and chlorite schists (some of which reveal the relic texture of pillow lavas) and a group of phyllites, banded ironstones and marbles. In the north, the surrounding rocks of the Dyke are even more intensely plutonized being the paragneisses and migmatites of the Zambezi mobile belt, younger in age than the Shield.

The Selukwe and Hartey belts abound with gold quartz-vein deposits, and the country's largest chromite deposit is near Selukwe, an irregular lenticular deposit in Archean serpentinite that seems to have no relationship to those in the Dyke itself.

The Great Dyke

The Great Dyke is not really a dyke in the ordinary sense, but a series of very elongated, boat-shaped basic-ultrabasic intrusions in a long line. Four separate intrusions have been identified, each of which contains a similar sequence of layered rocks. At the base are dunites and harzbergites, mostly serpentinized, and further up the sequence are distinct layers of pyroxenite and chromitite that are very persistent and constant in thickness. Above the highest pyroxenite layer is a thick gabbroic layer, hypersthene-bearing at the base, while in some of the better preserved parts of the Dyke a capping of quartz-gabbro is preserved. The bottom contact of the Dyke slopes gently inwards, and is always sharp with some contact metamorphism traceable for 50–100 m into the country rock. The layering shows that each separate intrusive complex has a synclinal form, and geophysical evidence shows that each as a 'root' at its centre. The Dyke is faulted along its length, and deformed at its northern end where it passes into the Zambezi mobile belt. It is intruded in a few places by dolerite dykes of younger age. Radiometric ages of the Dyke vary between 2·53 Ga and 2·8 Ga.

In addition to the chromitites, there is one asbestos deposit in the Dyke; the Ethel, at a place where east–west faults cut the serpentinite. A little magnesite occurs in veinlets. Just below the top of the uppermost pyroxenite band is a mottled, coarse-grained zone which contains platinum metals, nickel and copper, and the nickel content of some of the basal ultrabasic rocks is as high as 0·28%. None of these metals is of a grade to attract mining at the moment.

The Chromite Seams

The chromitite layers are quite distinct, and persistent, though less so than the pyroxenite layers. Seven are known in the northern, Musengezi complex, eleven in the larger Hartley, seven in the Selukwe and six in the southern Wedge complex. The corresponding seams numbered from the top are similar in each complex. Generally, each seam has a sharp footwall contact with the underlying rock, and often consists of massive chromite at the base becoming more mixed with serpentine higher up. Each of the seams has its characteristics of thickness, texture and composition. The upper seam is generally 200–250 mm thick with a low chromium content and high ferric-iron content. The second seam may be 450 mm thick and has a nodular texture, the third is rather thin but massive. Generally, the content of chromium increases downwards at the expense of ferric iron, aluminium remains quite constant, while magnesium and ferrous iron vary rather erratically.

The texture of the chromite as mined depends on the original texture of the chromitite, the degree of serpentinization, surface alteration and fracturing; the material produced being either friable or lumpy.

Extensive areas of the rather poor soil over the serpentinite outcrops contain grains of chromite, which are worked.

Size and Grade

One estimate based on detailed mapping of the Dykes gives the reserves of chromite as 326 Mt of the mineral, two-thirds of which is classified as potential chemical grade, and one-third as metallurgical grade. The seams are usually mined selectively, and only simple processing carried out, and the chromite content of the mined ore is high. The eluvial material can be mined at any grade down to 15% mineral. If the price justified deep mining, the reserves could be considerably extended, and the total resource of the mineral in the Dyke as a whole is probably well over ten times the measured reserves.

Geological Interpretations

Everyone agrees that the Great Dyke is an intrusive magmatic body, but there are many arguments about the detailed history of the intrusion, and there is still no explanation of its extraordinary shape beyond the suggestion that it occurs along a major crusted fracture. The sharp floors and gradational tops of the chromitite layers and some of the pyroxenites suggest to some people an intrusion in pulses, followed by crystallization and differentiation by gravity settling. It would seem that after each pulse, chromite crystallized and settled out, followed by silicates. There was an overlap in time between the two phases of crystallization so that some chromite becomes incorporated in the hanging-wall rocks of each seam. The form of the Dyke shown by the form of the chromite seams is attributed to the collapse of the floor of the intrusion to form a graben at some stage late in the intrusion process. The change in the composition of the Dyke and of the chromite it contains, is that well-known from other major layered bodies, and in this case presumably some differentiation took place in a magma chamber below the present site of the Dyke. Experimental work suggests that the gradual change in the composition of chromite is controlled by temperature and the activity (or fugacity) of oxygen, in a way that is consistent with the changes observed if conditions in the Dyke developed in a normal way.

The chromite deposits can properly be called magmatic, except for the material in soil, which can be classed as eluvial.

Mining

The various seams are mined by a variety of small syndicates and larger mines, by methods that vary from open cuts, to relatively simple underground methods. Because the seams are thin, generally a cut is made in the soft hanging-wall serpentinite and the chromite selectively blasted from the floor. Processing is normally confined to simple crushing, screening and washing to remove some of the serpentine, but more sophisticated treatment is given to some ores, particularly the low-grade eluvial ores where a combination of flotation and magnetic separation is used. The upper three seams in each complex can only be sold as chemical grade with about 51% Cr_2O_3 and a Cr/Fe ratio of less than 2·8, particularly if friable. The lower seams may have up to 55% Cr_2O_3 and Cr/Fe ratios over 2·8 and can be sold as metallurgical grade so long as the product is hard and lumpy, or briquetted. Most of the Great Dyke chromite can be sold as refractory grade when it is hard and lumpy, since it has a relatively high aluminium content.

Further Reading

WORST, B. G. (1960), The Great Dyke of Southern Rhodesia. *S. Rhodesian Geol. Surv. Bull.*, **47**, 239pp. [The standard work on the area with good maps and full description of the deposits.]

WORST, B. G. (1964), Chromite in the Great Dyke of Southern Rhodesia. In: *The Geology of some Ore Deposits in Southern Africa* (Haughton, S. H., ed.), Geol. Soc. S. Africa, Johannesburg, Vol.2, pp.209–24.

BICHAN, R. (1969), Chromite Seams in the Hartley Complex of the Great Dyke of Rhodesia. In: Magmatic Ore Deposits – A Symposium. (Wilson, H. D. B., ed.) *Econ. Geol. Mon.*, **4**, 95–113. [Two shorter and more compact accounts, with slightly different views of the origin of the deposits.]

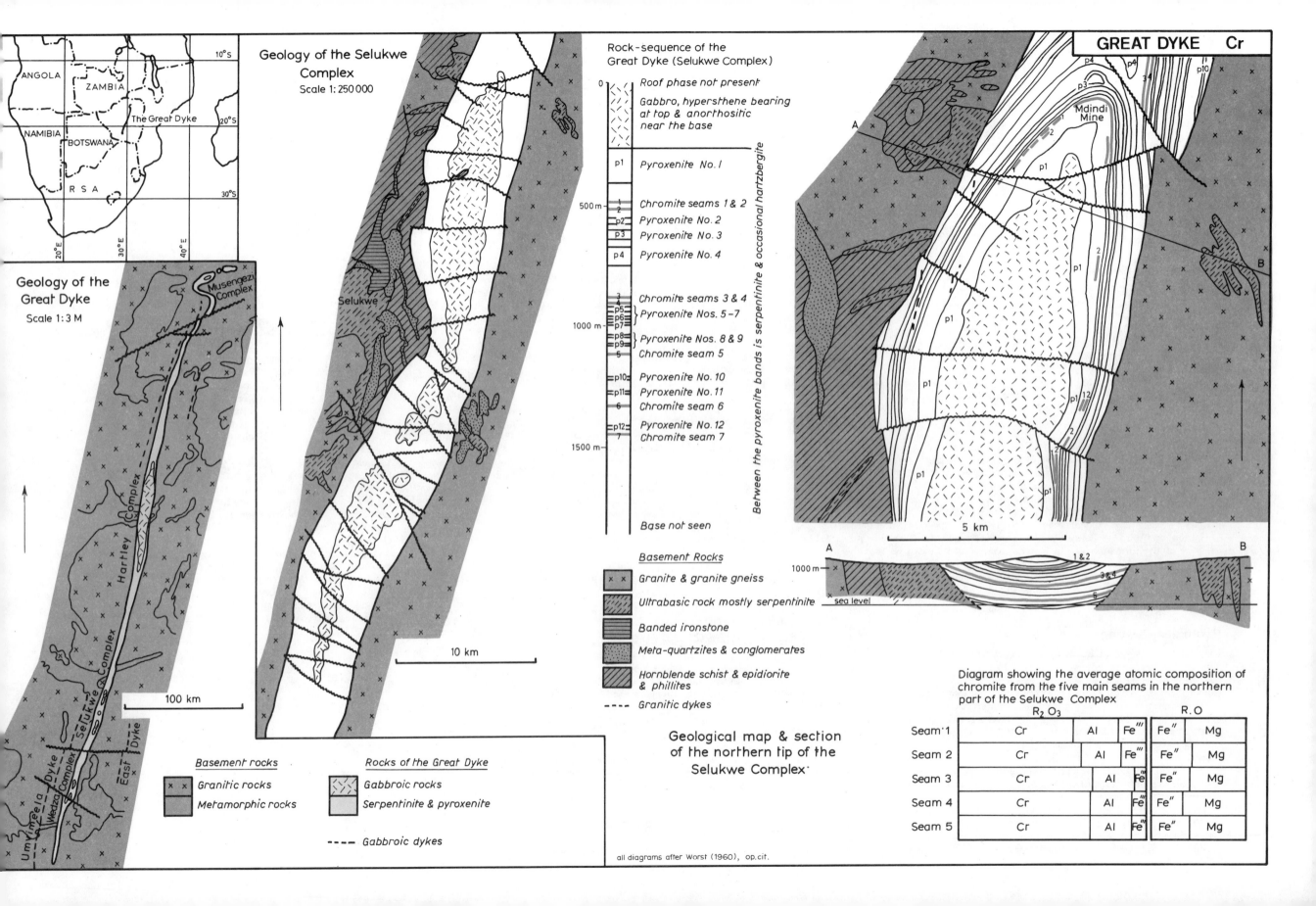

Geology of the Selukwe Complex
Scale 1: 250 000

Geology of the Great Dyke
Scale 1: 3 M

GREAT DYKE   Cr

Mdindi Mine

5 km

A   1 & 2   B
1000 m   3 & 4
sea level   5

Rock-sequence of the Great Dyke (Selukwe Complex)

0	Roof phase not present
	Gabbro, hypersthene bearing at top & anorthositic near the base
p1	Pyroxenite No. 1
500 m — 1	Chromite seams 1 & 2
2	
p2	Pyroxenite No. 2
p3	Pyroxenite No. 3
p4	Pyroxenite No. 4
3	Chromite seams 3 & 4
4	
p5	
p6	Pyroxenite Nos. 5–7
1000 m — p7	
p8	Pyroxenite Nos. 8 & 9
p9	
5	Chromite seam 5
p10	Pyroxenite No. 10
p11	Pyroxenite No. 11
6	Chromite seam 6
p12	Pyroxenite No. 12
1500 m — 7	Chromite seam 7
	Base not seen

Between the pyroxenite bands is serpentinite & occasional hartzbergite

Basement Rocks

- Granite & granite gneiss
- Ultrabasic rock mostly serpentinite
- Banded ironstone
- Meta-quartzites & conglomerates
- Hornblende schist & epidiorite & phillites
- - - - Granitic dykes

Geological map & section of the northern tip of the Selukwe Complex.

Basement rocks
- Granitic rocks
- Metamorphic rocks

Rocks of the Great Dyke
- Gabbroic rocks
- Serpentinite & pyroxenite
- - - - Gabbroic dykes

all diagrams after Worst (1960), op.cit.

Diagram showing the average atomic composition of chromite from the five main seams in the northern part of the Selukwe Complex

	R_2O_3			$R.O$	
Seam 1	Cr	Al	Fe'''	Fe''	Mg
Seam 2	Cr	Al	Fe'''	Fe''	Mg
Seam 3	Cr	Al	Fe'''	Fe''	Mg
Seam 4	Cr	Al	Fe'''	Fe''	Mg
Seam 5	Cr	Al	Fe'''	Fe''	Mg

100 km

10 km

The Sudbury Nickel Deposits – Canada

This is the largest nickel-mining district in the world. In outline the geology seems simple; nickel and copper sulphides segregated from a basic magma; but there are some puzzling features that have led to great controversy. It has even been suggested that the whole complex was formed by a gigantic volcanic eruption triggered off by the impact of a large meteorite.

Location

The Sudbury district is 60 km × 30 km, lying to the north-west of the town of Sudbury (latitude 46° 29′ N, longitude 81° 00′ W, 260 m in altitude) in south-eastern Ontario.

Geographical Setting

The Sudbury complex is a topographic feature; a rim of hills about 425 m high, almost enclosing a rugged, flattish area just over 300 m in altitude. There are numerous rock outcrops interspersed with lower, till-covered areas, small lakes and rivers and, originally, the area was thin boreal forest. The climate is harsh and continental, sub-zero winters, hot summers, some rain (or snow) in each month, totalling about 700 mm in a year. Sudbury is the main town; several towns have sprung up round the complex, largely centred on mines, and others, quite small, dependent on lumber and the very limited agriculture, are found in the centre.

History

In 1856, a government surveyor named Salter reported wild variations in his compass bearings but the report was forgotten until August 1883, when the Canadian Pacific Railway was driven north-west from Sudbury and, in a cutting, Thomas Flanagan, a blacksmith, found some gossan with traces of copper sulphides. Curiously, it was not until 1887 that it was realized that nickel was an even more important constituent. Within a few years of Flanagan's discovery, several other deposits were found, including the Creighton, on the site of Salter's original compass problems. It was in 1902, the year of the formation of the International Nickel Company, that the metallurgical problems of this complex ore began to be solved. By the First World War, Sudbury dominated the world nickel market and, by 1917, the problems caused by the rapid growth of the district led to the formation of the Royal Ontario Nickel Commission.

Geological Setting

The Sudbury lies on the boundary between a large area of Archean granite gneisses and somewhat younger Aphebian rocks of the Penokean Fold Belt, belonging to the Huronian Super-Group. Cutting these rocks on the south side of the complex are a series of granites with radiometric ages of between 2·06 Ga and 2·15 Ga. Also cutting this southern complex are zones of the extraordinary Sudbury Breccia, forming zones a few centimetres to several kilometres wide and containing assorted, often rounded fragments of rock ranging in size from over a kilometre to a few micrometres.

In the centre of the Sudbury Basin is the Whitewater Group, beginning with a quartzite breccia (similar to the Sudbury Breccia) overlain by some spectacular ignimbrites, tuffs and pumicites, overlain in turn by calcareous argillites and slates and sandstones.

The basic and acid igneous complex, with which the mineral deposits are associated, separates the Whitewater Group entirely from the floor of Archean and Aphebian rocks.

Since the intrusion of the igneous complex (dated at 1·72 Ga), some downwarping and relatively minor faulting has occurred, and a small swarm of dykes of olivine dolerite about 1·0 Ga in age has been intruded. Little else is known of the geological history of the area. Quaternary glaciation, which eroded the area, left behind a little till in the lower-lying land.

The Sudbury Igneous Complex

Frequently also known as the Sudbury Nickel Eruptive, this outcrops as a continuous band round the edge of the Basin. The outer part consists of hypersthene-gabbro (norite) with a few olivine-bearing patches at the base. The inner part, consists of granophyre. Between these two is a narrow zone of quartz diorite and similar rocks, are found as dykes leading off the main body.

Round most of the outcrop the norite contact dips inwards towards the centre of the Basin from 30° to near vertical but, in a few places, dips outwards. In places the contact is marked by a breccia.

The Nickel–Copper Deposits

Some fifty deposits are known round the edge of the complex. The deposits contain pyrrhotite intergrown with pentalandite, and chalcopyrite. In addition, small amounts of magnetite or ilmenite occur. Pyrite is present in some ore bodies and a large number of other minerals occur. The range of minor and trace-elements is large, including the platinum metals, gold, silver, cobalt, lead, zinc, bismuth, arsenic, antimony, selenium and tellurium.

The deposits consist of lenticular bodies or irregular zones of disseminated sulphides in norite, quartz diorite or other rocks; contact breccias with all or part of the matrix consisting of sulphides; and veinlet zones.

Some ore bodies occur along or close to the contact, or in contact with breccias, particularly where there are embayments or undulations in the floor of the main norite body. The so-called 'off-sets', where ore occurs outside the norite complex, are usually associated with quartz diorites, in the floor of the intrusion. Along the north side are small off-set deposits, situated in quartz diorite dykes up to 8 km outside the norite body.

Most of the ore bodies that outcrop have gossan cappings but, like most recently glaciated regions, this is not an important feature.

Size and Grade

The total declared reserves of the district from the time of the original discovery are of the order of 930 Mt containing an average of 1½% nickel and 1% copper, and of this about 500 Mt has been exploited, but new reserves are constantly being blocked out, almost keeping pace with production.

Geological Interpretations

The form of the Sudbury igneous complex has given rise to much speculation. Originally, it was interpreted as a gently folded sheet or sill. The balance of opinion today favours a lopolith with a deep root, for which there is a little geophysical evidence. The idea that the complex started life as an astrobleme is supported by the occurrence of high energy shatter-cones in the surrounding rocks and the Sudbury Breccia; a stony meteorite 4 km in diameter would be required, but many geologists are sceptical of such catastrophic theories.

Ore bodies in the simpler situations are interpreted as forming as a result of the unmixing of a sulphide melt during the cooling of the norite, some of which remained as droplets in the rock to form the disseminated ore, others coalesced to form larger masses in breccias along the contacts or depressions in the floor. It is important to realize that a sulphide-rich melt containing iron, nickel and copper, remains molten long after most silicates have crystallized, and it is not surprising that molten sulphide should move outside the intrusion into favourable reservoirs, such as breccia zones or dyke margins. Older literature describes the deposits as being hydrothermal and associated with granites, but the advent of reliable radiometric ages has rendered this interpretation unlikely.

Mining

INCO and Falconbridge are the two companies that between them operate some 23 mines in the district. A variety of open-pit and underground methods are used, ranging from square-set stoping, shrinkage, cut and fill (usually using mill-tailing sand as fill) and caving methods. The ore is hoisted and treated by selective flotation to produce pyrrhotite and chalcopyrite concentrates. There follows smelting to produce separate nickel or copper matte. In the case of nickel the matte is cooled slowly, and then treated by a combination of magnetic and flotation separation to produce high-grade nickel-sulphide concentrates for conversion to metal or oxide. The nickel is refined by electrolytic methods, or the carbonyl process. The district produces about 280 000 t of nickel each year with 180 000 t of copper, 50 000 kg of precious metals and a host of byproducts including sulphur chemicals from smelter gasses.

Further Reading

BOLTD, J. R. (1967), *The Winning of Nickel*. Methuen & Co. London, 487pp.[Comprehensive book for the general reader that contains a good survey of the geology of Sudbury as well as much other useful information.]

CARD, K. D. (1967), Geology of the Sudbury Area. In: *Geology of Parts of Eastern Ontario and Western Quebec*. Geol. Ass., Canadian Guidebook, pp.109–22. [Good summary.]

HAWLEY, J. E. (1962), *The Sudbury Ores; Their Mineralogy and Origin*. Min. Ass. Canada, Univ. Toronto Press, Toronto, p.207. [Thorough documentation and analysis of the ore petrology.]

Staff, International Nickel Company of Canada Ltd. (1946), The Operations and Plants of the International Nickel Company of Canada Ltd. Can. Mon. J., 67, (5), 322–31.

Staff, Falconbridge Nickel Mines Ltd. (1954), The Falconbridge Story: I Geology. Can. Min. J., 80, (6), 116–27.

ZURBRIGG, H. F. (1957), The Frood–Stobie Mine. In: *Structural Geology of Canadian Ore Deposits*, Vol.2. Canadian Inst. Min. Met., Montreal, pp.341–50. [A group of informative mine descriptions.]

Royal Ontario Nickel Commission (1917), *Report*; A. T. Wilguess, Toronto, 584pp. [Contains a good section on geology and is well worth reading as a case history of the impact of the discovery and development of a large mining district in a developing country.]

DIETZ, R. S. (1964), Sudbury Structure as an Astrobleme. *J. Geol.*, 72, 412–34. [Almost unequalled for its audacity and inventiveness, well worth reading; converts should also read French, B. M. (1967), *Science*, 156 (3778), 1094–8.]

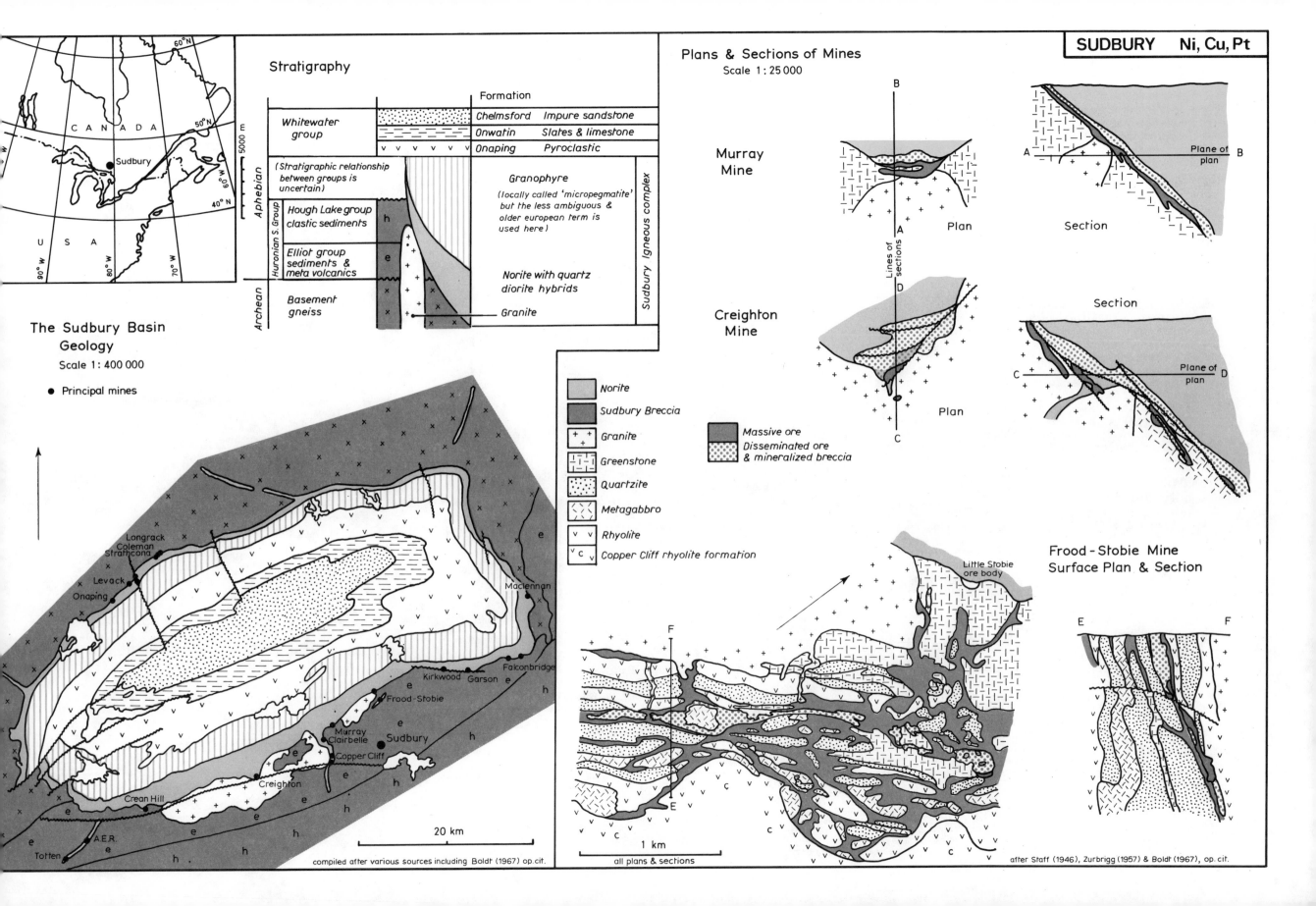

SUDBURY Ni, Cu, Pt

Stratigraphy

		Formation	
Whitewater group		Chelmsford	Impure sandstone
		Onwatin	Slates & limestone
		Onaping	Pyroclastic

Aphebian

(Stratigraphic relationship between groups is uncertain)

Huronian S. Group

Hough Lake group clastic sediments — h

Elliot group sediments & meta volcanics — e

Archean

Basement gneiss

Granophyre
(locally called 'micropegmatite' but the less ambiguous & older european term is used here)

Norite with quartz diorite hybrids

Granite

Sudbury Igneous complex

The Sudbury Basin Geology
Scale 1 : 400 000

● Principal mines

Longrack
Coleman
Strathcona
Levack
Onaping
Maclennan
Kirkwood Garson Falconbridge
Frood-Stobie
Murray
Clairbelle Sudbury
Copper Cliff
Creighton
Crean Hill
A.E.R.
Totten

20 km

compiled after various sources including Boldt (1967) op.cit.

Plans & Sections of Mines
Scale 1 : 25 000

Murray Mine

B
A
Lines of sections
D
C
Plan

Plane of plan
A B
Section

Creighton Mine

D
Plan
C

Section
C
Plane of plan
D

Legend

Norite	
Sudbury Breccia	
Granite	+ +
Greenstone	
Quartzite	
Metagabbro	x
Rhyolite	v v
Copper Cliff rhyolite formation	v c

Massive ore
Disseminated ore & mineralized breccia

Frood-Stobie Mine
Surface Plan & Section

Little Stobie ore body

F
E
1 km

all plans & sections

E
F

after Staff (1946), Zurbrigg (1957) & Boldt (1967), op. cit.

CANADA
Sudbury
USA
60°N
50°N
40°N
90°W 80°W 70°W 60°W

5000 m

The Tellnes Ilmenite Deposit – Norway

Although most ilmenite is recovered from breach-sands, an important quantity is mined from magmatic segregations in anorthosite complexes such as at Stillwater in Montana, U.S.A., Allard Lake in Canada and this example in Norway. Successful exploitation has depended on its location near a fjorded coastline close to the industrialized part of Europe.

Location

The Tellnes deposit lies under a lake, Tellnesvann, from which it derives its name. Tellnes is at latitude 58° 19′ N, longitude 16° 26′ E and at 225 m above sea level, 7 km from the south-west coast of Norway between Jøssingfjord and Åna Sira.

Geographical Setting

The deposit is situated in a remarkable landscape, rounded hills interspersed with small rivers and lakes, the hills almost devoid of vegetation; the valleys where a little glacial debris remains, tend to be swampy but may support a few pines and birches. The hills rise to over 300 m and fall away to a barren and rocky coast cut by a few deep fjords. One of these, Jøssingfjord, forms a sheltered harbour convenient for the export of ilmenite to Europe and, despite the mine and its waste products, lobsters grow at its mouth. The climate, though mild for Scandinavia, is wet, stormy seas and over 2000 mm of rain in a year, near zero temperatures with a little snow in winter, and summers averaging 15°C. Few people live in the area. The mine is half-way between the small towns of Egersund and Flekkefjord which depend on the sea, and on small but lush dairy farms in the bigger valleys.

History

Mining of ilmenite in the Egersund area is said to date from the eighteenth century though at that time worked as an iron ore; a British company worked mines to the west of Tellnes in the mid-nineteenth century. The Storgangen Mine, 6·5 km to the north-west of Tellnes, was opened in 1918 when the market for ilmenite grew as titanium dioxide began to be used as a pigment, and continued in a modest way into the 1960s. The Tellnes deposit, which, half hidden by the lake, remained undiscovered until 1954 when it was revealed by an aeromagnetic survey, came into production in 1957.

Geological Setting

The Egersund Complex is a major area of basic igneous rocks dominated by anorthosite, that outcrops over an area of 900 km². It is probable that only a part of the Complex is visible, being truncated by a fault-controlled coastline. The Complex forms part of and lies at the south-western tip of the Pre-Cambrian sector of Norway, itself one of the younger parts of the Baltic Shield. It is surrounded by gneisses and migmatites mostly of granitic composition, frequently hypersthene-bearing, of considerable structural complexity. The basic mass itself is rimmed by strongly banded migmatites, generally of noritic composition described by some geologists as charnockitic. The main body is anorthosite composed almost entirely of andesine. It is rather featureless in the west, but strongly porphyritic in the east with phenocrysts over 30 cm long, not uncommon. These phenocrysts and the accessory pyroxenes that do occur, are oriented to show a complex dome structure to the eastern and central parts of the mass. The other more mafic rocks of the complex appear to form a broad syncline overlying the anorthosite, beginning at the base with norite, passing through perthite-bearing

norite (locally called mangero-norite) into a granular monzonite. Quartz monzonite occurs as dykes cutting the complex and a swarm of east–west trending dolerite dykes, probably of much younger age, cut the whole complex and its surrounding gneisses. The complex has been dated at 960 Ma, which corresponds closely to the age of the main deformation of this part of the Shield.

There is no record of the subsequent history of the area until the Quaternary Glaciation formed the present topographic surface, and left minor debris in the valleys. At the end of the Ice Age the coastline was drowned forming the fjords that have proved such an asset to the mine.

The Geology of the Tellnes Deposit

The deposit consists of one large ore body in the form of a complex lens surrounded by porphyritic anorthosite. The contacts are everywhere steep and sharp, and a number of apophyses lead off the main lens into the country rock. The body is composed of ilmeno-norite but a more leucocratic norite occurs on the margins and at the south-eastern end, while at the opposite end the body tails off into a quartz-monzonite dyke. At the edges, a few xenoliths of anorthosite occur in the norite, together with norite-anorthosite breccias. The norite is not particularly banded except at the end and is, on the whole, a rather tough massive rock.

The main part of the body has the following composition:

Chemical		Modal		
SiO_2	30·4%	Plagioclase (An_{49})	36%
Al_2O_3	11·7%	Hypersthene	15%
Fe_2O_3	7·25%	Ilmenite	39%
FeO	17·4%	Magnetite	2%
MnO	0·18%	Biotite	3·5%
TiO_2	18·4%	Sulphides	0·6%
MgO	6·13%	Apatite	0·5%
CaO	4·39%		
Na_2O	2·40%		
K_2O	0·60%		
P	0·09%		
S	0·32%		

The plagioclase is slightly perthitic, and the hyperstehene carries exsolution lamellae of titan-augite.

The main ore mineral is ilmenite, which occurs as millimetre-sized crystals containing about 12% hematite as fine exsolution lamellae, too small to be separated. Magnetite is found throughout the ore as well-formed octahedra and is a recoverable by-product. It is slightly vanadium-bearing. Apatite is a serious impurity that occurs throughout, but much higher than average quantities are found on the margins and in the apophyses of the ore body. Very minor traces of a chromium spinel occur sporadically to the detriment of the ore quality (a few g/t of Cr imparts a green colour to TiO_2 pigment).

Size and Grade

The present mine was begun on ore reserves of 200 Mt containing 18% TiO_2 but the total reserve is probably much larger, although parts of the ilmeno-norite body are either too low in grade or too high in impurity content for present production.

Geological Interpretations

Like other anorthosite complexes, such as Allard Lake in Canada, the origin of the Egersund mass remains something of a mystery. A very general petrologic comparison may be made with the better understood layered masses, such as the Bushveldt Complex, but the overwhelming dominance of anorthosite and even the rich development of ilmeno-norite are striking differences. Undoubtedly the complex has been deformed, and some authors refer to large parts of the complex as charnockite or granulite. The ilmenite

deposits occur in banded norite entirely within the anorthosite, and are regarded by some as deformed remnants of a layered complex and by others as separate intrusions into the anorthosite. However, few would dispute the classification of the deposit as a magmatic segregation.

Mining

The Tellnes deposit is mined by Titania A/S who operate an open-pit (which has necessitated the partial draining of the lake) with a productive capacity of 1·5 Mt/a. The ore is crushed and ground at the mine, the magnetite separated by wet magnetic separation, and the ilmenite concentrated by flotation. The ilmenite concentrate is leached with sulphuric acid to remove apatite, and is then piped to Jøssingfjord where it is dried, and shipped elsewhere in Europe for the manufacture of TiO_2 pigment. The magnetite is sold as a dense medium to other mines, and the tailings and mine water dumped in the fjord with no serious environmental effects, except that dissolved apatite nourishes marine life. The annual production of ilmenite concentrates is about 500 000 t, representing 11% of the world's total.

Further Reading

DYBDAHL, I. (1960), Ilmenite Deposits of the Egersund Anorthosite Complex. In: *Mines in South and Central Norway.* (Vokes, F., ed.), Guide to Excursion No. C10. 21st Int. Geol. Congr., Norden 1960. [The best general description of the deposit so far published.]

BUGGE, J. A. W. (1953), En del Hovedtyper av Jern – og Titan-Malmer i Norge. *Kgl. Norsk Vid. Selsk. Forh.* **26**. [A general work that includes descriptions of the deposits in the Complex as a whole.]

MICHOT, J. (1955–8), Papers on the Anorthosites of the Egersund Area. *Ann. Soc. Geol. Belg.* **62, 63, 79, 80.** [A series of petrological works on the Complex, rather speculative, and with little to say about the ilmenite deposits.]

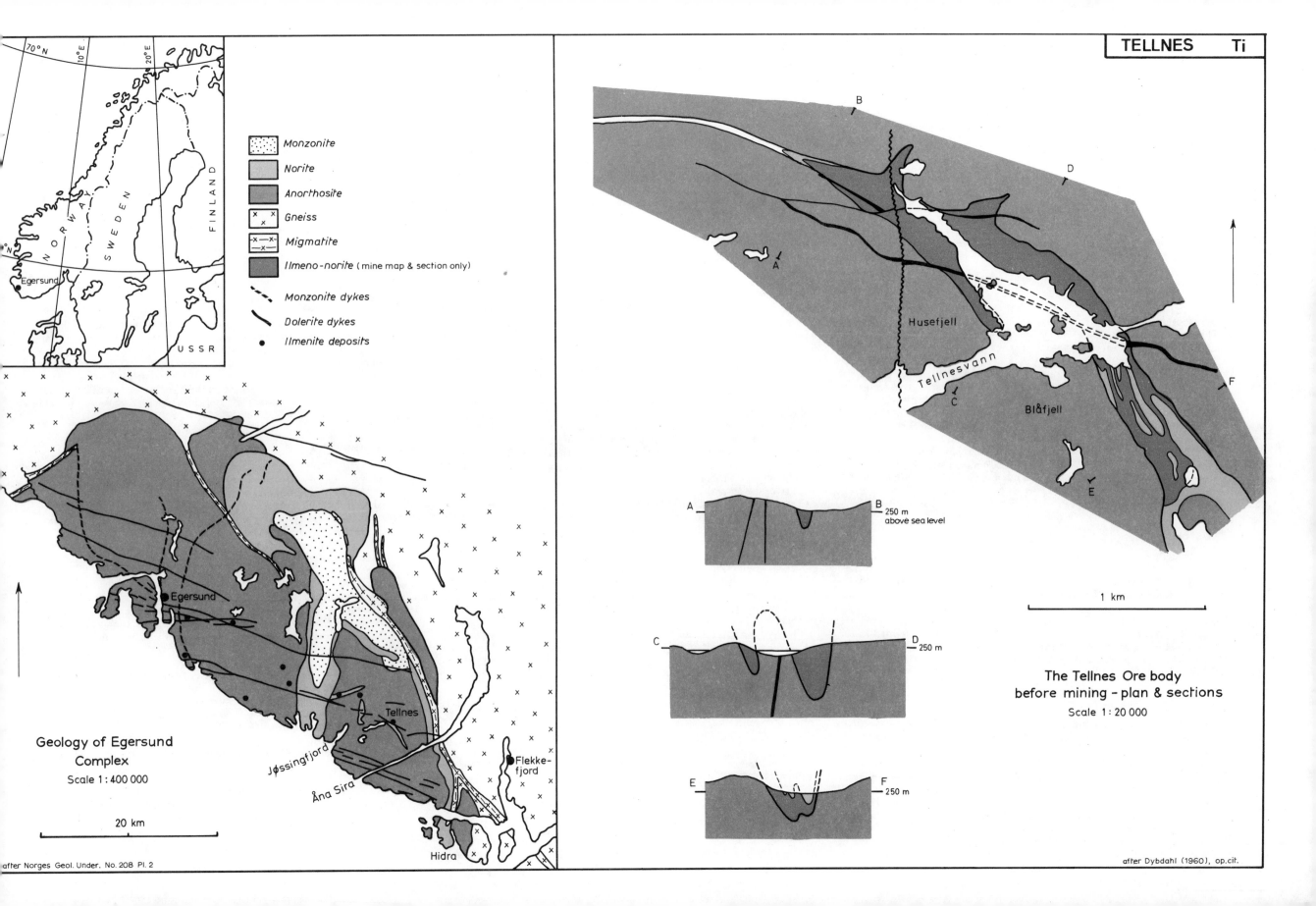

TELLNES   Ti

Monzonite

Norite

Anorthosite

Gneiss

Migmatite

Ilmeno-norite (mine map & section only)

Monzonite dykes

Dolerite dykes

Ilmenite deposits

Geology of Egersund
Complex

Scale 1 : 400 000

20 km

The Tellnes Ore body
before mining – plan & sections

Scale 1 : 20 000

1 km

A — B 250 m
above sea level

C — D 250 m

E — F 250 m

Husefjell

Tellnesvann

Blåfjell

Egersund

Tellnes

Jøssingfjord

Åna Sira

Flekke-
fjord

Hidra

NORWAY   SWEDEN   FINLAND   USSR

70° N   10° E   20° E

after Norges Geol. Under. No. 208 Pl. 2

after Dybdahl (1960), op.cit.

The Chromite Deposits of the Muğla District – Turkey

There are two types of chromite deposit; the stratiform or layered type characteristic of large layered intrusions such as the Great Dyke (q.v.) and the 'podiform' deposits such as these. Examples of this type of deposit are found in several other parts of Turkey, and in Greece, Albania and Yugoslavia, as well as on various of the islands of south-east Asia. The problem of their origin is closely related to that of their host rocks, sometimes known as the Alpine or 'ophiolitic' peridotites.

Location

The area is 150 km by 80 km and lies to the south-east of the town of Muğla (latitude 37° 12′ N and longitude 28° 22′ E) in the province of that name and the adjoining province of Denizli.

Geographical Setting

The southern part of Turkey is a region of dramatic and varied topography. Deep valleys running down to the Aegean Sea separate ranges of mountains up to and over 2000 m in altitude. Over the large areas of limestone, pronounced karst topography has developed, and this contrasts sharply with the rolling wooded country over the peridotites, and with the fertile alluvial plains on the valley floors. The climate is Mediterranean, with a temperate wet winter and a warm summer. Half a metre of rain falls during the year and the winter and summer average temperatures are 8°C and 25°C. The vegetation cover is the usual mixture of Mediterranean pines, evergreen oaks and scrub, and only the alluvial valleys can be cultivated to any extent. Mixed farming is practised with citrus, olives, figs and vines, and grazing of sheep and goats in the hills. Muğla and the chromite district is rather remote from the rest of Turkey because of the mountains, but there is a reasonable network of roads and several good ports at Fethiye, Gokova and Marmaris in bays sheltered from the Aegean storms.

History

Before the middle of the last century chromite was not used to any great extent, but it became important with the discovery of electro-plating, and the development of stainless steel. In Turkey, the mineral was first reported by an American geologist, Lawrence Smith, in the Bursa area in 1848 and was produced from small surface showings from about 1860. There followed a period of active prospecting and small-scale mining which induced the discovery and development of mines around the port of Fethiye in the south-east of the Muğla area. During this century Turkey has grown to be a major chromite producer only exceeded in importance by Rhodesia and South Africa.

Geological Setting

The Muğla area forms the south-western part of the Taurus Mountain Range, a zone of almost baffling tectonic complexity, which formed during the middle Tertiary. To the north of the area are metamorphic rocks of the Menderes Massif. The core of this massif is composed of paragneisses and schists, mantled by a series of metamorphic rocks among which marbles predominate. From the beginning of the Mesozoic, great thicknesses of sediments were laid down in a trough on the southern side of the Menderes Massif (which at that time was probably an island in the Tethys Ocean). On the present land area most of the rocks are thrust sheets, displaced by some distance from the original site of deposition, plus a post-tectonic cover of continental clastic sediments.

There are three main units to the thrust sheets. Firstly there is a thick (approximately 3 km) sequence of Mesozoic limestones. These form most of the mountains of the area. The second group of rocks are mid-Tertiary clastic sediments classified as 'flysch' deposits. The third group are the ophiolite complexes, which consist of great areas (apparently thin slabs) of peridotite with small intrusive bodies of gabbro. The ultrabasic rocks are mostly hartzbergites and dunites, often very broken and faulted and serpentinized, although only in a few areas is the serpentinization pervasive. Round the edges of some of the peridotite masses are found small wedges of pillow-lavas associated with radiolarites, and other pelagic sediments, and often associated with dolerite dyke complexes.

Along the boundary between the thrust sheets and the Menderes Massif is a zone of strong disturbance, in which Mesozoic rocks and ophiolites are mixed up with the marbles and schists of the basement. In several places along this zone are found metamorhic rocks of blue-schist facies. The thrust sheets seem to have been emplaced by the beginning of the Miocene, after which uplift and block faulting began to occur, accompanied by the formation of a number of basins which became filled (and continue to be filled) with continental clastic sediments. Several of these sediment-filled valleys are fault valleys or grabens and many of them remain seismically active.

Chromite is the main mineral that occurs in the area, although there are also some sedimentary manganese deposits. Some lignite seams occur in the Neogene basins, and in the marbles of the Menderes Massif are the famous emery deposits.

Geology of the Chromite Deposits

Apart from some locally derived eluvial material from chromite outcrops, all the deposits occur in the peridotites. There is considerable variation in form. Some deposits occur as layers or lenses, often a sequence of them parallel to one another (see Merlütla Mine) others are irregular pods or flat-lying pipes. In some cases, they are of a form that looks like a fold structure (see Kandak Mine). Where there is any layering to be seen in the host rock, the deposits seem to be conformable to it, although there are exceptions. The host-rocks are relatively fresh, but the rock is normally serpentinized in the immediate vicinity of the chromite.

The ore itself varies from solid massive chromite to zones of disseminated crystals in a matrix of serpentine. The texture of the ore often looks as if the chromite crystals were sedimented along with olivine but, in some cases, the lenses have a central core of solid chromite entirely surrounded by disseminated ore. When parallel lenses occur, there is a tendency for the chromite to change composition from one lens to the next, with higher aluminium contents higher in the sequence and an increase upwards in the FeO/MgO ratio. The chromite contains between 48 and 56% Cr_2O_3; 16 to 14% FeO; 20 to 16% Al_2O_3 and 14 to 16% MgO (Cr/Fe ratio 2:8 on average).

The chromite deposits are badly affected by deformation at all scales. The ore bodies are often faulted by very large numbers of small faults. The larger chromite crystals in the massive ores are frequently affected by dilation, crystal being cracked, and the cracks infilled with serpentine (sometimes known as 'pull apart' texture).

Size and Grade

This is an area of numerous small deposits and reserve estimates are unreliable. The ore bodies usually range from a few thousand tonnes up to 200 000 tonnes and the area has been quoted as containing 16 Mt in total. Massive ore grades about 48% Cr_2O_3 and disseminated ore is worked down to 30% Cr_2O_3.

Geological Interpretations

The processes by which chromite becomes concentrated during the cooling of a large mass of ultrabasic igneous magma are fairly well understood (see the Great Dyke). There cannot be much doubt that a broadly similar process leads to the segregation of chromite in smaller bodies of similar magma. The evidence of systematic changes in chromite composition in a series of parallel ore bodies indicates the similarity.

But the geometrical irregularity of the podform deposits is difficult to understand. It is perhaps due to the fact that smaller bodies of magma are subjected to movements, and changed during cooling. The main problem in interpreting these deposits is, where and when their host-rocks, were formed, because in all cases it is clear that their present-day location is tectonic. They may have been intrusions formed in ocean trenches during the intensive downwarping, or they may represent sections of oceanic crust. It is envisaged that the Western Taurus Mountains were formed by the forcing together of the Menderes Massif with the ocean plate that formed the Tethys Ocean; and that the thrust-sheets formed as a result. During their tectonic emplacement the peridotites became very faulted and broken, and partially serpentinized.

Mining

There are a number of mines in the Muğla area worked by small private companies, and by the State Mining Corporation (ETIBANK). The mining methods are small open-pit, simple underground, open-stoping or cut-and-fill. The chromite ore is usually selectively mined, and may be upgraded by hand picking, or simple screening and washing to remove serpentinite. Heavy-media separation is used in some of the more modern plants, and chromite concentrates are also produced from lower-grade ores by gravity separation methods. Most of the chromite produced at Muğla is metallurgical grade, and is sent to ferro-chrome smelters, but some is sold for refractory use. Turkey produces about 1 Mt/a of chromite (15% of Western world total) and of this the mines in the Muğla area produce about one-third.

Further Reading

M.T.A. (1966), *The Chromite Deposits of Turkey*. Publication No.132, Maden Tetkik Ve Arama Enstitüsii Yayinlarindan. Ankara. [A systematic documentation of the known deposits, but with only the briefest note on each; contains good bibliography.]

KOVENKO, V. (1945), Fethiye ve Dagardi Bolgeleri Kromit Yalaklari. (Chromite Deposits of the Fethiye and Dagari Regions), *M.T.A. Bull.* 10 1/33, 42–75. [A short review of the geology of deposits in the western part of the country – in Turkish with French translation.] See also:

KOVENKO, V. (1949), Gites de Chromite et Roches Chromeifères de l'Asie Mineure *Mem. Soc. Geol. France*, **28**, 61.

PAMIR, H. N. (1974), Denizli – 1: 500 000 Olcekli Turkiye Jeoloji Haritasi. (*Explanatory text of Geological Map of Turkey*, 1: 500 000 Denizli Sheet), M.T.A. Ankara. [A general summary of the geology of the area with maps and sections.]

CENTO, (1960), *Symposium on Chrome Ore*. Central Treaty Organisation/Maden Tetkik Arama – Ankara. [An interesting volume of papers including several papers on the area with good review of the geology by Kaaden and a review of the deposits of Borchert.]

BORCHERT, H. (1960), *Die Chromiterorkommen in Peridotitmassiv Westlich von Acipayam – Denizli*. Publication of Maden Tetkik Arama, No.106. [One of a series of detailed studied by this author on chromite deposits of the country, this particular one referring to the north-east part of the Muğla area.]

THAYER, T. P. (1964), Principal Features and Origin of Podiform Chromite Deposits and some Observations on the Gulema-Soridäg District, Turkey. *Econ. Geol.*, **59**, 1497–1524. [This is probably the best and most accessible genetic essay on the subject which refers to part of the Muğla area.]

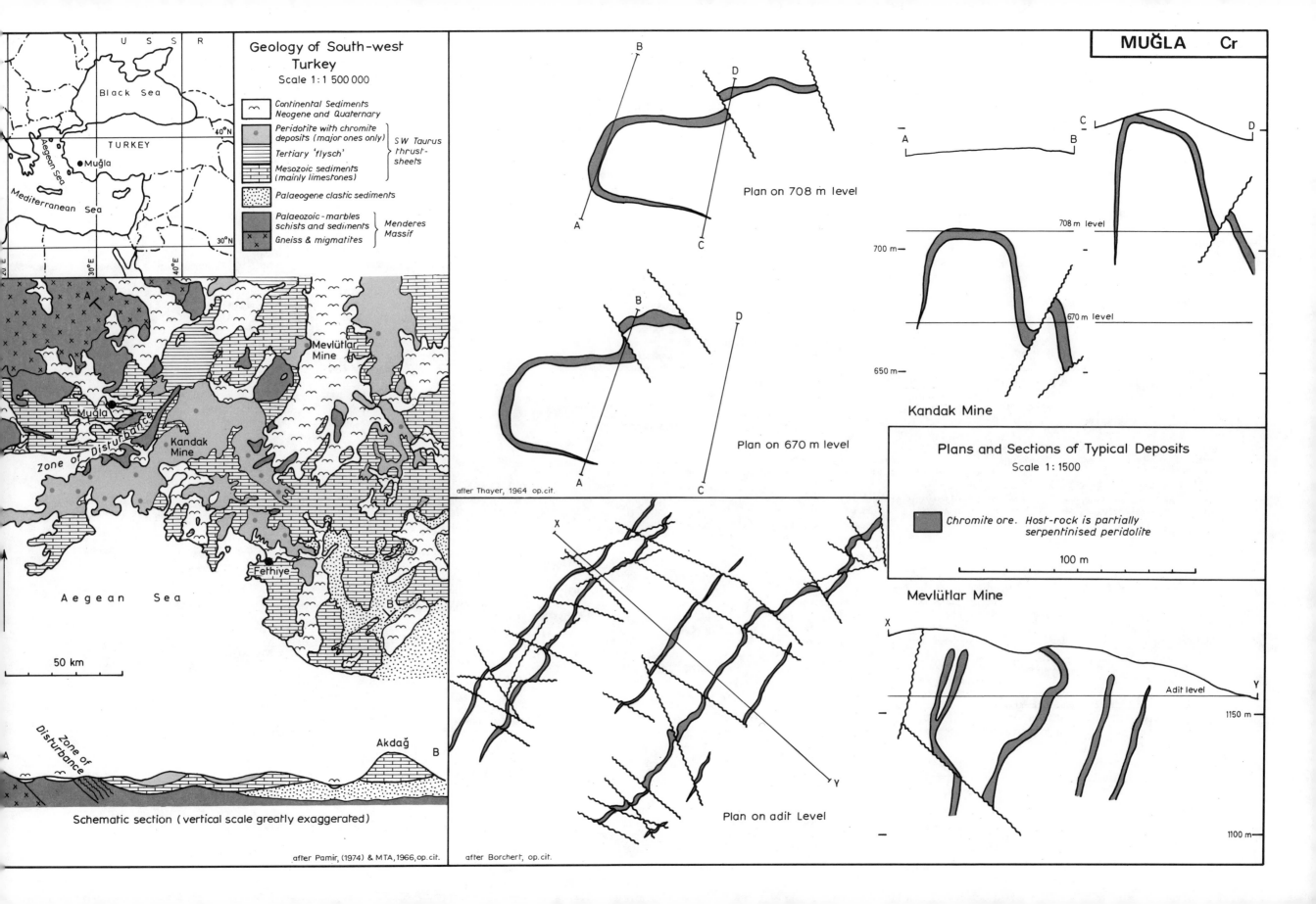

Geology of South-west Turkey
Scale 1:1 500 000

Continental Sediments Neogene and Quaternary

Peridotite with chromite deposits (major ones only)
Tertiary 'flysch'
Mesozoic sediments (mainly limestones)
} SW Taurus thrust-sheets

Palaeogene clastic sediments

Palaeozoic - marbles schists and sediments
Gneiss & migmatites
} Menderes Massif

U S S R

Black Sea

TURKEY

40°N

Aegean Sea

Mediterranean Sea

30°N

Muğla

20°E   30°E   40°E

Mevlütlar Mine

Muğla

Zone of Disturbance

Kandak Mine

Aegean Sea

Fethiye

50 km

Zone of Disturbance

Akdağ

A   B

Schematic section (vertical scale greatly exaggerated)

after Pamir, (1974) & MTA, 1966, op.cit.

B   D

Plan on 708 m level

A   B   C   D

708 m level

700 m

670 m level

650 m

Kandak Mine

B   D

Plan on 670 m level

A   C

after Thayer, 1964 op.cit.

Plans and Sections of Typical Deposits
Scale 1:1500

Chromite ore. Host-rock is partially serpentinised peridolite

100 m

Mevlütlar Mine

X

Y

X

Adit level   Y

1150 m

Plan on adit Level

1100 m

after Borchert, op.cit.

The Asbestos Deposits of the Thetford District – Canada

There are two kinds of asbestos. There are asbestoform varieties of certain minerals (such as the amphiboles) that find rather specialized uses in industry; and there is chrysotile, which is a much more abundant, cheaper and more generally used. Among deposits of the latter, those of Eastern Canada are the largest. The deposits have some geological interest, because they are found in ultrabasic rocks occurring near a major thrust zone at which folded rocks of the Appalachian Belt overlap the St. Lawrence Platform. In recent years, the interpretation of the origin of these rocks has become critical to the understanding of orogenic belts.

Location

The deposits occur in a belt over 100 km long, centred on the town of Thetford Mines (latitude 46° 06′ N, longitude 71° 18′ W and 300 m altitude) 80 km from the state capital in the part of Quebec known as Eastern Townships.

Geological Setting

South-eastwards from the River St. Lawrence is a gently rising plain and as one approaches the asbestos belt, the land rises to 300 m and over. To a large extent the drainage is super-imposed on the country; rivers drain north-westwards to the St. Lawrence, and southward to the Hudson River Basin, in contrast to the north-easterly trend of the geology.

The climate is cool, continental, with sub-zero winters and warm summers; a metre of precipitation falls each year, carried by storms from the south in summer and as snow from the north-west in winter. The area was originally mixed forest, some of which persists on the higher ground and, despite only one-third of the year being free of frost, mixed agriculture is possible on the lower ground. Towns have grown up with the mines, of which Thetford is the largest, but other important towns are Black Lake and (appropriately) Asbestos. These are connected by road and rail to Quebec City and the St. Lawrence.

History

Asbestos was discovered by local farmers and woodcutters in 1862 some 40 km north-east of Thetford, and such was the importance of the find that other deposits were searched for and found; those at Asbestos in 1870 and Thetford in 1877. Working by hand methods soon started but was limited because of the difficulty of separating the fibre from its rock matrix. It was in 1895 that a satisfactory milling process was devised, and production at several mines began on a larger scale. The post-war slump and competition from other mines in south and central Africa forced the mines to merge into larger units in the 1920s, when large-scale open-pit and underground mining began. Even then, there were problems, because the Thetford asbestos proved to be less suitable as an electrical insulator than the African, a disadvantage overcome by the Canadian mines by pioneering the manufacture of asbestos cement goods for building. In a few years they created a new market of unprecedented proportion.

Geological Setting

The Eastern Townships region of Quebec is for the most part underlain by deformed Palaeozoic rocks of the so-called Arcadian–Taconian fold belt, which represents the outer zone of the Appalachian orogenic belt. The most significant tectonic feature is Logan's Line; a line of faults along which the folded Palaeozoic rocks are thrust north-westwards, on to the Pre-Cambrian rocks of the Canadian Shield, or on to Lower Palaeozoic rocks of shelf and basinal facies. The core of the belt consists of late Pre-Cambrian schists, on either side of which are elongated belts of clastic sediments. Various groups have been named, in particular the Caldwell, an unfossiliferous sequence of quartzites, slates and volcanic fragmental rocks, probably of Cambrian or Ordovician age. Another, the Bennett Group, consisting of schists, is probably the metamorphic equivalent of the Caldwell. Rather better understood is the Beauceville Group that begins with a sequence of basic volcanic rocks, pillow lavas, tuffs and associated dyke rocks, overlain by conglomerates, quartzites and slates. Thin limetones yield a mid-Ordovician fauna and the volcanics have a radiometric age of 441 Ma. It is in the Beauceville volcanics and the underlying Caldwell slates that a line of basic and ultrabasic rock masses occur which contain the asbestos deposits. There are two groups of these rocks, peridotites with dunitic patches, moderately serpentinized; and a complex of dunite, pyroxenite and gabbro. Small dykes of granite and syenitic rocks cut them. The mode of emplacement of these rocks is not clear. They display some intrusive features but no contact metamorphism, and the contacts are frequently obscured by strong shear zones. The granitic dykes cut all rocks of the complex and have radiometric ages of 477 to 481 Ma. A significant feature is a strong positive Bouguer gravity anomaly along the line of ultrabasic rocks.

Devonian rocks outcrop on the higher ground to the south-east of the ultrabasic belt. These are undeformed and represent the period when this part of the fold belt became stabilized. The area was probably eroded down to a peneplain during the Tertiary, and then heavily glaciated in the Quaternary, large terminal moraines being left in the area.

The serpentinized peridotites contain what is probably the greatest concentration of chrysotile asbestos deposits in the world, as well as minor deposits of chromite and talc rock.

Geology of the Asbestos Deposits

A line of ultrabasic bodies follows roughly the Caldwell–Beauceville contact, but within them the asbestos deposits are concentrated in three groups; at Asbestos in the south-west, Thetford–Black Lake in the centre, and along the so-called Pennington dyke to the north-east. The host rock of the asbestos is normally a hartzbergite with patches of dunite which, although sometimes fresh, is generally traversed by a network of serpentine veinlets or, in places, pervasively altered to serpentinite. Chrysotile asbestos is found both as cross-fibre and slip-fibre veins from 150 mm in width down to a millimetre or less. The asbestos veins are surrounded by a zone of serpentinized rock averaging six times the width of the vein; but the veins show no preferential concentration in peridotite of differing degrees of serpentinization. The dunite, pyroxenite and gabbro, although altered, do not contain asbestos.

A very marked feature of the deposits is the presence of large zones of shearing in which almost all evidence of original rock type is lacking, and which are frequently composed of talc rock. Where the content of talc is high enough, these have been mined, although not on a large scale. In addition to the shear zones, the ore is interrupted by granitic dykes.

Size and Grade

There are over a dozen major deposits in the belt, and figures of reserves only given sporadically. The size of workable deposits is between 50 and 300 Mt, and the whole area contains about 700–800 Mt in all. Grades worked vary according to mining methods and cost, some zones are worked at 12% fibre, others down to 2 or 3%; the average is probably 5 to 6%.

Geological Interpretations

Most of the writers on these deposits have concluded that the ultrabasic bodies were intruded into the sediments and volcanics (finding their way up along the Caldwell–Beauceville contact) and that the various rocks were differentiates of some gabbroic parent magma. However, ultrabasic and basic rocks (including pillow lavas) in association with argillaceous sediments over a strong positive gravity anomaly and in proximity to a major zone of overthrusting, suggests that the Beauceville (and the ultrabasic rocks) are a slice of oceanic crust thrust over a flysch-wedge (the Caldwell) during the destruction of some pre-Appalachian ocean or marginal ocean basin. Serpentinization clearly took place more than once, and the formation of asbestos was the latest phase of this alteration. The relationship of the peridotite bodies to structures in the surrounding rocks, and the structures that may be seen within them (revealed by chromite seams) suggests that the asbestos bodies are zones subjected to tension during deformation. Failure of the rock permitted the serpentinization and growth of asbestos fibres in the cracks. Where shearing, rather than tension, was the dominant deformation style, the alteration is to talc. Just how and why asbestos grows in such circumstances is a subject of much controversy.

Mining

The major mines are controlled by Asbestos Corporation and Canadian Johns Manville, who have several underground mines (most of which work by block caving methods) and a number of large open-pits. Some of the mines are 400 m deep. These mines were among the pioneers of block caving, and large open-pit mining. Underground operations are much influenced by the rock-mechanical conditions, particularly where talcose shear zones adjoin hard peridotite. The ore is dried, crushed and screened, so that the rock matrix is gradually removed, leaving the fibres to be blown off and sorted into different length-groups. Much of the Thetford product is used to make asbestos cement and asbestos-plaster goods for the building industry. The heat-insulating qualities of these products, valuable in the cold Canadian winters, and their resistance to atmospheric corrosion in urban areas, make them superior to galvanized corrugated steel. The area produces about 1 Mt of asbestos products per year, approximately half the world's production.

Further Reading

COOKE, H. C. (1937), Thetford, Disraeli and Eastern Half of Warwick Map-Areas, Quebec. *Geol. Surv. Can., Mem.*, **211**. [The basic work on the area but rather old; a better account is to be found in Dresser and Denis, 1944.]

DRESSER, J. A. and DENIS, T. C. (1944), *Geology of Quebec*, Vol. II. Quebec Dept. Mines Geol. Rep.20, 544pp. [Particularly pp.365–454.]

RIORDON, P. H. (1957), The Asbestos Belt of Southeastern Quebec.

RIORDON, P. H. (1957), The Asbestos Deposits of Thetford Mines, Quebec.

RIORDON, P. H. (1957), The British Canadian Mine.

RIORDON, P. H. (1957), Normadic and Vimy Ridge Mines.

BOURASSA, P. J. (1957), The Asbestos Mine of Nicolet Asbestos Mines Ltd.

ALLEN, C. C., GILL, I. C., KOSKI, J. S. *et al.* (1957), The Jeffrey Mine of Canadian Johns-Manville Co. Ltd. All in: *The Geology of Canadian Industrial Mineral Deposits.* (Goudge, M. F., ed.), Canadian Inst. Min. Met. Montreal, 247pp. [A fine group of papers covering the important features of the whole area.]

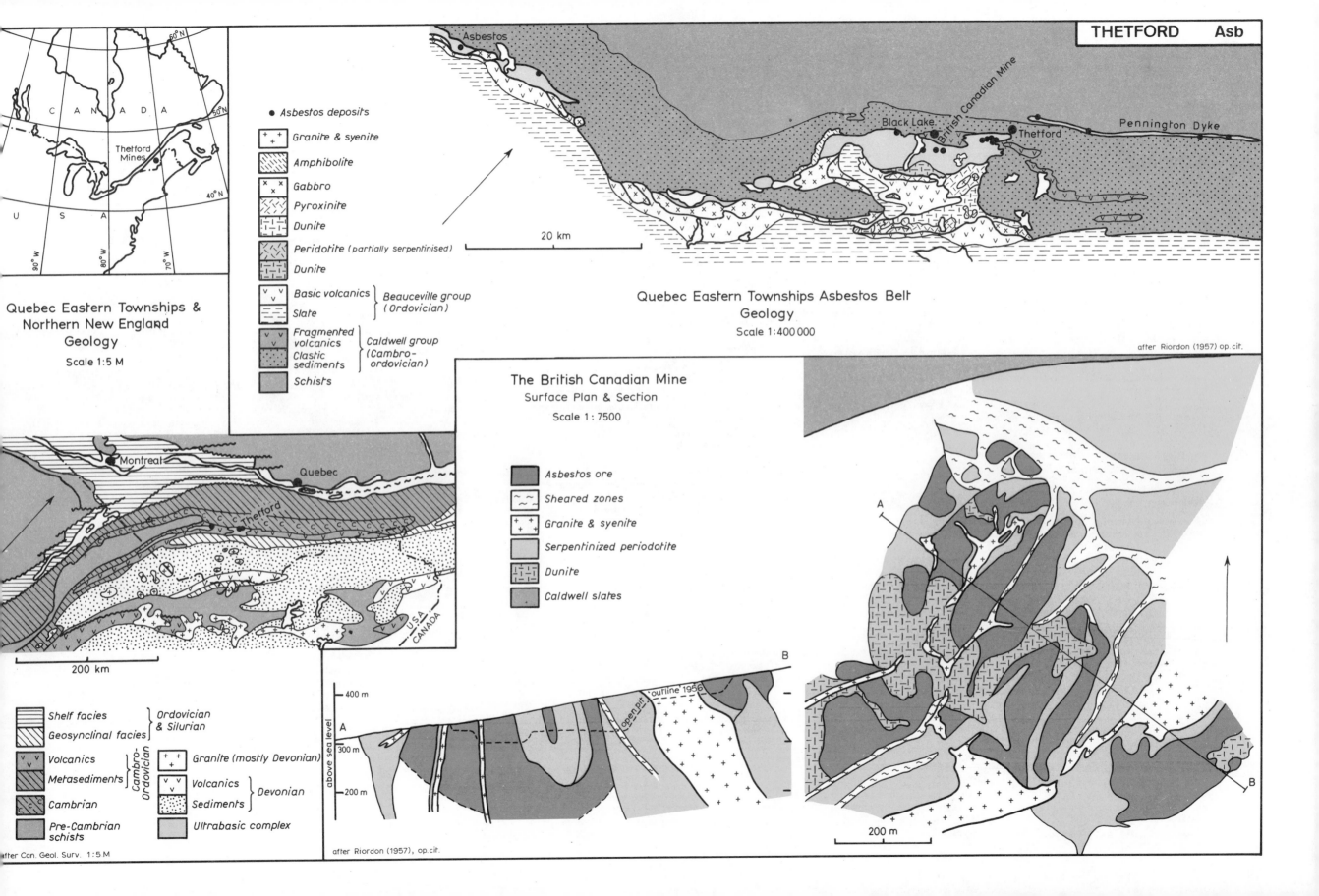

Quebec Eastern Townships & Northern New England Geology

Scale 1:5 M

Quebec Eastern Townships Asbestos Belt Geology

Scale 1:400 000

after Riordon (1957) op.cit.

The British Canadian Mine

Surface Plan & Section

Scale 1:7500

Legend (centre):

- Asbestos deposits
- Granite & syenite
- Amphibolite
- Gabbro
- Pyroxinite
- Dunite
- Peridotite (partially serpentinised)
- Dunite
- Basic volcanics } Beauceville group (Ordovician)
- Slate
- Fragmented volcanics } Caldwell group (Cambro-ordovician)
- Clastic sediments
- Schists

Legend (British Canadian Mine):

- Asbestos ore
- Sheared zones
- Granite & syenite
- Serpentinized periodotite
- Dunite
- Caldwell slates

Legend (lower left):

- Shelf facies } Ordovician & Silurian
- Geosynclinal facies
- Volcanics } Cambro-Ordovician
- Metasediments
- Cambrian
- Pre-Cambrian schists
- Granite (mostly Devonian) } Devonian
- Volcanics
- Sediments
- Ultrabasic complex

after Can. Geol. Surv. 1:5 M

after Riordon (1957), op.cit.

Map labels: Asbestos, Black Lake, British Canadian Mine, Thetford, Pennington Dyke, Montreal, Quebec, Thetford Mines, CANADA, U S A

Scale bars: 20 km, 200 km, 200 m

Section: 400 m, 300 m, 200 m (above sea level); open pit outline 1956; A, B

The Palabora Complex – South Africa

Palabora is the largest known alkaline ring complex with a carbonatite core. In addition to the apatite-rich rocks, iron ores, and traces of radioactive minerals that these complexes normally include, Palabora is unique for containing a major copper sulphide ore body in the carbonatite, and for a substantial zone of vermiculite-rich rock. It may not be the most typical carbonatite complex, but it is certainly the most important economically.

Location

The centre of the Complex is the peak of Loolekop, located at latitude 24° 00′ S, longitude 31° 08′ E and 478 m in altitude in the Letaba district of the Transvaal in the Republic of South Africa.

Geographical Setting

Palabora lies on the Lowveldt, the lower ground that lies between the elevated plateau that forms much of southern Africa, and the coastal plains of Mozambique. From the peak of Loolekop at 478 m the land descends to under 400 m in the south, where small seasonal rivers drain down to Olifants River, a tributary of the Limpopo. The climate is subtropical, with warm, wet summers, and temperate, dry winters. The area is generally covered by short grass with a few trees, and makes good farming country, particularly for cattle, even though much of the Complex itself is rocky outcrop. On the northern edge of the Complex a sizeable town has grown up since mining began, called Phalaborwa, connected by modern highway to Johannesburg 550 km away. Palabora is on the edge of the Kruger National Game Park, and even has a substantial 'passing tourist' industry. Unlike many other countries in the world, South Africa was shrewd, and stopped the boundary of the Park short of the mineralized Complex.

History

There are remains of copper workings on Loolekop that are probably 2000 years old, and the area was mined for iron by the Bantu in more recent times. In the modern era, Palabora was first studied by Carl Mauch between 1868 and 1871, and he reported the copper, iron and other mineral occurrences in the area. In the 1930s C. H. Cleveland, a prospector, noticed the exfoliation of 'rotten mica' after a grass fire, and by the mid-1930s vermiculite was being produced from the northern part of the complex. About the same time, open-pit mining of phosphate-rich rock began from outcrops below Loolekop, and a thorough study of the mineral potential of the Complex was made by Hans Merensky. In 1952 the carbonatite core was tested for radioactive minerals; traces were found but not enough to warrant mining. But the survey did find a considerable amount of copper mineralization, below the surface outcrops that had not previously been taken seriously (though known for a long time). After years of drilling and bulk sampling underground, a large open-pit operation for copper, iron and apatite was started in 1968.

Geological Setting

The Palabora Complex is situated in an area of Archean granitoid rocks that form the eastern exposed edge of the Transvaal Shield. The Murchison Range greenstone belt lies to the north and, although most of the country is gneiss, there are some granites such as the Mashishimala Granite 20 km to the west. 45 km to the east there is an abrupt change where Karoo sediments overlie the Shield (mostly basaltic lavas), while 70 km to the south-west the Shield disappears beneath Proterozoic sediments.

All round the Complex are bodies of syenite or syenite intrusion breccia that probably belong to the Complex itself, and round part of the perimeter there are fenites clearly formed by felspathization of the surrounding gneiss. Most of the Complex consists of pyroxenite, feldspathic round the outer rim, within which are three centres of different rocks. The northern centre consists of a very coarse-grained olivine: phlogopite diopside rock that becomes richer in phlogopite towards the centre. A not dissimilar, but less well-developed centre occurs in the south; while, in the middle, is the Loolekop Complex which has a similar outer zone but passes to an olivine, magnetite, apatite rock locally called 'phoscorite', and in the centre is finally traversed by a complex of carbonatites. In a few places there are massive bodies of a biotite-rich rock (glimmerite) and small carbonatite dykes traverse the pyroxenite and other rocks outside the main Loolekop mass.

Near the surface there are substantial alteration effects. In the northern centre, vermiculite takes the place of phlogopite, and most of the olivine is serpentinized. It is this rock that is worked for vermiculite, the other mineral wealth coming from the phoscorite and carbonatite of Loolekop.

The form of the outcrop of the Complex, and its internal structure known from drilling suggests a large vertical pipe, and this is substantiated by the gravity anomaly that coincides with it. The accepted radiometric age of the Complex is 2·06 Ga, much younger than the surrounding gneiss, but much older than the majority of carbonatite in Southern Africa. The complex is traversed by numerous dolerite dykes of Karoo age (c. 200 Ma).

The Loolekop Mineralization

The limit of the mineralized zone is marked by the inner limit of the phlogopite-biotite pyroxenite. Within this is the main mass of phoscorite, consisting of coarse-grained serpentinized olivine, titaniferous, magnetite, fluorapatite with accessory mica, calcite and baddeleyite. The proportions of the minerals are very variable, but generally there is about 35% magnetite, 25% apatite, and 18% carbonates; locally there are quite large lenses of pure serpentine and calcite. The olivine is magnesium-rich (Fo_{91}). The phoscorite mass interdigitates with the surrounding pyroxenite, and has zoning and banding which is vertical and concentric with its outline. Small quantities of copper sulphide minerals occur, but rarely yield more than 0·3% Cu. In the centre of the phoscorite body is a strongly banded vertical pipe of carbonatite. It is a søvite, the main carbonate being (Cal_{93} Mag_7) with 0·3% Sr and 300 µg/g Ba, and there are bands of other minerals such as magnetite, apatite and phlogopite, more or less concentric with the outline of the pipe. There are sulphides, including bornite, but the copper content is not much higher than in the phoscorite. The banded carbonatite is cut by a swarm of dykes, and one large elongated body of a massive granular to sugary-textured søvite, slightly richer in magnesium (Cal_{86} Mag_{14}) with a similar assemblage of accessory minerals (except that the magnetite is relatively low in titanium, and the sulphide content is much higher). Pyrrhotite and pentlandite occur, with amounts of chalcopyrite that gives a copper content of 1% or more. There is a wealth of minor minerals, including olivine, slightly more iron-rich than the general (Fo_{87}), dolomite in intergrowths with calcite, chondrodite, biotite, baddeleyite, spinel, ilmenite, uranothorianite, fluorite and a variety of sulphides. The maximum uranium content found is 150 g/t.

Weathering of the rocks reaches a few metres on Loolekop, and up to 70 m on the lower slopes. Olivine is serpentinized, copper sulphides altered to carbonates and the phoscorite is rendered somewhat friable.

Size and Grade

The mine at Loolekop was started on reserves of 300 Mt containing 0·69% copper and 27% magnetite, since which time the reserves have been increased to over 400 Mt. The reserves of phosphate in the phoscorite are very large, but much depends on how deep it could be mined; one estimate is 11 Mt for each metre of depth containing about 6% P_2O_5. The reserves of vermiculite in the northern part of the Complex are also very large.

Geological Interpretations

The problem of the formation of these ring complexes with their carbonatite cores is a general and highly controversial one. There was even a period, decades ago, when geologists were studying these complexes in Africa at the same time as theoretical petrologists were saying that they had proved that they could not exist. However, the post-Second World War uranium boom led to much more detailed study, and it became generally accepted that these complexes are highly differentiated intrusions originally from magmas generated in the upper mantle. One of the persistent problems has been the source of the carbonate, but even that is now regarded as primary. Quite why they form remains a mystery. They seem to be associated with fracture systems of continental or global scale, which also seem to control the occurrence of kimberlites and major intrusions of basic rock. Palabora differs in only a few respects from the typical carbonatite complex. It is larger than most, its main rocks contain no felspathoids, the phoscorite, though not unique, is unusually well-developed and, above all, its carbonatite is rich in copper and other sulphides. No explanation has been found for these peculiarities, particularly the occurrence of copper.

Mining

Most of the mining is now in the hands of the Palabora Mining Company which operates the Loolekop pit, which is 1·5 km long, 1·1 km wide and 430 m deep with a 45° slope. Production is 53 000 t per day of ore, and slightly less waste. By careful geological mapping and sampling, the rock is selectively mined in several categories according to the copper, phosphorus and titanium content. Sulphides and apatite are recovered by flotation, and the copper sulphide and oxide concentrates smelted to blister copper, some of which is electrolytically refined and formed into copper rod on site. The low titanium magnetite, recovered by magnetic separation, is sold as iron ore; much of the titanium-rich magnetite is stockpiled against the day when it can be used. Sulphuric acid is made from the smelter gasses. Palabora is one of the best examples of an integrated mineral-industrial complex developed in modern times.

Further Reading

HANEKOM, H. J. et al. (1965), The Geology of the Palabora Igneous Complex. S. African Geol. Surv. Mem., 54, 185pp. [The standard work on the geology of the complex as a whole.]

VERWOERD, W. J. (1967), Loolekop at Palabora. In: The Carbonatites of South Africa and South West Africa. S. Africa Geol. Surv. Hndbk., 6, 15–25. [More detailed description of the carbonatite core. The volume also includes an interesting petrogenetic essay on carbonatites, pp. 307–35.]

LOMBAARD, A. F. et al. (1964), The Exploration and Main Geological Features of the Copper Deposit in Carbonatite at Loolekop, Palabora Complex. In: The Geology of Some Ore Deposits in South Africa. (Haughton, S. H., ed.), Geol. Soc. S. Africa, Johannesburg, Vol.2, pp.315–37. [Title explains itself, written by some of the team that took part in the evaluation at the time it was finished; well-illustrated.]

KUSCHKE, O. M. and TONKING, M. J. H. (1971), Geology and Mining Operations at Palabora Mining Company Limited, Phalabora, N.E. Transvaal. J. S. Africa Inst. Min. Metall., 72, 12–23. [Good summary of the geology of Loolekop and description of the mining operation.]

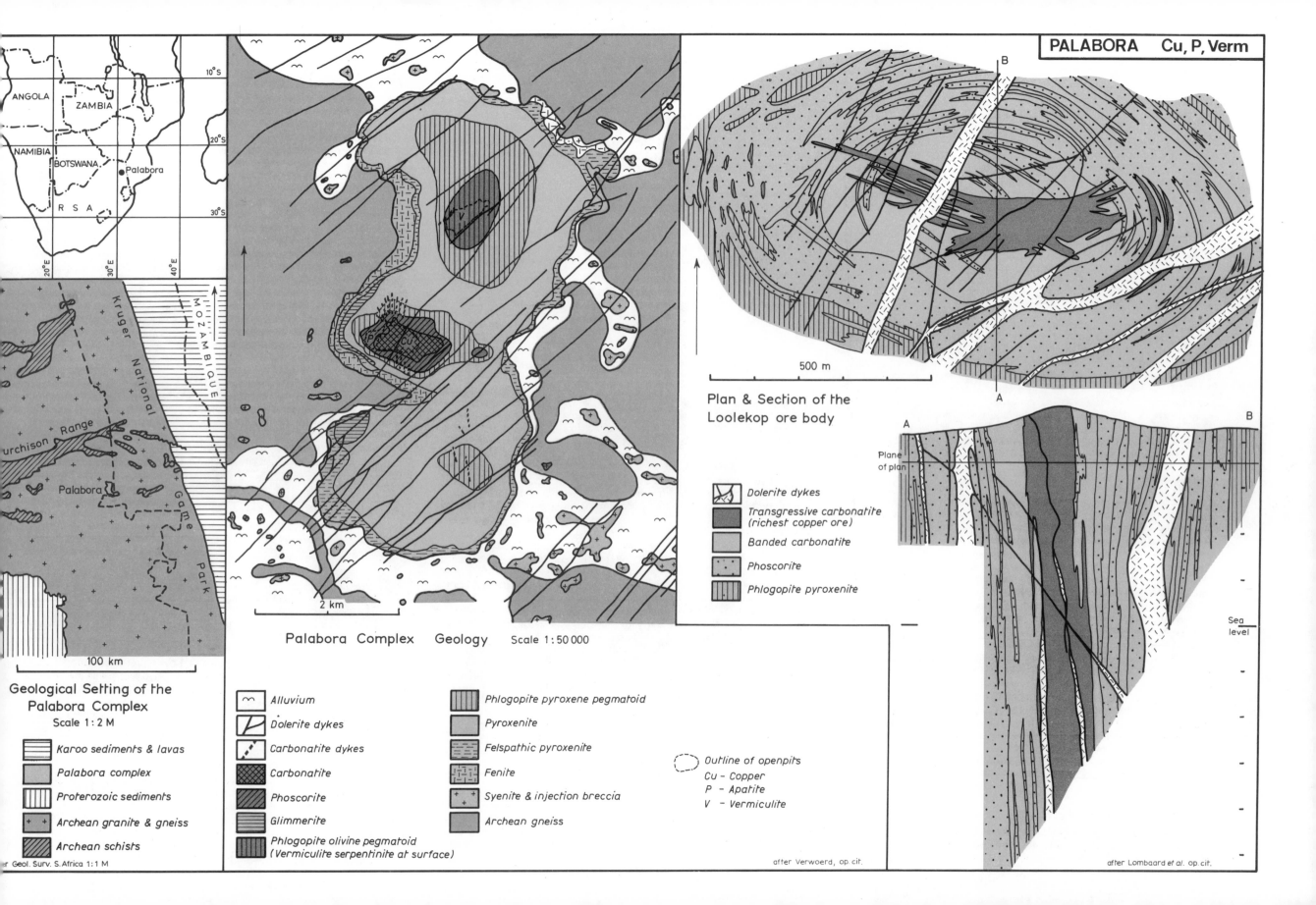

PALABORA   Cu, P, Verm

Geological Setting of the Palabora Complex
Scale 1 : 2 M

- Karoo sediments & lavas
- Palabora complex
- Proterozoic sediments
- Archean granite & gneiss
- Archean schists

100 km

er Geol. Surv. S.Africa 1:1 M

Palabora Complex   Geology   Scale 1 : 50 000

2 km

- Alluvium
- Dolerite dykes
- Carbonatite dykes
- Carbonatite
- Phoscorite
- Glimmerite
- Phlogopite olivine pegmatoid (Vermiculite serpentinite at surface)

- Phlogopite pyroxene pegmatoid
- Pyroxenite
- Felspathic pyroxenite
- Fenite
- Syenite & injection breccia
- Archean gneiss

- Outline of openpits
  - Cu – Copper
  - P – Apatite
  - V – Vermiculite

after Verwoerd, op.cit.

Plan & Section of the Loolekop ore body

500 m

- Dolerite dykes
- Transgressive carbonatite (richest copper ore)
- Banded carbonatite
- Phoscorite
- Phlogopite pyroxenite

Plane of plan

Sea level

after Lombaard et al. op.cit.

The Mwadui Diamond Pipe – Tanzania

The discovery of the Kimberley diamond pipe in South Africa in 1871 was an event of great importance for the diamond industry and for geology, because it revealed for the first time the source of the stone up till then only known from alluvial gravels. Several generations of geologists have been baffled by the origin of the strange rock, 'kimberlite', found in these pipes. A small proportion of kimberlite pipes contain very small quantities of diamond and the largest of these, and the one displaying the most interesting range of features, is Mwadui.

Location

Mwadui is situated at latitude 3° 32′ S, longitude 33° 36′ E and 1190 m altitude, 110 km south of Lake Victoria in the Shinyanga district of Tanzania.

Geographical Setting

The northern part of Tanzania is a high plateau with a seasonal drainage system that gently descends to Lake Victoria. The land is gently undulating around 1200 m in altitude, with a few hills rising up to 1400 m. The climate is tropical and fairly dry, the annual rainfall being about 600 mm, mostly falling during the autumn. Temperatures are rather moderate for such a low latitude because of the altitude, and average in the low twenties (°C), but with extreme daily variations. The land is tall grass savannah, with wooded areas where there is more water. The area is not densely populated, but there are scattered villages and farms. Mwadui is 23 km from the provincial town of Shinyanaga, and the main road and railway that connects Tabora with Mwanza on Lake Victoria passes within a few kilometres of the mine.

History

The discovery and development of Mwadui is one of the great one-man successes of the mining industry. Diamonds were discovered at Mabuki during the German colonial period, but no serious work was done on them until a South African named Blignault began systematic prospecting in 1921. By 1925 a diamond industry was established and a number of kimberlite pipes were discovered below the alluvial diamond-bearing gravels. Among the geologists and prospectors who came to the district between the two world wars was Dr. J. T. Williamson, a Canadian, who in 1930 decided to take over several blocks of ground, and rework some of the tailings and small patches of unworked gravels. With the proceeds of this and money borrowed, he set about a systematic study of all the Shinyanga occurrences, and the numerous outcrops of kimberlite. Previous work had shown that none of the known kimberlite pipes contained economically workable diamonds, but Williamson was convinced that there must be a source of the alluvial stones in the area. In 1940, almost out of money, he had chosen the most likely area of the source to exist by identifying the trend lines of kimberlite showings, and the probable provenance areas of the alluvium. He had apparently almost given up when by chance he struck rich diamond-bearing sediments of a type that was rather unusual. Test pits revealed the top of the Mwadui pipe, and two years later Williamson formed his own company and began production. When Williamson died in 1958 (a rich man) the mine was sold to Anglo-American Corporation, but was later acquired by the Tanzanian State.

Geological Setting

Mwadui is at the northern end of the Tanzanian Shield, an area of Archean granitoid rocks consisting in the main of gneisses and granulites. There are a few fragmentary schist belts, one of which occurs around the Shinyanga diamond area. The younger granite plutons of the area that cut the schists, have been dated at 2.5 Ga.

Apart from dolerite dykes which cut the basement rocks, believed to be Mesozoic in age, there is almost no record of the geological history of this area between 2.5 Ma and the Neogene. At that time a number of small basins developed that were filled with locally derived sediments, and this continued almost to the present day. Many of these sediments are cemented by silica or calcite during a period of residual, chemical weathering. Part way through the formation of these sediments, the area was affected by faulting which modified the sedimentation. This faulting was part of the tectonic events which led to the development of the great East African rift valley system, and the rift with its associated volcanic cones dominates the landscape 150 km to the east of Mwadui.

Pipes and dykes of kimberlites are found in clusters, arranged in lines over an area of 60 000 km² south and east of Lake Victoria. Some hundreds of occurrences are known but only Mwadui contains significant amounts of diamond. They are younger than the dolerite dykes (and the basement) but older than the Neogene sediments, and are generally thought to be Cretaceous in age.

Geology of the Mwadui Pipe

The Mwadui kimberlite pipe is roughly elliptical in cross-section and where it is actually filled with kimberlite at about 400 m below the surface, it measures 1 km by 700 m. Above this it widens out into a debris-filled crater which is 1600 km by 1100 km in size. The kimberlite itself is the usual fragmental rock composed largely of serpentine. It contains recognizable, and quite large phenocrysts of serpentinized olivine and fragments of granitic rocks, up to 25 mm across, that display a wide range of angularity and alteration. The matrix between the more coherent fragments of serpentinite and the granitic xenoliths is a mixture of soft serpentine, chlorite and carbonate with an assortment of crystals of garnet, ilmenite, chrome-diopside and phlogopite. (The first three of these minerals are important because they survive weathering, and are used as indicator minerals of the possible presence of diamonds in prospecting.)

The contact between the kimberlite and the surrounding gneiss is sharp, and is sometimes marked by a zone of strong shearing forming a clay-like gouge. There is almost no evidence of contact alteration, except that in places the gneiss is extensively shattered and stained green along the fractures for some distance from the pipe.

The crater filling is very variable. At the base overlying the kimberlite are fine-grained kimberlite tuffs, above which is a thick mass of breccia, consisting of large and small fragments of gneiss with a kimberlite matrix. The top 200 m of the pipe filling is composed of laucustrine sediments. At the edges these are a relative coarse-grained and poorly sorted mixture of material, in part derived from the gneiss, but with some kimberlite material and there is much evidence that these sediments were washed into the crater from the sides. These sediments also show slump structures trending inwards to the centre of the crater. At the centre are much finer-grained greenish and yellowish clays and grits. Towards the top of the crater the sediments are better sorted and there is a partial cover, of more normal alluvial gravels, some of which are cemented by silica. Diamonds occur throughout the pipe and crater, most numerous in the alluvial gravels at the top. Below, their frequency of occurrence seems to depend on the proportion of non-kimberlite material, being lowest where the proportion of gneiss fragments is highest.

Size and Grade

Like most diamond deposits, not all of the rock contains enough diamonds to warrant treatment. The total material available down to the depth of development is about 1 Gt and it is said that the grade varies between 10 and 14 carats per 100 t. The pipe is at present selectively mined at a grade of 20 carats per 100 t (equal to 40 mg/t). The diamonds are variable in colour, often twinned, and average 0.24 carats in weight. The grade and quality of the diamonds in the surface gravels was much higher.

Geological Interpretations

There is a vast literature on the origins of kimberlite and diamond, and it is a subject of great speculation. Opinion now favours the origins of the kimberlite from the melting of a very impure garnet lhertzolite deep in the mantle. Quite how it comes to contain a high content of volatile material is still uncertain. The arrangement of kimberlite clusters strongly suggests that the magma rises along fractures that penetrate through the crust, although quite how they form is another problem. Kimberlite magma seems to rise first in the form of dykes, from which circular diatremes penetrate up to the surface. As the magma nears the surface, it cools and becomes fluidized by its enormous content of gas. At Mwadui the most plausible idea is that the magma advanced upwards, the gases under pressure fracturing the wallrock and forming a crater of relatively cool but fluidized rock fragments and kimberlite, eventually blowing out a mass of material on to the rim of the crater. After the 'eruption' this material was gradually washed back into a crater-lake, forming the upper part of the crater filling. People have always argued about the origins of the diamonds. Some think that they formed in the pipe during the eruption, but most people now follow the theory that they form deep in the mantle where diamond is the stable form of carbon, and are carried up in the magma with such rapidity that they can remain in a meta-stable condition and do not revert to graphite.

Mining

The Mwadui pipe is mined by a company that bears its discoverer's name, now controlled by the State. There is a shaft down to 400 m below surface, originally sunk for exploration and sampling. The rock is extracted from an open-pit and underground workings, and sent to two large heavy medium separation plants where a slurry of magnetite and water is used to float off the lighter material. What remains is a concentration of the more dense minerals and rock fragments that includes the garnet, ilmenite and the diamonds. Diamond, along with several other minerals is hydrophobic and oleophilic, and is recovered by passing the concentrate over the grease tables. Periodically the grease is removed, and the mineral concentrate separated from which the diamonds are picked by hand. The mine is capable of treating 10 000 t of ore per day and produces between 500 000–700 000 carats per year.

Further Reading

TREMBLAY, M. (1956), *Geology of the Williamson Diamond Mines, Mwadui, Tanganyika.* Ph.D. Thesis, McGill University, Canada. [The main descriptive work on the deposit but difficult to get hold of.]

EDWARDS, C. B. and HOWKINS, J. B. (1966), Kimberlites in Tanganyika with Special Reference to the Mwadui Occurrence. *Econ. Geol.*, **61**, 537–54. [A good general description written at the end of an extensive prospecting campaign of the whole area that followed the Anglo-American purchase.]

WILLIAMS, G. J. (1938), The Kimberlite Province and Associated Diamond Deposits of Tanganyika Territory. *Tanganyika Geol. Surv. Bull.*, **12**, 38pp.

WILLIAMS, G. J. and GADES, N. W. (1938), Exploration of the Geology of Degree Sheet No. 18 (Shinyanga). *Tanganyika Geol. Surv. Bull.*, **13**, 20pp. with coloured map. [These two works are the standard ones on the area and are interesting as they were written two years before Williamson's discovery.]

DAWSON, J. B. (1971), A Review of the Geology of Kimberlites. *Earth Sci. Rev.*, **7**, 187–214. [Of the many, this is probably the best up-to-date review on the subject and contains a very comprehensive bibliography to the vast literature.]

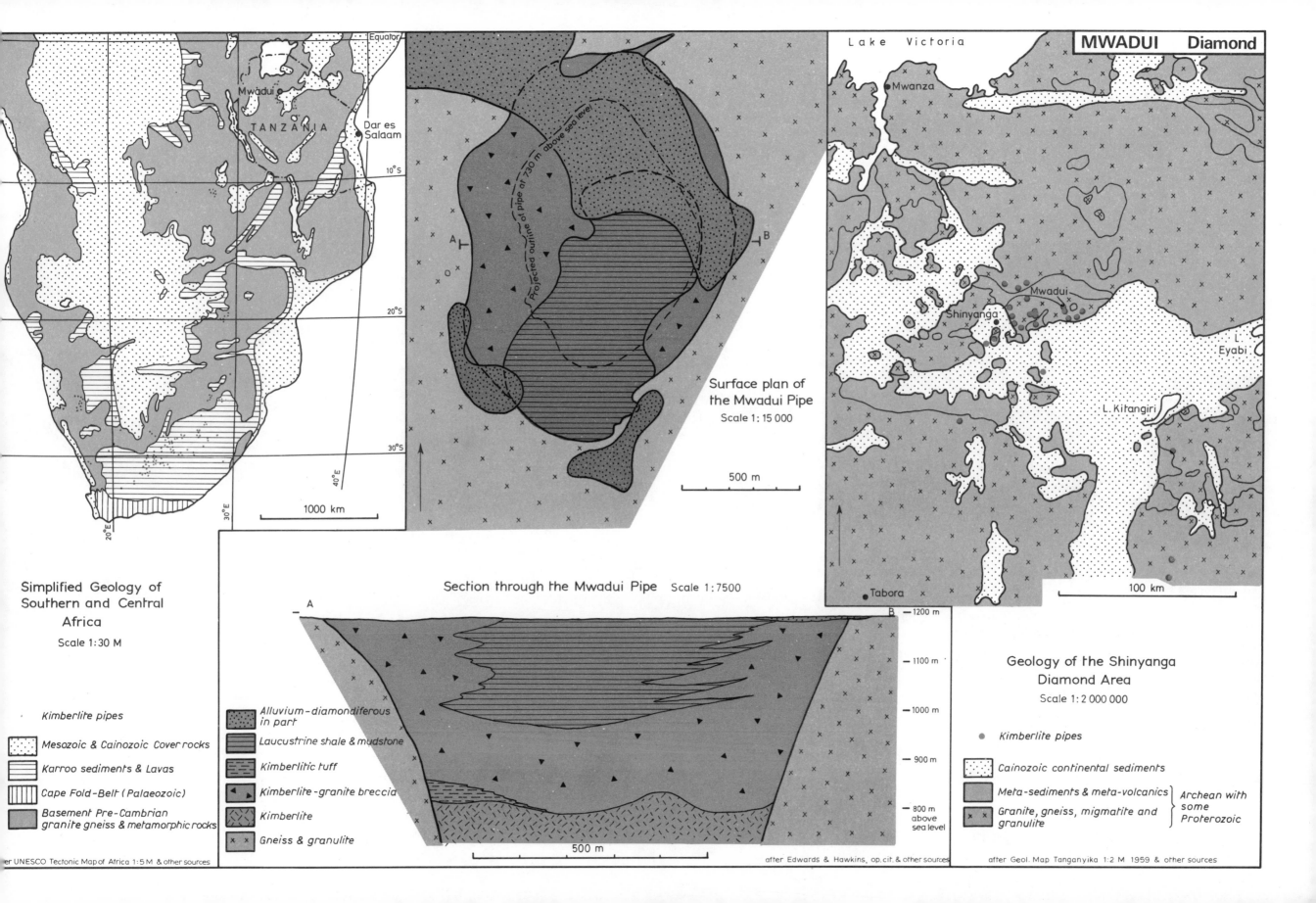

MWADUI Diamond

Lake Victoria

Surface plan of
the Mwadui Pipe
Scale 1:15 000

Projected outline of pipe at 730 m above sea level

A

B

O

500 m

Mwanza

Mwadui

Shinyanga

L. Eyabi

L. Kitangiri

Tabora

100 km

Simplified Geology of
Southern and Central
Africa

Scale 1:30 M

Equator

Mwadui

TANZANIA

Dar es
Salaam

10°S

20°S

30°S

20°E

30°E

40°E

1000 km

Section through the Mwadui Pipe   Scale 1:7500

A

B

1200 m

1100 m

1000 m

900 m

800 m
above
sea level

500 m

Geology of the Shinyanga
Diamond Area

Scale 1:2 000 000

Kimberlite pipes

Mesozoic & Cainozoic Cover rocks

Karroo sediments & Lavas

Cape Fold-Belt (Palaeozoic)

Basement Pre-Cambrian
granite gneiss & metamorphic rocks

*Alluvium-diamondiferous
in part*

Laucustrine shale & mudstone

Kimberlitic tuff

Kimberlite-granite breccia

Kimberlite

Gneiss & granulite

● *Kimberlite pipes*

Cainozoic continental sediments

Meta-sediments & meta-volcanics } Archean with
some
*Granite, gneiss, migmatite and
granulite* } Proterozoic

after UNESCO Tectonic Map of Africa 1:5 M & other sources

after Edwards & Hawkins, op.cit. & other sources

after Geol. Map Tanganyika 1:2 M 1959 & other sources

Section Five: The World Distribution of Mineral Deposits

Introduction

The series of world distribution maps presented in this section show the important economic deposits of a selection of the most important commodities. The deposits are shown by different symbols, based on a simple classification into geological types against a background that shows the essential tectonic architecture of the earth's crust. Only the land areas are dealt with here, because, for the foreseeable future it is likely that most of the non-fuel mineral raw materials will come from the land.

It is notoriously difficult to define a single mineral deposit, because on the one hand it is a geological feature, but on the other, it is an object of potential or actual economic interest. Most of the localities indicated on the following maps are mines. Some, however, are deposits defined geologically, that are exploited by several mining operations; others are the converse; that is to say, groups of deposits which can be regarded as independent geological entities, but exploited by a single mining operation. Included also are some well-known deposits that for reasons of local economics, are not mined, even though they may come to be exploited in the not too distant future. But in general, the assemblage of points shown on the maps represents the earth's mineral wealth which has been both discovered, and shown to be technologically exploitable. From this discussion it is clear that the question 'what is a mineral deposit?' is only answerable both in terms of geology and human-imposed factors; and it is worth while to discuss some of these latter factors before examining the geological ones that govern the distribution of mineral deposits.

Human civilization depends and always has depended, on mineral resources, and from the time that groups of human beings began to specialize in the production of minerals, they have been rewarded by their fellows for so doing, most notably in the Western industrial capitalist societies through the medium of money, according to a price set by the interaction of supply and demand. It is however important to note that in the present-day world there are many examples of mineral exploitation operations that are rewarded by society in a way that is not strictly controlled by a market-determined price. Wars, alternative political ideologies (or the desire to avoid their spread) have motivated rulers and governments to intervene in the market place in order to assure the mineral raw material base for their survival (or furtherance of their political ambitions). Whatever the socio-economic system within which a mineral deposit is exploited, there is always some criterion by which the operators can judge the threshold of exploitability of their deposit. The result of this complex set of controls on exploitability is that the world distribution of mineral deposits is certain to depend on socio-political factors, which are only tenuously related to geology.

It is pertinent to discuss the reason why the rewards for the production of different minerals vary. The most obvious is the scarcity. For example, gold and platinum command a high price because they are rare; iron is common and commands a much lower price. The total quantity of a metal or mineral that has been identified as exploitable in mineral deposits is known as the 'reserve' of that substance. It has been observed in the case of many substances, that if the known reserves at a certain time are R tonnes, then:

$$R = k\,C$$

where C is the so-called 'Clark' (or the average concentration of the substance in the earth's crust as a whole) and k is a factor that changes from 10^{12} to 10^{13} depending on the substance. What this seems to indicate is that in the earth's crust there is a certain amount of each substance, in the case of each there are concentrations in certain rocks which become exploitable by human civilization, and that there is a similar spectrum of concentration for each substance. The degree of concentration necessary for economic exploitability varies from one substance to another. The proportion of a valuable substance in a rock is known as its 'grade'. A rock which has a high enough grade to satisfy the appropriate economic criterion of exploitability, is known generally as 'ore'. The ratio between the average grade of ore worked for a given substance, and the Clark is called the 'concentration clark', a factor that varies from one substance to another, being about 10^1 for iron, and 10^3 for gold.

If we examine the amount of an economic mineral in a single mineral deposit we find that it varies considerably from place to place, and we must, therefore, define an 'ore deposit' as being that part of a chosen rock-mass which contains, on average, enough of the mineral to satisfy the economic criterion of exploitability. This quantity of exploitable ore is known as the 'ore reserve' of the deposit. In many types of mineral deposit we find that if we estimate the quantity (Q) of ore available at differing grades (g) then the following relationship holds:

$$Q = e^{a\,-\,bg}$$

where a and b are positive constants for any single mineral deposit, and almost constants for a single geological type of deposit. This phenomenon, known as Lasky's Law, has definite validity over well-defined volumes of rock, but is also valid when one considers the proportion of the earth's crust that contains a high enough quantity of a given substance in proportion to the Clark of that substance.

There is another factor that affects the exploitability of rocks for substances they contain. Two thousand years ago a person would have to have spent a lot of time and effort in generating enough disposable wealth to acquire a few kilos of iron in the form of a sword or cooking pot. Now a substantial portion of the world's population can, with less effort, acquire a motor car which contains upwards of a tonne of the same substance; the difference is a matter of technology. The effect of such technological developments is largely to reduce the grade of ore that can be exploited with sufficient reward. And while it is possible that there are limitations to these technological developments (evidence for which as emerged from the recent debates on the environmental impact of mineral extraction), so far there are no signs that technological limits are being reached in the case of most mineral commodities.

From all this discussion of the factors that control the exploitability of certain rocks for their mineral content, it should be evident that, to some extent, the world distribution of economic mineral deposits must be a function of human effort as well as the geology of the earth's crust. And there is yet another aspect of the distribution of deposits that warrants our attention. A mineral deposit is a body of rock of actual or potential value to humanity, but it is also a geological entity. To be either a subject of human economic or scientific interest, its existence must be known. Mineral deposits are not easy to find; many are found by chance, sometimes in quite unexpected places.

Most mineral deposits are discovered because someone with the requisite expertise recognizes it by some direct or indirect surface expression. Some deposits outcrop; others are disguised by a thin cover of superficial material, or are affected by weathering processes so that they have a leached and oxidized outcrop. Other deposits are found because they have anomalous responses to geophysical or geochemical tests, or because their presence can be inferred by the extrapolation from geological observations made at, or below, the surface. Very few deposits have been found that are buried deeper than a few metres, the only cases are rather special ones like the Orange Free State goldfield in South Africa which was found by a combination of geological deduction and geophysical responses, beneath hundreds of metres of cover. This was only possible because the deposits were of a type that was well known in other areas from outcrops and mine workings and reasonable extrapolations could be made.

So, of the factors controlling the discoverability of mineral deposits, there are the geological and geographical ones that control the likelihood that recognizable surface expressions will be present; and also human factors that influence the likelihood that someone will recognize these expressions.

Topography has some effect, because on steeper ground more rock is exposed, and being more rapidly eroded, superficial modifications form less easily. Perhaps the most important controls are the presence of superficial deposits. Some mineral deposits are found in accumulations of superficial deposits, but the majority occur in the bedrock beneath. Ice is one example that totally obscures most of Antarctica and Greenland, and a substantial area of the Northern Hemisphere for part of the year; water, particularly in the form of lakes, covers large areas in northern latitudes which were covered by ice during parts of the Quaternary period. This Ice Age also left behind considerable accumulations of glacial till (and related surface sediments) that not only obscure the bedrock, but, being transported products of erosion, make the detection of sub-surface mineralization by geochemical responses difficult. In both high and low latitude regions, thick accumulations of vegetable matter occur, such as peat and tropical forest debris. Not only do these obscure the bedrock, but they are electrically conducting and have a rather specialized chemistry, rendering the detection of physical and chemical responses almost impossible. Where dry climates exist (or existed in the recent geological past) the surface tends to be covered by wind-blown material; sand in hot deserts and loess in cold sub-arctic ones. Mineral deposits have few if any detectable expressions through such transported materials and, in hot desert regions, sand tends to contain saline moisture which is electrically conducting. Even rock outcrops in hot deserts may be obscured by a few millimetres of 'desert varnish' making all rocks look alike.

In tropical and subtropical areas rocks may be covered by one of a number of materials that are residues of the chemical processes of weathering. Materials such as calcrete and silcrete are examples, but perhaps the most bothersome is laterite. Although some mineral deposits form in laterite profiles (such as bauxite), in general the laterite-forming process changes the appearance of rock outcrops, and leaches them even of traces of minerals and substances that would indicate the presence of mineralization, leaving merely a hard irregular mass of iron and aluminium hydroxides.

Fortunately erosion continues all the time, and valleys and gullies are eroded into superficial materials; although the immediate products of erosion; eluvium, alluvium, etc. also obscure bedrock outcrops. Soil, also obscures bedrock, but fortunately, these are dynamic materials. The chemistry of soil frequently reflects that of the underlying bedrock, and there is a continual, but slow movement of material in soil by the action of organisms, including most notably, human beings with ploughs. Many mineral deposits have been found by the discovery of mineralized rock fragments in soil, either by direct observation, or by means of geochemical surveys.

The human factors that affect the discoverability of mineral deposits include motivation and technology. People will find deposits if they receive some reward for so doing, and if they have the knowledge and the tools with which to recognize their surface expressions. Population density and development of civilization have effects. In hostile climates few people live, and few deposits are discovered; the availability of the right knowledge and tools tends to be found in those cultures that are accustomed to using mineral raw materials on a large scale. The classification of cultures in history and pre-history has a nomenclature based on minerals; Stone Age, áceramic-ceramic, Bronze Age, Iron Age, etc. A great deal of the mineral wealth known today, and the majority of that which is indicated on the following maps, was discovered as a result of the spread of industrial civilization from Europe during the 'colonial' era. Many of the major deposits of the world were discovered by people who, while being part of that industrial civilization, were for some reason out on the ground. A very large number of deposits were discovered by shepherds, cowboys, fur-trappers, farmers, foresters and people with similar land-based occupations. Quite a few deposits were, and are, found during the building of railways and roads. These are built to improve transport and, in some cases, to open up the possibility of mining, and while being built, mineralized outcrops are often revealed in cuttings and tunnels. Regrettably, it is the most undesirable aspect of industrial civilization, high-technology warfare, that is the greatest stimulus to mineral discovery; and the distribution of mineral deposits sometimes has an effect on the choice of battle-ground.

One of the main features that emerges from the study of mineral deposit distribution is that their discovery so far keeps pace with demand. This is largely due to the fact that the technology of exploration and exploitation develops in response to demand. In that part of the world where price is the main control of exploitability, we find that the real prices of mineral raw materials have been either stable or declining relative to other things, for a long time. At the same time the *per capita* consumption increases each year. In the case of most mineral raw materials, the rate of new discoveries of exploitable ore are similar in magnitude to, or exceed, the rate of consumption. The periods during which this is not true are temporary, and are times when there are sudden increases in consumption. It has been concluded that if the whole population of the world used mineral raw materials at the rate per head that pertains in the most industrialized countries, then we would soon be finding real shortages of supply. This may seem to have some validity, but it is somewhat hypothetical because such a situation would depend on socio-political changes that are very slow to take place. The most important restriction on the supply of mineral raw materials is probably the availability of energy to turn the raw material into a usable product. To liberate a useful mineral from a tonne of rock-matrix, consumes on average 15 kWh of energy, irrespective of the amount recovered. There is some hope that this figure might be reduced by technological developments, but it remains a barrier to the development of lower and lower grade materials for their mineral content.

Two important conclusions may be drawn from this analysis. Firstly, there are many influences on the discovery of mineral deposits that derive from the activities of human civilization, and have little to do with geology; therefore the distribution of deposits that we see is in part man-made, and we must be cautious lest we see geological controls where there are in fact only human ones. For example, we often find mineral deposits in clusters, but before we call one a 'metallogenic province' we must note that exploration tends to be most intensive in areas around previous discoveries; that is to say, an apparently geological province may be man-made.

Secondly, changing technology and changes in the pattern of industrial uses of mineral raw materials may convert what was formerly regarded as waste rock into ore; and uneconomic resources into exploitable reserves. Therefore, metallogeny as a science is the study of a group of objects and phenomena that are constantly changing. Changes in metallogenic theory derive in part from increased knowledge and understanding of mineral deposits, but also as much from the discovery of new objects to study, in the form of new kinds of rock that become interesting as economic mineral deposits.

In spite of these difficulties it is possible to say something about the geological controls of mineral deposit distribution and, to begin with, an explanation is needed for the background against which the deposits are shown on the following maps. There are many ways of classifying the rocks we see at the surface of the earth. First of all one can divide the surface into the oceans, areas of relatively thin crust largely covered by water, and continents, where the crust is much thicker and is, at least for the most part, not covered by water. We have also concluded that the crust is divided into a series of 'plates' which are relatively rigid and with a low thermal flux, which meet along zones where heat flow is high, vulcanicity and earthquaks common. These latter are the 'mobile zones' of today.

Continental areas that are stable can often be divided vertically into a lower layer or 'basement' which is dense, compact, rigid and relatively impermeable to water. Overlying this is a 'cover' of sediments and volcanic products which are softer, less dense, much more permeable and, to a large extent, saturated with water. The older, more deeply eroded areas of basement that have no significant cover are known as 'shields'.

Unlike the neighbouring bodies in the solar system, the earth's surface constantly renews itself by a continuous process, because of the presence of an atmosphere and hydrosphere. What seems to happen is that the plates split along rift systems which become mobile zones along which thermal activities (such as magmatic, volcanic and metamorphic) take place, building up a thicker crust which becomes added to that formed in earlier times. Contemporaneously, other parts of the crust are eroded down, first to form the cover, and later to become deformed rock masses in the deep crust. Thus the structure of basement contains a record of past eras of mobility.

All mineral deposits can be classified into groups that form as part of the cover, or are formed in the zones of mobility, and are thus found either in present-day mobile zones, or preserved in the basement. The background of the following maps shows this broad classification of rocks. The mobile zones and former mobile zones, at least of the later (and better understood) eras of geological history, are distinguished from the cover. Knowledge and cartographic limitations do not allow the older basement areas to be subdivided, and these are simply shown as shield areas.

There are many ways of grouping mineral deposits in relation to the geological features of the crust, and three important aspects will be dealt with here. There are patterns of distribution that are related to the spacial distribution of major crustal features; others which are related to changes which have taken place during geological time, and a number of features which are difficult to explain in either of the two previous ways.

Mineral deposits occur in groups either as clusters, or arranged in more or less linear belts; and in many cases these can be explained because they correspond to areas where the various geological features that are associated with their formation also tend to be more common. Thus in mobile zones we find belts of intrusive or volcanic centres along crustal weaknesses, and with them the associated deposits. For example, the so-called 'Copper Belt' of Central Africa corresponds to the ancient zone of tectonic activity which controlled erosion and the development of shorelines. Provinces of bauxite deposits correspond to areas of flat topography indicating crustal stability. Such areas can properly be called 'metallogenic provinces', albeit of a straightforward kind, being clusters or groups of deposits with a common origin associated with geological features that are fairly easy to understand.

Examination of the ages of mineralization shows that there were periods in geological history when over wide areas, and including those that can be ascribed to several different metallogenic provinces, deposits were formed during a relatively narrow range of time. These so-called 'metallogenic epochs' vary both in their duration, and area of the crust affected. There is some risk of false interpretation here because of the variable precision of geochronological measurement; the uncertainty of which increases with age. We may be tempted to lump together a series of metallogenic events of great age as an epoch, which we would not do if they were younger and the ages of individual deposits more precisely measurable. Despite these reservations, there are examples of age groupings that warrant our attention.

We must regard the concentration of the large taconite and itabirite basins and their associated iron deposits, in a narrow range of age around 2 Ga, as being of some significance. The grouping of iron deposits in rocks of this age is primarily due to the preponderance of a certain kind of iron-rich chemical sediment, and some attempts have been made to explain this phenomenon in terms of a major change in the geochemistry of iron that took place in response to a change in the oxidizing power of the atmosphere and hydrosphere brought about by the evolution of plant life.

All over Europe, the Permian (or a period, at least, extending from the late Carboniferous into the early Triassic) is a metallogenic epoch. Provinces of hydrothermal deposits, and other late magmatic types are found in abundance, associated with intrusions of about this age. These are deposits which had their origin during the period of mobility that formed the basement that lies beneath a large part of the cover rocks of western and central Europe. At the same time, several kinds of deposits were forming in the cover itself; including the late Carboniferous coals, stratiform metal-rich shales, many provinces of lead–zinc–fluorite deposits in carbonate sequences, and substantial accumulations of salt. (Some of these deposits cannot be classed as metalliferous, but this does not invalidate the conclusion.) If one compares the mineralization of this epoch with, say, the Devonian or Cretaceous, the contrast is extreme; but these two ages were very important, metallogenically, in other parts of the world.

Very often metallogenic epochs are interpretable in much the same way as metallogenic provinces because of the widespread action, at a particular time, of a particular kind of geological activity. For instance, zones of mobility at the edges of continental plates (at least in recent geological history) are active in various parts of the globe at the same time.

There is some evidence that the types of mineral deposit that form, change with geological time. The case of the large basins of iron deposits has already been mentioned, which could be regarded as a major global metallogenic epoch during which deposits were formed of a type poorly represented in younger rocks. And there are other sorts of change. If we examine gold deposits (and deposits in which gold is found in significant amounts) throughout geological time, we first find them as vein-deposits with certain particularities which seem to have formed during the deformation of the Archean greenstone belts; they are widespread, and despite some difference in detail, have considerable similarities and probably range in age from over 3 Ga to just over 2 Ga. In a few parts of the world, notably in the Capvaal craton of South Africa, the era of greenstone belt formation seems to terminate with the production of large sedimentary basins (that is to say, the first true cover rocks). These sediments contain quite a different kind of deposit: gold in quartz-pebble conglomerates and associated carbon seams. Gold deposits in the Proterozoic are not particularly common, although several groups of polymetallic sulphide deposits of the volcanic-exhalative type contain gold in fair quantities; for example, those of the Skellefteå district in Sweden. Gold in quartz veins, somewhat different from the Archean ones, occurs in deformed terrigenous clastic sediments of Palaeozoic age, and subvolcanic vein deposits and gold-bearing porphyry–copper deposits, appear in the late Mesozoic and Tertiary. In Tertiary and Quaternary rocks, gold is found in aluvial deposits. Thus, if one traces the geological history of deposits, of a selected commodity, one seems to see an 'evolution' from one type to another.

Some part of this type of change can be attributed to erosion, because as times goes on there is a tendency for higher-level rocks to be removed and become less commonly preserved, and at the same time more of the deeper crustal levels to be exposed. But this cannot account for all such changes, and we are left with the conclusion that certain mineralizing processes are either confined to certain periods in the history of the crust, or change their nature with time. Several kinds of mineral deposit do not seem to have formed at all before and others not after, a certain critical time. The characteristic greenstone belt vein-gold deposits are hardly found at all in rocks younger than 2 Ga; copper sulphide deposits in clastic sediments are not found earlier than 1 Ga; neither are lead–zinc deposits in carbonate rocks. The latter feature is relatively easy to explain because such deposits first appear at the same time as their usual host rocks, corresponding to the evolution of calcium carbonate-secreting organisms on a large scale.

Many possible groupings of mineral deposits into provinces and into epochs have no easy explanation, and it poses problems of interpretation why the occurrence of a particular set of geological features is mineralized in one place, and not in another. Sometimes the answer is clearly lack of exploration, or more difficult exploration conditions. Neither is it easy to explain why a metallogenic epoch occurred in one part of the crust, and not in another sensibly similar in geology at the time. There are a number of examples of metallogenic provinces or belts in which deposits are found of a wide variety of types, formed over a wide range of age and containing several different assemblages of minerals. Some of these are belts with an orientation that is quite different from that of the prevailing structure of the rocks that can be seen at surface. In some cases these belts or provinces have a position, or orientation, that corresponds to the deep structure of the crust which may be observed indirectly from the evidence of regional geophysical interpretations, particularly regional magnetic and gravity anomalies. Boundaries between major geochronological provinces may also be correlated with such belts. The best-known examples of this latter phenomenon are to be found in North America, where despite the general north–south orientation of the major tectonic provinces such as the Appalachian Belt, the West Coast belts, and Rocky Mountains, there seem to be a predominance of metallogenic belts oriented in a north-easterly direction, parallel in fact to major tectonic features of the Canadian Shield (such as the Grenville Front).

Explanations for these types of phenomena have been proposed which involve features of the deep crust of the underlying mantle. One such explanation is the so-called Principal of Inheritance which supposes that during the formation of continental crust, discontinuities are built in that remain influential for a very long time. This means that after the stabilization of a region of the crust, certain zones go on being active for a long time controlling magmatic activity, sedimentation and a variety of other processes, and so as each age passes, the cover rocks and other near-surface phenomena inherit the tendency to become the locus of mineralization.

It has also been proposed that there are differences in the mantle from place to place that control the distribution of mineral deposits in the overlying crust. This has been particularly popular in the explanation of zones of diamondiferous kimberlite pipes and carbonatites. Variations in the heat-flow from the mantle have also been evoked as an explanation of metallogenic provinces and epochs, on the assumption that it takes energy to facilitate many mineralizing processes. It has been suggested, for example, that areas of high heat-flow in the mantle remain static for long periods, but the crustal plates are continually moving in relation to them. This could provide an explanation for mineralized belts, although only in a few such belts can one actually see a progressive change in age of deposit along the length.

Some, but not all, of the features of the distribution of deposits can be explained in terms of observable geology at the earth's surface. It is often the detailed features, the type, morphology, mineralogy, etc. of the deposits that are controlled by the geological processes that we conclude, from observation, to have taken place. The reason why deposits of a particular, substance are where they are many require explanations that are both more fundamental and much less easy to study.

On the succeeding pages are shown the major deposits of certain important mineral commodities. They are shown, classified into broad groups according to type and geological environment and, to some extent, age. In most cases, examples of each type have been previously described in the first four sections of the book, and the names of the described examples are underlined.

Copper Deposits of the World

Copper was the first metal to be used by mankind, and was essential in the development of the Bronze Age. When electric power was harnessed in the last century, it became even more important. Deposits of copper minerals are concentrated in the mountainous belts that surround the Pacific Ocean, although quite a number are found in Central Africa, Asia and the Canadian Shield.

Geochemistry and Mineralogy

Copper is a rare metal, the rocks of the earth's crust containing about 47 μg/g on average; but to be worked as an ore a rock needs to contain over 100 times this quantity. Copper is classified as a chalcophile element; that is to say it has a strong affinity for sulphur in its reduced state, commonly forming sulphide minerals. Much of the world's production comes in the form of the mineral chalcopyrite, although other sulphides, chalcocite, bornite, enargite and tetrahedrite-tennantite are important. In the zone of weathering it is often found as one or more of a number of oxides, carbonates and silicates, among which are cuprite, tenorite, malachite, azurite, chrysocolla; or simply adsorbed on to the lattice of clay-like siliates such as vermiculite.

Being a chalcophile element, copper tends to accumulate where sulphur exists in its reduced state, that is, in igneous and volcanic environments and in those sediments where reducing conditions are generated.

The 'Porphyry Coppers'

Although not always used strictly as a geological term, there is a group of copper deposits the earlier worked examples of which are found in porphyritic igneous rocks – hence the name. Their main characteristic is large size, relatively low grade, with copper sulphides occurring in veinlets and disseminations in a zone of alteration derived from an igneous rock, frequently granodiorite. They are often found in younger fold belts along the edges of the major continents, and frequently the parent intrusive cuts sequences of andesitic volcanics. Some cut sediments and, where these are carbonate rocks, disseminations of sulphide minerals may be found in the wallrocks in the zone of contact metasomatism.

Many of the best known 'porphyry coppers' contain commercially important quantities of molybdenite, as well as copper sulphides, while others contain gold and silver. Tungsten minerals occur in some, recoverable in a few and others contain the minerals enargite, tetrahedrite and tennantite.

Quite a number of examples have been modified by weathering near the surface to produce 'oxide ore' in which occur oxides, carbonates and silicates. Rich chalcocite ore, apparently derived by the reprecipitation of copper derived from the weathered zone, is important in many cases.

The porphyry coppers, which include the largest known deposits, make up 52% of the world inventory of the metal and account for a large proportion of the world's resources of molybdenum. The examples for which data is available contain 191 Mt of copper metal, of which 140 Mt remain as reserves. The smaller porphyry coppers contain about 2 Mt and the largest around 35 Mt metal with grades from 0·5% to 1·5%.

The 'Volcanic Group'

Although the degree of association varies, many deposits containing copper are associated with volcanic rocks. Massive pyrite deposits containing copper sulphides are sometimes found in association with sequences of basaltic lavas, while others are in intermediate or acid volcanic rocks. Many of these deposits are situated on the boundary between the volcanic series and sediments, some are wholly contained in volcanic or volcanoclastic rocks, while others are found in sediments, containing a substantial amount of volcanic material. Some of the best-known examples, such as those of Pre-Cambrian age in Canada and Scandinavia, contain zinc as well as copper; others contain useful amounts of the precious metals.

Many massive pyrite deposits have zones of disseminated copper sulphide ore associated with them and there are a number of examples of disseminated copper deposits in volcanic rocks that have no massive ore with them, although pyrite is usually present.

About 13% of the known inventory of copper is contained in this type of deposit amounting to over 50 Mt of metal, although probably only half of this exists today as reserves. Individual deposits vary from quite small ones to those containing 3·7 Mt of copper with grades averaging 1·5% Cu.

Copper–Nickel Deposits

Although better known as nickel deposits, appreciable quantities of copper come from massive or disseminated ores containing pyrrhotite and chalcopyrite associated with basic or ultrabasic rocks. Sudbury in Canada and Noril'sk in Russia are of spectacular size, but others are important.

Sediment Group

Very locally, sediments such as black shales contain copper sulphides. The best-known examples of this group in Central Africa where copper occurs in carbonaceous and dolomitic shales as well as a variety of coarser-grained rocks. Not infrequently these deposits show a zonal arrangement of minerals around an original island or topographic high present when the sediments were deposited. Like the porphyry coppers, weathering has formed oxide and enriched ore where the sediments outcrop. Many of the best-known examples have undergone deformation and metamorphism, some even to the extent that one may dispute their inclusion in this group at all. Cobalt is found in some of these deposits, and most of the world's supply of this metal comes from this type.

Of the known world inventory, 105 Mt of copper belongs to this group, 27% of the total and of this perhaps 80 Mt remains as unexploited reserves. Due in part to the remote location of the largest examples in Central Africa, these deposits are worked at higher grades averaging around 2·3% Cu.

Inventory, Reserves and Production

Available statistics show a world inventory of 405 Mt of metal of which about 188 Mt has been exploited, leaving 217 Mt reserves. The current rate of production is approaching 5 Mt/a but new discoveries of ore are at present keeping pace with this rate of exploitation.

Mining and Processing

Some of the largest mining operations in the world today are copper mines, particularly the large conical open-pits working the porphyry coppers and deposits of the sediment group. Daily productions of 40 000 t are relatively common. The whole range of underground mining techniques are used for steeper and deeper deposits.

In the case of copper sulphide-bearing deposits, almost invariably the minerals are recovered by crushing, grinding and flotation. A few sulphides and most oxide ores are treated by acid leaching, often by sulphuric acid produced locally from smelter gases; others are allowed to weather so that sulphide minerals oxidize rendering the copper soluble, in which form it can be recovered by precipitation with scrap iron. A recently developed process known as TORCO is now coming into use, in which salt and a fuel are added to silicate-bearing ore that forms sodium silicate and volatile copper chloride, which is then reduced to the metal and hydrogen chloride.

Concentrates of copper minerals are normally smelted in a furnace to a slag and a substance known as 'matte' with the composition Cu_2S. This is then converted to copper by blowing air through the melt. The process is made reasonably efficient by heat derived from the burning of sulphur, some of which is usually recovered. The crude copper is refined by further furnace treatment, or electrolytically in acid solution. Much copper is cast into special long ingots known as wire bars, that can be rolled and drawn into wire for electric cable manufacture, one of the principal uses of the metal.

Further Reading

PELISSONNIER, H. (1972), *Les Dimensions des Gisements de Cuivre du Monde*. Memoires du BRGM, No.57.

McMAHON, A. D. (1965), *Copper, A Materials Survey*. U.S. Bur. Mines, IC 8225.

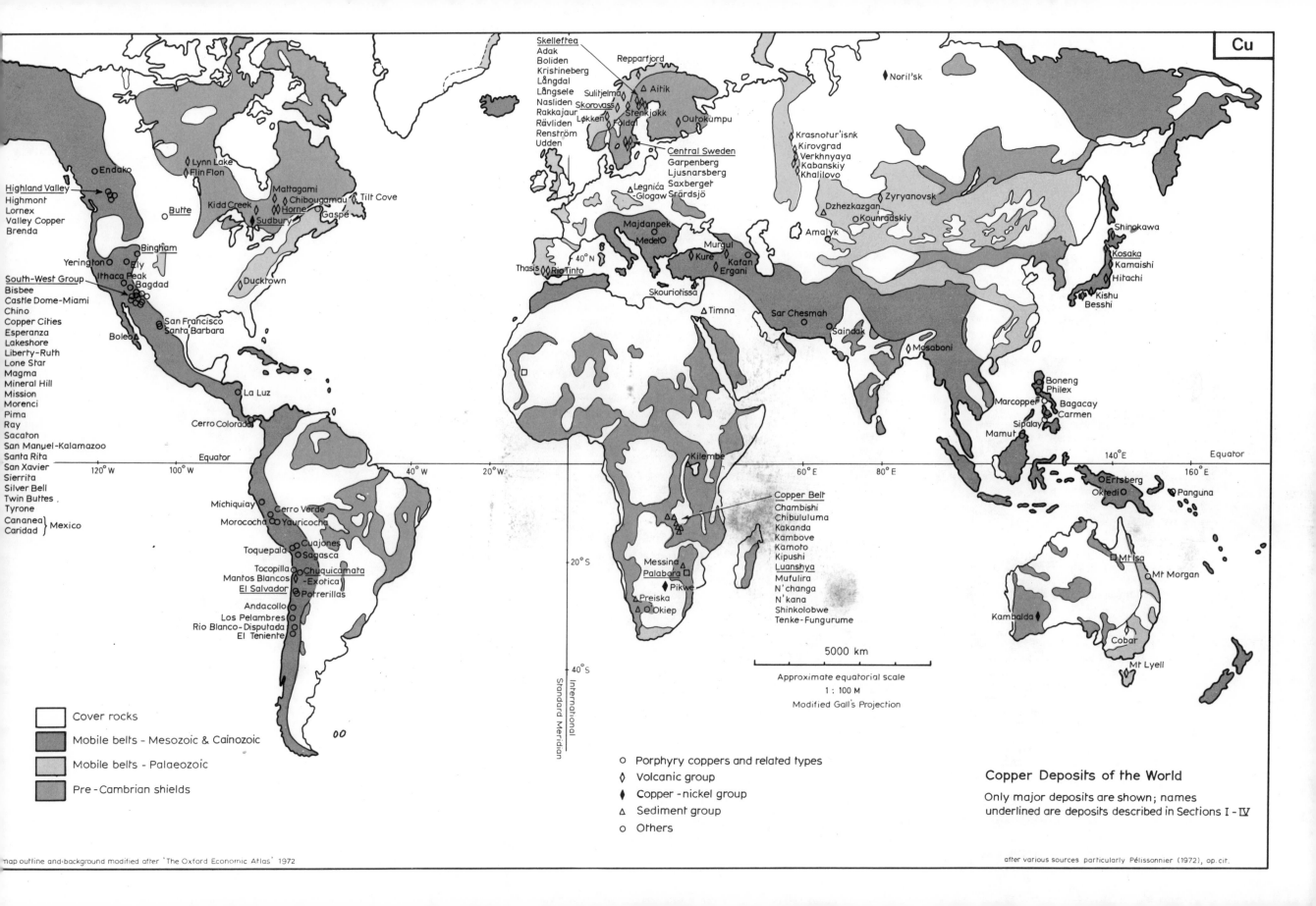

Cu

Skellefteå
Adak
Boliden
Kristineberg
Långdal
Långsele
Nasliden
Rakkajaur
Rävliden
Renström
Udden

Repparfjord

Aitik

Noril'sk

Sulitjelma
Skorovass
Løkken
Foldal
Stenkjokk
Outokumpu

Krasnotur'isnk
Kirovgrad
Verkhnyaya
Kabanskiy
Khalilovo

Central Sweden
Garpenberg
Ljusnarsberg
Saxberget
Srärdsjö

Lynn Lake
Flin Flon

Endako

Highland Valley
Highmont
Lornex
Valley Copper
Brenda

Butte

Kidd Creek

Mattagami
Chibougamau
Horne
Gaspé

Tilt Cove

Sudbury

Zyryanovsk

Dzhezkazgan
Kounradskiy

Shinokawa

Bingham

Yerington
Ithaca Peak
Ely
Bagdad

Legnića
-Glogaw

Majdanpek
Medet

Murgul
Kure
Kafan
Ergani

Amalyk

Kosaka
Kamaishi
Hitachi
Kishu
Besshi

40°N

Thasis
Rio Tinto

South-West Group
Bisbee
Castle Dome-Miami
Chino
Copper Cities
Esperanza
Lakeshore
Liberty-Ruth
Lone Star
Magma
Mineral Hill
Mission
Morenci
Pima
Ray
Sacaton
San Manuel-Kalamazoo
Santa Rita
San Xavier
Sierrita
Silver Bell
Twin Buttes
Tyrone
Cananea
Caridad } Mexico

Ducktown

San Francisco
Santa Barbara

Boleo

Skouriotissa

Timna

Sar Chesmah

Saindak

Mosaboni

Boneng
Philex
Marcopper
Bagacay
Carmen
Sipalay
Mamut

La Luz

Cerro Colorado

Equator

120°W   100°W   40°W   20°W   60°E   80°E

Ertsberg
Oktedi
Panguna

Michiquiay
Morococha
Yauricocha
Cerro Verde

Kilembe

Copper Belt
Chambishi
Chibululuma
Kakanda
Kambove
Kamoto
Kipushi
Luanshya
Mufulira
N'changa
N'kana
Shinkolobwe
Tenke-Fungurume

140°E   Equator   160°E

Cuajones
Sagasca
Toquepala
Tocopilla
Mantos Blancos
Chuquicamata
-Exotica
El Salvador
Potrerillas

20°S

Messina
Palabora
Pikwe

Mt Isa
Mt Morgan

Andacollo
Los Pelambres
Rio Blanco-Disputada
El Teniente

Preiska
Okiep

Kambalda

40°S

5000 km

Approximate equatorial scale

1 : 100 M

Modified Gall's Projection

Cobar

Mt Lyell

International Standard Meridian

Cover rocks

Mobile belts - Mesozoic & Cainozoic

Mobile belts - Palaeozoic

Pre-Cambrian shields

○ Porphyry coppers and related types

◇ Volcanic group

◆ Copper-nickel group

△ Sediment group

○ Others

Copper Deposits of the World

Only major deposits are shown; names
underlined are deposits described in Sections I - IV

map outline and background modified after 'The Oxford Economic Atlas' 1972

after various sources particularly Pélissonnier (1972), op.cit.

Lead and Zinc Deposits of the World

Lead and zinc usually occur together, and a number of other elements and minerals are commonly associated with them. Silver and cadmium are both found, as minor but often economically important by-products in lead–zinc ores and fluorite and barite often occur as important constituents.

Geochemistry and Mineralogy

In the crust as a whole, lead and zinc are rare, the Clarke for lead being about 16 μg/g and for zinc 80 μg/g. Both metals readily form sulphide minerals, galena and sphalerite being the most common lead and zinc ore minerals. Galena commonly contains silver in amounts varying between 10 mg/g and 100 mg/g, exceptionally as high as 1%. Silver-rich galena (or argentiferous galena) is normally an unmixed intergrowth of galena and silver sulphide minerals, and these are usually accompanied by some arsenic or antimony. Cadmium is found in sphalerite up to 2%; there being a cadmium isomorph of the mineral, greenockite. Several other elements substitute for zinc in sphalerite, and it is unusual to find the mineral with its theoretical content of the metal. Both galena and sphalerite can contain small amounts of tellurium and selenium in place of sulphur.

Both metals can be transported in chloride solutions even at low temperatures, and it is probable that some deposits occur where such solutions met with sulphide. The two metals can equally well be transported as polysulphide complexes in solutions, from which the two sulphide minerals may be precipitated by cooling.

In the zone of weathering, zinc is normally very soluble, but is sometimes preserved as the carbonate, smithsonite or the silicate, hemimorphite; a mixture of these and other minerals derived from sphalerite by oxidation is known as 'calamine ore'. Lead too may be preserved as the carbonate, cerussite or the sulphate, anglesite and in some localities as complex minerals containing vanadium, phosphorus and arsenic.

Stratiform Deposits

Several large stratiform deposits of lead and zinc occur, all of which differ from one another and can only loosely be regarded as a type. The world's largest worked deposit at Broken Hill in Australia is a highly deformed series of conformable lenses, in a series of high-grade metamorphic rocks that include amphibolites, sillimanite-garnet gneisses and meta-taconites. The metamorphism and deformation make the understanding of this deposit very difficult, but most of the evidence suggests that it was originally a stratiform deposit in a sequence of sediments. Somewhat less modified is the Mt. Isa deposit in Queensland that, in addition to zones of stratiform, bedded lead–zinc sulphide mineralization, has a large zone of silicious dolomite rich in copper. Even larger than Broken Hill is the MacArthur River deposit in north-western Australia, again stratiform in a series of tuffaceous dolomite shales, but cannot at present be exploited because of its fine grain size and mineralogical complexity. Sullivan in British Columbia is another large lead–zinc deposit rich in silver and copper. It too is stratiform (with modifications) but is surrounded by alteration phenomena of the wallrocks which suggests a magmatic parentage. It occurs in a somewhat unusual environment; an immensely thick sequence of shallow-water clastic sediments. A few other examples of this type occur, but not many.

Strata-bound and Related Deposits in Sediments

An ill-defined group of deposits that are for the most part found in shelf facies carbonate sediments, and are widely distributed in the platform areas of the world. Some of these are bordering on stratiform, but most are irregular zones of zinc and lead sulphide mineralization confined to specific beds or rock units. The most important area is the platform region in the centre of North America, including the great districts of Pine Point, Tri-State and East Missouri. The other important area is western Europe and North Africa, where several districts occur in Ireland, France, Spain and Morocco. Almost all of these deposits occur in rocks of uppermost Pre-Cambrian to late Mesozoic age and in warm climate carbonate rocks often associated with evaporites. Some are associated with volcanic provinces and most show a strong relationship to structures in cover rocks related to basement tectonic movements. However, the morphology, size, mineralogy and detailed relationship to the host rocks is very variable. Both barite and fluorite occur in some of these deposits, and major deposits of these two minerals, usually with a little zinc and lead, occur in similar environments.

Deposits in Volcanic Environments

Quite a number of deposits of lead and zinc, usually with copper and iron sulphides, occur in volcanic sequences, and there is a complete range between these and the copper-zinc or copper-pyrite type. Probably the best-known group of these are the Kuroko deposits in the Tertiary submarine, island-arc volcanic province of Japan. These deposits seem to be characteristic of the outer zones of the island-arc environment.

Hydrothermal Vein and Replacement Deposits

Quite large quantities of lead–zinc ore are found in veins or replacements around igneous stocks in the younger Cordilleras. Several porphyry copper deposits have outer zones of lead–zinc mineralization. Included in this class are the pipe and manto deposits best known in Mexico, and a number of contact replacement deposits.

Resources and Reserves

Identified reserves of lead and zinc exceed 300 Mt of metal. The handful of large stratiform deposits account for most of this, followed by the larger deposits in carbonate sediments and some of the pipe and manto deposits.

Mining and Processing

Lead–zinc deposits are frequently mined underground at grades of anything from 5 to over 20% combined metals, although normally below 8% open-pit methods are more common. Some open-pit mines work ores down to under 4%. Almost universally galena sphalerite ores are concentrated by selective flotation to produce separate lead and zinc concentrates. In the case of lead, the concentrate is smelted direct to the metal. One complex smelting process produces lead and zinc from a single plant. In the case of zinc, smelting of roasted concentrates (the metal being recovered from condensers) is practised but, increasingly, the roasted concentrate is dissolved in sulphuric acid (generally made on site from the roaster gases) and the metal recovered by electrolysis of the purified solution.

Further Reading

DUNHAM, K. C., ed. (1950), *The Geology, Paragenesis and Reserves of the Ores of Lead and Zinc.* 18th Int. Geol. Cong. Part VII, 400pp. [Although somewhat old, this symposium volume contains the best collection of papers on the subject extant.]

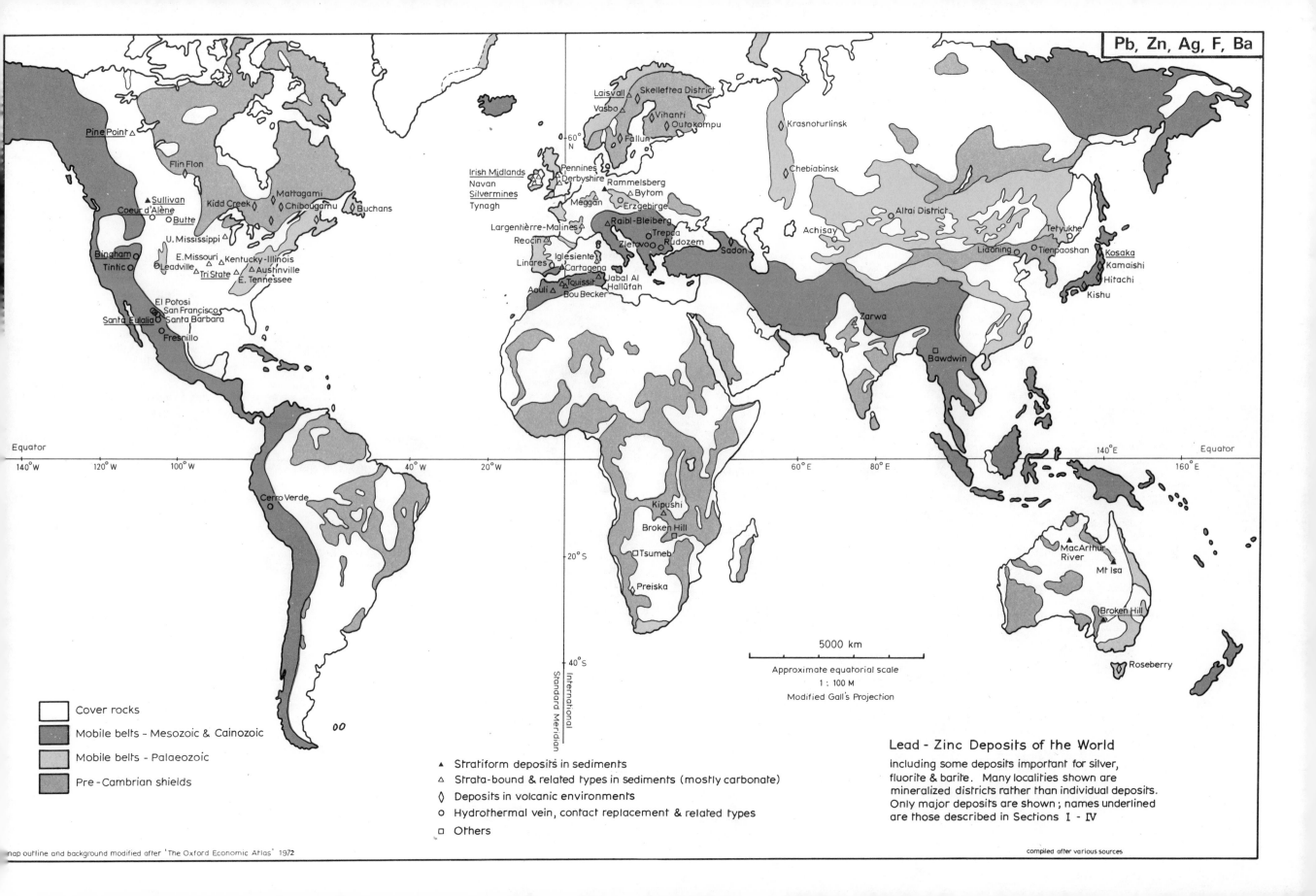

Pb, Zn, Ag, F, Ba

Labels on map:

Pine Point
Flin Flon
Sullivan
Coeur d'Alène
Butte
Kidd Creek
Mattagami
Chibougamu
Buchans
U. Mississippi
Bingham
Tintic
E. Missouri
Leadville
Kentucky-Illinois
Austinville
Tri State
E. Tennessee
El Potosi
San Francisco
Santa Eulalia
Santa Barbara
Fresnillo
Cerro Verde

Laisvall
Skellefrea District
Vasbo
Vihanti
Outokompu
Fallun
Krasnoturlinsk
Irish Midlands
Pennines
Derbyshire
Navan
Rammelsberg
Chebiabinsk
Silvermines
Bytom
Tynagh
Meggan
Erzgebirge
Largentière-Malines
Raibl-Bleiberg
Altai District
Reocin
Trepca
Zletovo
Rudozem
Tetyukhe
Linares
Iglesiente
Sadon
Achisay
Liaoning
Tienpaoshan
Cartagena
Jabal Al
Aouli
Hallûfah
Kosaka
Touissit
Kamaishi
Bou Becker
Hitachi
Zarwa
Kishu
Bawdwin

Kipushi
Broken Hill
Tsumeb
Preiska
MacArthur River
Mt Isa
Broken Hill
Roseberry

60° N

Equator
140° W   120° W   100° W   40° W   20° W   00   20° S   40° S
60° E   80° E   140° E   160° E   Equator

International Standard Meridian

5000 km

Approximate equatorial scale

1 : 100 M

Modified Gall's Projection

Legend:

- Cover rocks
- Mobile belts - Mesozoic & Cainozoic
- Mobile belts - Palaeozoic
- Pre-Cambrian shields

▲ Stratiform deposits in sediments
△ Strata-bound & related types in sediments (mostly carbonate)
◊ Deposits in volcanic environments
○ Hydrothermal vein, contact replacement & related types
□ Others

Lead - Zinc Deposits of the World

including some deposits important for silver, fluorite & barite. Many localities shown are mineralized districts rather than individual deposits. Only major deposits are shown; names underlined are those described in Sections I - IV

map outline and background modified after 'The Oxford Economic Atlas' 1972

compiled after various sources

Iron and Ferro-alloy Metal Deposits of the World

Iron is the predominant structural metal used in engineering, and modern industrial society is still very much part of the 'Iron Age'. Although an abundant element in the earth's crust, the majority of the iron produced comes from deposits in iron-enriched sediments, and largely from certain basins of sedimentation that formed about two thousand million years ago. Chromium, nickel and manganese are often used to produce alloy steels, and the major deposits of these are included.

Geochemistry and Mineralogy

The rocks of the earth's crust contain about 5% iron and a workable iron ore may contain anything from 25 to 70%. A particularly important feature of the metal is that it exists in two valency states. In its divalent or ferrous state, it is soluble in natural waters, though readily precipitated as silicates, carbonates and sulphides. In its oxidized or trivalent (ferric) state it is only soluble in very acid waters, and otherwise remains as oxide or hydroxide minerals. Because of the pervasive effect of the atmosphere, exceptional conditions are necessary for iron to remain long in the ferrous state near the earth's surface. However, deep in the earth's crust, iron often exists in a form equivalent to the ferrous and, as a rule, the ferric/ferrous ratio of magmatic rocks increases towards the earth's surface.

The principal ore minerals of iron are the oxides hematite and magnetite, but a variety of other minerals can be important; such as the hydrated oxide goethite, the carbonate siderite, and the silicates greenalite and chamosite.

Stratiform Iron Deposits

Most of the world's deposits fall into this category and they have been classified into three types; the Algoma, Superior and Minette. These three types of deposit have one thing in common, they are iron-rich facies of, or concentrations of iron minerals from, a sequence of sediments that is already iron-rich.

The Algoma Type. These are found in sequences of cherts that carry substantial amounts of iron oxides, carbonates and sulphides. The iron deposits themselves are either concentrations of siderite, banded cherts containing hematite or magnetite, or hematite/goethite ores formed by weathering from the cherty sediments.

The best examples of the type are found in the Archean greenstone belts associated with intermediate and acid volcanic rocks. The form and situation of these deposits suggests that they are subaqueous volcanic exhalative in origin.

The Superior Type. The vast majority of the world's iron resources falls into this category. These deposits, named after those found in the Lake Superior region of America, are found in sequences of iron-rich sediments or meta-sediments that form extensive basins associated with other sediments. The sequences are generally devoid, or contain only minor amounts, of volcanic rocks or intrusions of similar age to the sediments. The iron-rich sediments may be classified into two kinds. Taconites are granular massive, or laminated, cherts containing iron in the form of oxides, carbonates and silicates. There are usually gradual facies changes between the different types. The other kind, perhaps best called 'itabirites' (after the type locality in Brazil's Quadrilátero Ferrifero), are essentially quartz-rich rocks containing iron oxides with some carbonates and iron silicates. It is generally held that itabirites are metamorphosed taconites.

The iron ore bodies occur in rocks of either type and are of several kinds and origins. Many are concentrations of iron-oxide minerals formed by residual weathering that involves the removal of silica in solution. However, many variations occur because the weathering may take place at various stages in geological history and not necessarily just once. Some ore bodies may be original concentrations in the sediment or formed by contemporaneous desilication, others are zones that are workable because metamorphism has changed the iron minerals to magnetite or coarse-grained hematite.

A number of very large basins of this type of sediment occur in Protoerozoic rocks generally around 2 Ga in age, but smaller examples occur in both younger and older rocks.

The Minette Type. The principal characteristic of the type is the predominance of oolitic textures. These are generally associated with nearitic or shelf-facies sediments in which carbonates are usually an important element. The typical rock types are siderite, chamosite or hematite-rich sediments containing quartz, and other minerals of detrital origin.

The ore bodies themselves are either facies that were originally much richer in the iron minerals, or are concentrations derived from them by weathering. The typical ores are chamosite-siderite oolites, hematite oolites, pure siderite ores or soft brown goethite-rich ores. In examples that have not suffered subsequent weathering, it is apparent that the iron minerals were formed during the diagenesis of the sediments; probably from detrital iron minerals. Examples are found of all ages, but the largest and best known are in the Lower Palaeozoic of eastern North America, and Jurassic and Cretaceous of western Europe.

Other Deposits

Many other types of iron deposit are found, and although they may be locally important, they do not compare, on a world scale, with the Superior type. Most of the other types are closely associated with volcanic or magmatic rocks. A type of some importance is that found in north and central Sweden; the Kiruna type. These are found in a volcanic environment and are magnetite/apatite ores. Both hematite and magnetite ores occur in zones of contact metasomatism adjacent to intrusives, and concentrations of magnetite can occur in ultrabasic rocks. Large bodies of titanium-rich magnetite occur in layered basic intrusions and anorthosites; but are not always workable ores, being too rich in titanium to be economic as iron ore, or too poor in titanium for its economic recovery.

Inventory, Reserves and Production

It is very difficult to give reliable estimates for the world reserves of iron because the classification of an iron-rich rock as an iron ore is so dependent on a complex interplay of economic, technological, environmental and other factors. Identified reserves of iron ore are usually sufficient to satisfy current demand rates for several decades ahead, and any stimulation to demand usually results in the identification of new reserves in excess of the need. Identified reserves of the metal certainly exceed 100 Gt. Present production is about 480 Mt/a of ore and concentrates and the productive capacity of mines over the world is well in excess of this.

Mining and Processing

Iron-ore mines vary in size from those producing a million or so tonnes, to ones that produce tens of millions of tonnes per year. Most are open-cast, but a few are underground. Generally, however, the value of the product does not warrant the production by open-pit mining beneath thick overburden, or underground mining of narrow complex shaped ore bodies. Some deposits can sustain production of ore which can be shipped to the smelters after only simple crushing, screening, and perhaps washing. These 'direct shipping' ores generally come from the residually concentrated zones of iron deposits or from the large ore bodies of magmatic affinity. Lower grade deposits are used as feed to a concentrator that by various methods (including magnetic, gravity and electrostatic methods) produces iron-ore concentrates. An increasing tendency is to form these into 'pellets' for shipping to the smelter.

Most iron is still smelted in a blast furnace, a device which requires coke as a fuel and reducing agent, and limestone as a flux. But some of the ores, particularly of the minettes and Algoma types, contain some calcium carbonate and are 'self-fluxing'. Other processes for producing iron are used, including solid or semi-solid state reduction in rotary kilns. In all cases, the product is iron which is either refined and used as such; or more usually, made into steel by further furnace treatment, and the addition of carbon, manganese and other metals such as nickel, chromium, molybdenum or cobalt. In most steel-making processes, certain elements such as sulphur or phosphorus are troublesome impurities, and ores or concentrates low in these elements are of particular value.

Nickel

There are two types of nickel deposit, sulphide deposits and lateritic deposits. The nickel–iron sulphide deposits, usually containing some copper, occur as massive bodies or disseminations in basic and ultrabasic magmatic rocks. These are widely regarded as being formed by the segregation of a sulphide melt from magma at an early stage in crystallization. They are of very restricted distribution, the bulk of the world's reserves occurring in just two magmatic complexes at Sudbury in the Canadian state of Ontario, and Noril'sk in north-west Siberia. The second and more widely distributed type of deposits are the nickel–iron laterites found as weathering residues over bodies of ultrabasic rock. The parent rock is usually relatively rich in nickel, but the overlying weathered mantle is enriched. The nickel occurs in two forms; as accumulations of nickel silicates and carbonates, and hydrated iron oxide with absorbed nickel.

The metallurgy of nickel is very complicated, and both pure nickel and ferro-nickel are produced as raw materials for the production of alloy steels, principally stainless steel.

Chromium

Almost all chromium is obtained from chromite, an oxide mineral that segregates early during the crystallization of basic and ultrabasic magmatic rocks. One can, however, distinguish between deposits that are found as layers in large basic intrusive complexes such as the Bushveld in South Africa, and the less regular segregations that are found in serpentinites.

Chromite is smelted to produce ferro-chrome as raw material for the making of special steels. However, chromite has two other uses. Very pure chromite is used to produce chromium compounds for use in the chemical industry, and somewhat impure but mechanically hard chromite is used as a component of certain refractories.

Manganese

Manganese is an essential metal in the making of steel. The chemistry of the metal is similar to iron but with slight differences in chemical properties, and it is deposited in sediments separately from iron. Most manganese is mined as oxide ore from sedimentary deposits or residually concentrated derivations from them, often associated with iron-rich sediments. Manganese, as well as being used as an alloy metal in steel, is widely used (in its oxide form) as an oxidizing agent in steel-making in the chemical industry, and in the manufacture of dry-cell batteries.

Further Reading

BLONDEL, F. and MARVIER, L., Eds (1952), *Symposium sur les Gisements de Fer du Monde.* 19th Int. Geol. Cong. 1952 Algiers. 2 vols. 638pp and 594pp. [Although old, this is the best collection of papers on the subject covering the whole world, and all types of deposit.]

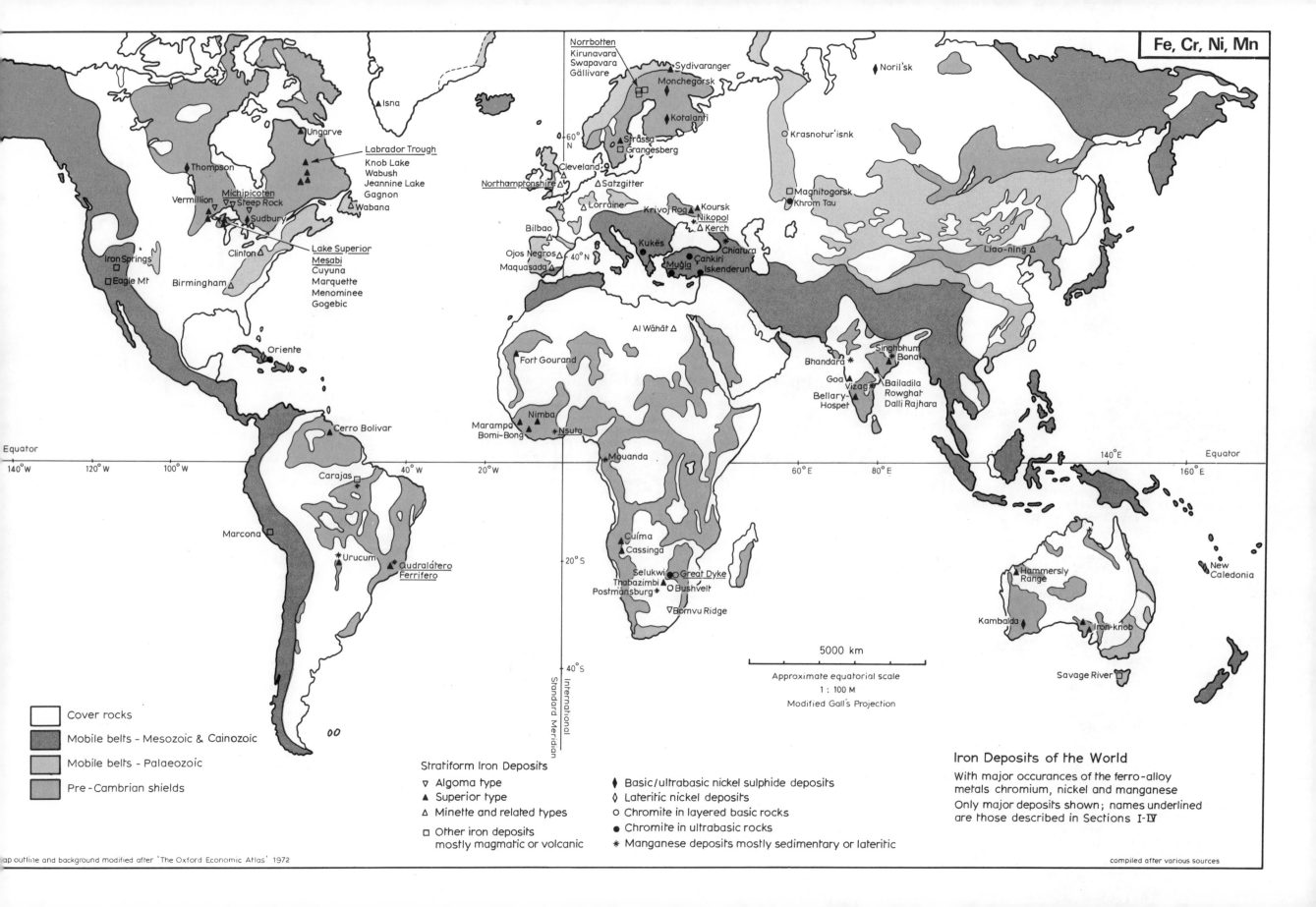

Fe, Cr, Ni, Mn

Norrbotten Kirunavara Swapavara Gällivare

Sydivaranger

Monchegorsk

Noril'sk

Isna

Ungarve

Thompson

Labrador Trough Knob Lake Wabush Jeannine Lake Gagnon Wabana

Kotalanti

Krasnotur'isnk

Vermillion

Michipicoten Steep Rock

Sudbury

Clinton

Strässa
Grangesberg

Cleveland

Magnitogorsk
Khrom Tau

Northamptonshire

Satzgitter

Krivoj Rog

Koursk
Nikopol
Kerch

Lake Superior Mesabi Cuyuna Marquette Menominee Gogebic

Lorraine

Liao-ñing

Iron Springs

Eagle Mt

Birmingham

Bilbao

Ojos Negros

Maquasada

Kukës

Muğla
Çankiri
Iskenderun

Chiatura

Oriente

Al Wāhāt

Singhbhum
Bonai

Bhandara

Fort Gourand

Goa
Bellary-Hospet

Vizag

Bailadila
Rowghat
Dalli Rajhara

Cerro Bolivar

Nimba

Marampa
Bomi-Bong

Nsuta

Carajas

Mouanda

Marcona

Urucum

Oudralátero Ferrifero

Cuíma

Cassinga

Selukwi
Thabazimbi
Postmansburg

Great Dyke

Bushvelt

Bomvu Ridge

Hammersly Range

New Caledonia

Kambalda

Iron-knob

Savage River

5000 km

Approximate equatorial scale

1 : 100 M

Modified Gall's Projection

Stratiform Iron Deposits

∇ Algoma type

▲ Superior type

△ Minette and related types

□ Other iron deposits mostly magmatic or volcanic

◆ Basic/ultrabasic nickel sulphide deposits

◇ Lateritic nickel deposits

○ Chromite in layered basic rocks

● Chromite in ultrabasic rocks

✳ Manganese deposits mostly sedimentary or lateritic

Iron Deposits of the World

With major occurances of the ferro-alloy metals chromium, nickel and manganese

Only major deposits shown; names underlined are those described in Sections I-IV

Legend

Cover rocks

Mobile belts - Mesozoic & Cainozoic

Mobile belts - Palaeozoic

Pre-Cambrian shields

Equator

140°W 120°W 100°W 40°W 20°W 60°E 80°E 140°E 160°E

60°N 40°N 20°S 40°S

Standard Meridian

International

Light Metal Deposits of the World

The less dense or 'light' metals such as aluminium, magnesium, titanium, lithium and beryllium have considerable strength, weight for weight, compared with iron or steel, but have never seriously challenged the latter as principal structural metal for many reasons. Aluminium, titanium and magnesium are abundant, but require much more energy to isolate the metal than is the case for iron. Lithium and beryllium are relatively rare, and have only limited uses. Magnesium is not discussed further here. It largely comes from sea water, or brines associated with evaporites.

Aluminium Deposits

Aluminium is third to oxygen and silicon in its crustal abundance, averaging about 8%. However, by far the largest proportion of the element occurs as complex silicates from which it can only be extracted at great cost. Substantial deposits occur of bauxite, a mixture of the aluminium hydroxide minerals gibbsite, boehemite and diaspore, with impurities of hematite, goethite, silica and clay minerals. Aluminium is only soluble in natural waters under extremely acid or alkaline conditions, and over the normal range of natural conditions at the earth's surface, remains as an insoluble residue. By contrast, the other major elements in rocks, the alkalies and alkaline earths, are readily leached, as is silica, although more slowly. Iron and maganese are readily leached from the surface, providing reducing agents are present; and these are generated by decomposing organic matter. Thus, given the right set of chemical conditions, bauxite may form from any aluminium-bearing rock. There are two major geological environments in which bauxite is found.

Lateritic Bauxite Deposits

Laterite is the name given to a class of surface residual materials largely composed of iron and aluminium hydroxide minerals. It forms an essential part of the so-called lateritic weathering profile that is frequently encountered in areas of low relief within 20° of the Equator. The profile may begin at the surface with a thin, rather infertile soil, followed by an iron-rich crust, capping an irregular porous but rather impermeable layer of laterite. Below this is a zone of softer pale coloured aluminium-rich material generally referred to as the 'pallid zone' that normally contains numerous veinlets of poorly crystalline silica or silica-gel, and this grades into a zone of weathered rock. The bedrock may be anything, and the composition of the laterite reflects that of the bedrock in a general way. Bauxites of this type are aluminium-rich varieties of the laterite layer, and may be so for a number of reasons. The bedrock may be relatively rich in aluminium as compared with iron or manganese; feldspar-rich igneous or metamorphic rocks, for example. But bauxites are found over comparatively iron-rich rocks, and in these cases the iron and aluminium tend to be separate as an upper and lower layer respectively, probably as a result of upward leaching of iron under the influence of dense forest cover which provides the reducing agents necessary for the mobility of the iron. Most lateritic bauxite deposits are found lying on erosion surfaces of relatively young age in South America, West and Central Africa, India and Australia. In every case it can be shown that the erosion surfaces developed over a long period of time during the Tertiary, of the order of 5–10 Ma in duration. Generally, they are not found in the mountainous mobile belts of the Tropics, such as the Andes or South-East Asia. Neither are they found in arid regions, presumably because reasonable rainfall is needed for the leaching processes. Lateritic bauxites are not always *in situ*. Large quantities are found as sediment, albeit locally derived. During transport there may be some sorting that will deplete the bauxite in iron-rich concretions and hence improve its quality, but it may also be contaminated with clay or sand. A few examples of palaeo-bauxites occur, lateritic zones covered by later sediments. Lateritic bauxites provide the larger part of the world's reserves, although they are sometimes of relatively low quality. The important areas are north-east South America, West Africa, Northern Territory and Queensland in Australia.

Other Bauxite Deposits

Most of the other bauxite deposits of the world form by broadly the same process as the lateritic type, but more rapidly and under rather special geological conditions. The so-called terra rosa type, for instance, is found in or associated with karst topography in limestone regions.

These tend to be rather small, irregular, but rich deposits. Terra rosa is an iron, aluminium-rich, clay-like material that accumulates as a residue in karst sink holes as the carbonate of the limestone is removed in solution. Given the right conditions (a moist sub-tropical climate for instance) terra rosa may be desilicated to form bauxite. Karst topography also acts as a trap for other materials such as fine-grained sediment or volcanic ash, which may contribute to the parent material of the bauxite. When compared with the lateritic type, these deposits tend to form relatively quickly. A series of palaeo-bauxites occurring in the south of France, for instance, seemed to have formed within individual stages of the Jurassic, sometimes during a period of no more than 1 Ma.

Bauxite also occurs over highly aluminous rocks such as undersaturated magmatic ones. While not dissimilar from the lateritic type, these seem to form much more rapidly and under sub-tropical conditions.

Mining and Processing of Aluminium

Almost all bauxite is produced from open-pits (in the case of the large lateritic deposits, from very large strip mines). The material is easy to mine, often requiring no blasting, although the irregular nature of the material makes quality control a problem in some cases. Bauxite is treated chemically by one of two processes. The most commonly used is the alkali process in which the bauxite is dissolved in sodium aluminate solution under pressure, then filtered and the pure aluminium hydroxide precipitated by lowering the pressure and temperature. Iron is removed as a red mud from the filters. The other process uses acid to dissolve the alumina and is more suitable for high-silica, low-iron types. The product of either process is pure aluminium oxide, which is smelted to the metal by hot electrolysis in a graphite bath containing the alumina dissolved in a melt of sodium aluminium fluoride. Rarely can this process be carried out near the bauxite deposits, because large quantities of electric power are required, and the smelters are usually sited alongside hydroelectric stations, or thermal power stations based on nuclear reactors, or cheap coal. To smelt a tonne of the metal from 2 tonnes of refined alumina takes about 22 MWh.

Not all bauxite is used for metal production. Either raw bauxite or refined alumina is used for many other purposes, such as alumina cement, refractories and ceramics.

Titanium Deposits of the World

Titanium is a relatively abundant element in the crust, over 4% on average, but like aluminium much of it occurs in silicates. There are two source minerals, the oxides ilmenite and rutile. Both are commonly found as accessory minerals in igneous rocks and both are durable or 'resistate' minerals, and occur as grains in sediments derived from magmatic rocks.

There are two quite different types of titanium deposit, one magmatic, the other sedimentary. The magmatic type consists of ilmenite-rich lenses or layers in basic igneous rock, generally norite or gabbro. Characteristically, these occur in large basic intrusions dominated by thick accumulations of anorthosite associated with charnockites, or similar rocks formed in a deep-seated environment. There are many examples in the world, but in many cases the oxide minerals consist of an intergrowth of magnetite or hematite with the ilmenite, referred to as 'titaniferous iron ores'. For the most part, these are not ore at all, it being too expensive to separate the iron and titanium. However, a few such deposits are rich enough in ilmenite to be worked. The major commercial deposits are found in the younger Pre-Cambrian shield areas of North America and Scandinavia.

The sedimentary deposits are beach sands, generally consisting of quartz sand with a small proportion of ilmenite, rutile, zircon, monazite and other resistate minerals. The deposits seem to form on stormy beaches facing the major oceans, where wave power concentrates the heavy minerals on the berm, which later become preserved under sand dunes. Accessory minerals are not easily liberated in pure form from their parent rocks during erosion, and it is probable that the ultimate source of ilmenite and rutile grains is in laterite, or similar chemical weathering profiles where the total destruction of the silicate matrix liberates the grains. But the grains may go through several phases of sedimentation, erosion and transport, before ending on the right kind of beach. Almost all beach sand deposits occur along long depositial coastlines, bordering stable shield or platform areas, in the climatic zones on either side of the Tropics where the trade winds blow, and hurricanes or monsoon storms are frequent.

Reserves of ilmenite from both kinds of deposit are large, as are those of rutile sands, probably totalling in excess of 200 Mt of metal. There are very large resources of sub-economic ilmenite, both in the form of titaniferous iron deposits in igneous rocks, and impure or low-grade beach sands.

Mining and Processing of Titanium Minerals

The magmatic type of ilmenite ore is usually worked by open-pit methods at grades of about 30% TiO_2. The ilmenite is recovered from the ground ore by flotation, gravity and magnetic methods, often combined with acid leaching to remove impurities such as apatite. Beach sands are worked by strip mining, dredging or pumps, and treated by gravity methods to produce a crude concentrate of heavy minerals. The concentrate is then separated into its constituents by a combination of high-intensity magnetic and electro-static separation.

Relatively little titanium metal is produced as it has only a few specialized uses and is costly. The metal is produced by solid state reduction from rutile. All ilmenite and some rutile is used to produce various forms of titanium oxide for use in the ceramic, glass and pigment industries. The oxide is produced either by a sulphuric acid or chloride process, and chemical separation of the iron from titanium oxides. For this process it is important that the concentrate is free from colouring impurities, particularly chromium. There is also a pyrometallurgical process that involves smelting the concentrate in a special blast furnace to produce pig iron, and a high titanium slag, which then becomes feed-stock for the chemical production of the oxide.

Lithium

Lithium is a widespread trace element in the crust, occurring largely in feldspars and micas. The average crustal content is about 65 μg/g. There are a number of lithium-rich minerals from which the metal can be recovered. The principal one is lepidolite, a mica containing up to 3·3% of the metal. Others are amblygonite, petallite and spodumene. All these minerals occur in complex pegmatites, usually in zones, and are recovered by selective mining and hand picking. Certain brines from continental player-lake evaporites contain lithium which can be recovered by chemical treatment. The metal is produced by fusion and solution of the raw mineral (different minerals require separate treatments) followed by chemical separation and electrolysis.

Beryllium

Beryllium is a rare element averaging only 6 μg/g in the crust. It is found as beryl in complex pegmatites, and as a trace element in certain sediments. Commercially, it is for the most part produced from beryl recovered from pegmatites, but recovery from certain beryllium-rich sediments is feasible.

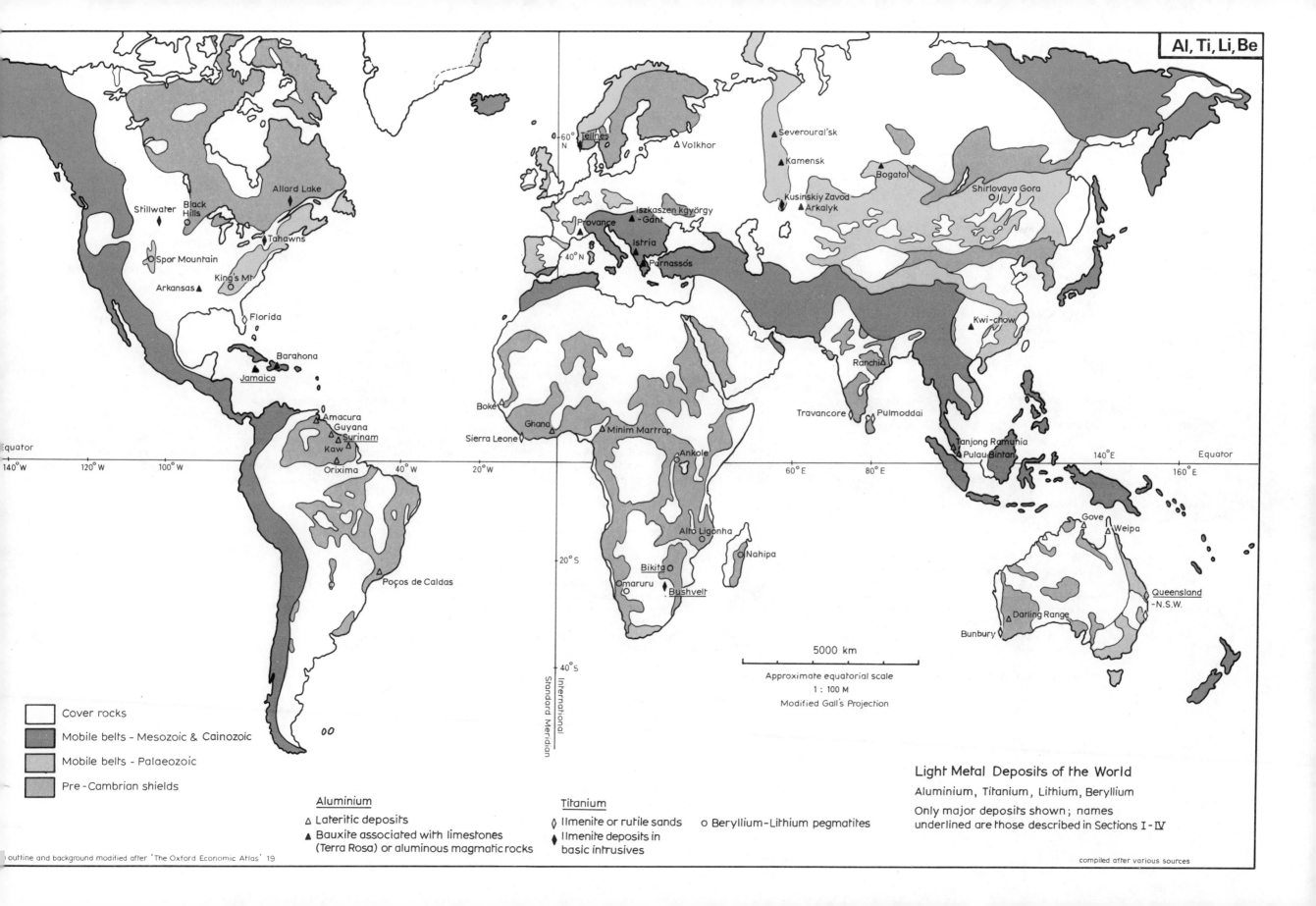

Light Metal Deposits of the World

Aluminium, Titanium, Lithium, Beryllium

Only major deposits shown; names
underlined are those described in Sections I - IV

Al, Ti, Li, Be

Cover rocks

Mobile belts - Mesozoic & Cainozoic

Mobile belts - Palaeozoic

Pre-Cambrian shields

Aluminium

△ Lateritic deposits

▲ Bauxite associated with limestones
(Terra Rosa) or aluminous magmatic rocks

Titanium

◊ Ilmenite or rutile sands

◆ Ilmenite deposits in
basic intrusives

○ Beryllium-Lithium pegmatites

5000 km

Approximate equatorial scale
1 : 100 M

Modified Gall's Projection

outline and background modified after 'The Oxford Economic Atlas' 19

compiled after various sources

Precious Metal Deposits of the World

The precious metals; gold, silver, platinum, palladium, rhodium, ruthenium, osmium and iridium, are extremely rare, but have chemical properties that enable them to exist in a wide variety of environments as native metal or simple compounds with sulphur, arsenic, selenium or tellurium. Deposits exist with concentration clarkes of 10^3 to 10^4. The metals are dense, durable and attractive to look at and have, in the case of gold and silver, maintained their value as a basis for currency, from the dawn of metallurgical cultures.

Geochemistry and Mineralogy

The average contents of these metals in the crust is difficult to measure, but they are of the order of 1 ng/g, except for that of silver, which is about 20 ng/g. The metals are rather unreactive and all exist as native metal both at the surface and in deep-seated rocks. Silver is rather more reactive than the others, and will go into sulphate solution. All of these metals react with the elements, sulphur, arsenic, selenium and tellurium, forming a variety of minerals, and there are many examples of these metals occurring as minor sulphide phases in other more common minerals.

Gold Deposits

Gold is found as a trace mineral in certain clastic sediments, and in a series of deposits of deep-seated origin that are somewhat difficult to classify. The grouping used here is a general one with many borderline cases.

Placer and Palaeo-Placer Gold Deposits

Gold occurs widely in gravels. Typically, these are fluviatile deposits near the headwaters of fast-flowing rivers where gold particles are trapped between pebbles or in bedrock irregularities. Gold, along with other heavy minerals, is concentrated where there is a velocity gradient in the waterflow, such as on bends, or changes in gradient of the water course. The major placer deposits like the famous Klondyke district in north-west Canada are thick and extensive, accumulations of alluvium formed by a series of rivers over a period of time. Placer deposits tend to be destroyed by erosion almost as fast as they are formed, and the biggest deposits are those which were preserved by some accident of geological history. There are two types of such accident, one is the abandonment at high level by crustal uplift or eustatic sea level change (or burial by the reverse process), and the other is preservation beneath lava flows. Most gold placers are no older than Tertiary, but in a few parts of the world occur what are regarded as palaeo-placers.

The classic and largest examples of palaeo-placer deposits are those found in the Witwatersrand Basin in South Africa. These early Pre-Cambrian deposits are of conjectural origin, because there are no modern parallels, or similar deposits in Proterozoic or Phanerozoic rocks. They are now regarded as being nearitic marine or esturine lag-gravels lying on low-angle outwash fans. They are particularly remarkable because of their association with uranium minerals, carbonaceous materials (which show evidence of organic origin) pebbles or concretions of pyrite, and the almost total absence of the usual assemblage of heavy minerals found in modern placers.

Young placers are found in proximity to the major bedrock gold deposit areas with important examples occurring in the high latitude zones of Canada, Alaska and Siberia. Apart from the Witwatersrand Basin, similar though much smaller, palaeo-placers are found in Brazil, Ghana and India. The uranium deposits of Blind River in Canada are essentially of the same type, although they do not contain workable gold.

Quartz-vein Gold Deposits

A large number of comparatively rich quartz-vein and related deposits occur. By far the greater proportion of them are found in the Archean greenstone belts. They usually consist of irregular pods, lenses, shear zones and other forms, filled with quartz and an association of gold, arsenopyrite, pyrrhotite, graphite and many other minerals. In some areas they are found cutting lavas or other volcanic rocks, others are found in close association with taconites or related iron-rich meta-sediments. They are usually structurally complex, and the veins can mostly be seen to be related to one phase of deformation of the host rocks. These deposits are found in many places; quite large districts occur in the Abitibi Belt of Canada, the Yulgarn Block of Western Australia, the Ashanti District of Ghana and the Rhodesian Shield. There are several large individual deposits such as Homestake in South Dakota, Morro Velho in Brazil and Kolar in Southern India.

Several groups of deposits of similar type occur in younger rocks and, although they may have been locally important in the past, they are overshadowed by the Archean ones. The Balarat–Bendigo districts of Australia and the Mother Lode district of California are examples.

The Magmatic and Sub-volcanic Gold Deposits

Largely occurring in the younger continental mobile belts are a whole variety of gold deposits that are broadly related to the hydrothermal phases of calc-alkaline and alkaline igneous rocks. The range of types and styles of mineralization is vast. Examples are Cripple Creek in Colorado, U.S.A. an alkaline volcanic pipe shot through with rich veins and sheeted zones containing gold tellurides. The Comstock district of Nevada is another, a vein zone in volcanic rocks associated with which are pockets of gold and silver-sulphosalt mineralization of great richness. By contrast, one can include in this group the Carlin deposit in Nevada which is a dissemination of extremely fine-grained gold in an altered shale, probably formed under the influence of sub-volcanic or hypabyssal intrusions. Many porphyry copper deposits contain important amounts of gold, and most have an alluvial dispersion of gold around their outcrops.

Gold in Sulphide Deposits

Quite a number (but by no means all) of sulphide deposits, both of the volcanic exhalative type and the magmatic type, contain gold. Generally the pyritic copper and copper–zinc volcanogenic and the Kuroko types contain a little, and some have concentrations of gold in the weathered zones, or gossans, or in surrounding alluvial deposits. Many of these deposits contain good grades of silver. Some related deposits do contain high gold values, notably the Boliden deposit in Sweden and Noranda in Canada in which the gold is largely associated with zones of rather more complex mineralogy than is usual in other pyritic deposits. Gold is usually present in the nickel–copper pyrrhotite ores such as those of Sudbury in Canada.

Silver Deposits

There are a number of distinct silver-rich deposits, but for the most part silver occurs either associated with lead or as an accessory metal to copper or gold.

Silver–Lead Deposits

There is great variation in the silver content of lead deposits. The silver is almost always found as inclusions of silver sulphosalts in galena, although this is sometimes augmented by copper–silver minerals such as tetrahedrite and tennantite. In general, the zinc-rich deposits that occur as strata-bound ore zones in shelf carbonate rocks do not contain much silver, but the large stratiform lead–zinc deposits such as Sullivan in Canada and Broken Hill in Australia are very rich in the metal. So, too, are the lead–zinc deposits that

are associated with magmatic rocks; the contact replacement deposits and the 'Kuroko' type.

Vein and Related Silver Deposits

A number of silver-rich vein deposits occur, the best known being in the Cœur d'Alene district of Idaho, U.S.A. and the Cobalt district of Ontario, Canada. Silver is also a very important constituent of the hydrothermal deposits associated with the calc-alkaline intrusions and volcanic centres of the American Cordillera, particularly the pipe and manto deposits of Mexico.

Platinum Deposits

These are of very restricted distribution and fall into three categories. Most important are the large basic intrusions like the Bushveld Complex of South Africa which contains the Merensky Reef, the resources of which overshadow all other deposits. Platinum metals are found and recovered from the nickel–copper sulphide deposits, and quite substantial alluvial deposits occur in Colombia and the Ural Mountains.

Mining and Processing

One cannot generalize about mining methods for these metals because there is too much variation. Small-scale underground mining is normally possible for gold ores containing 25 g/t or more and large-scale underground methods can be based on ores down to 10 g/t or less. Bedrock open-pit mining can successfully work down to a few grammes per tonne, and modern alluvial deposits can be worked down to as low as 150 mg/t.

All the precious metals have one important chemical property; they dissolve to form complex ions in cyanide solution, and this forms the basis for recovery of these metals in many cases. Other methods are used to pre-concentrate the ore, and it may have to be roasted to remove sulphur and arsenic before treatment. Gold is precipitated from cyanide solution by another metal, usually zinc, and the precipitate smelted to a crude bullion. This is then further refined, usually chemically, to separate the individual metals. Silver is recovered from lead ores after smelting, usually by extraction with molten zinc. Precious metals that are worked as by-products of copper, nickel and zinc mining are usually recovered from the anode-slimes produced during electrolytic refining.

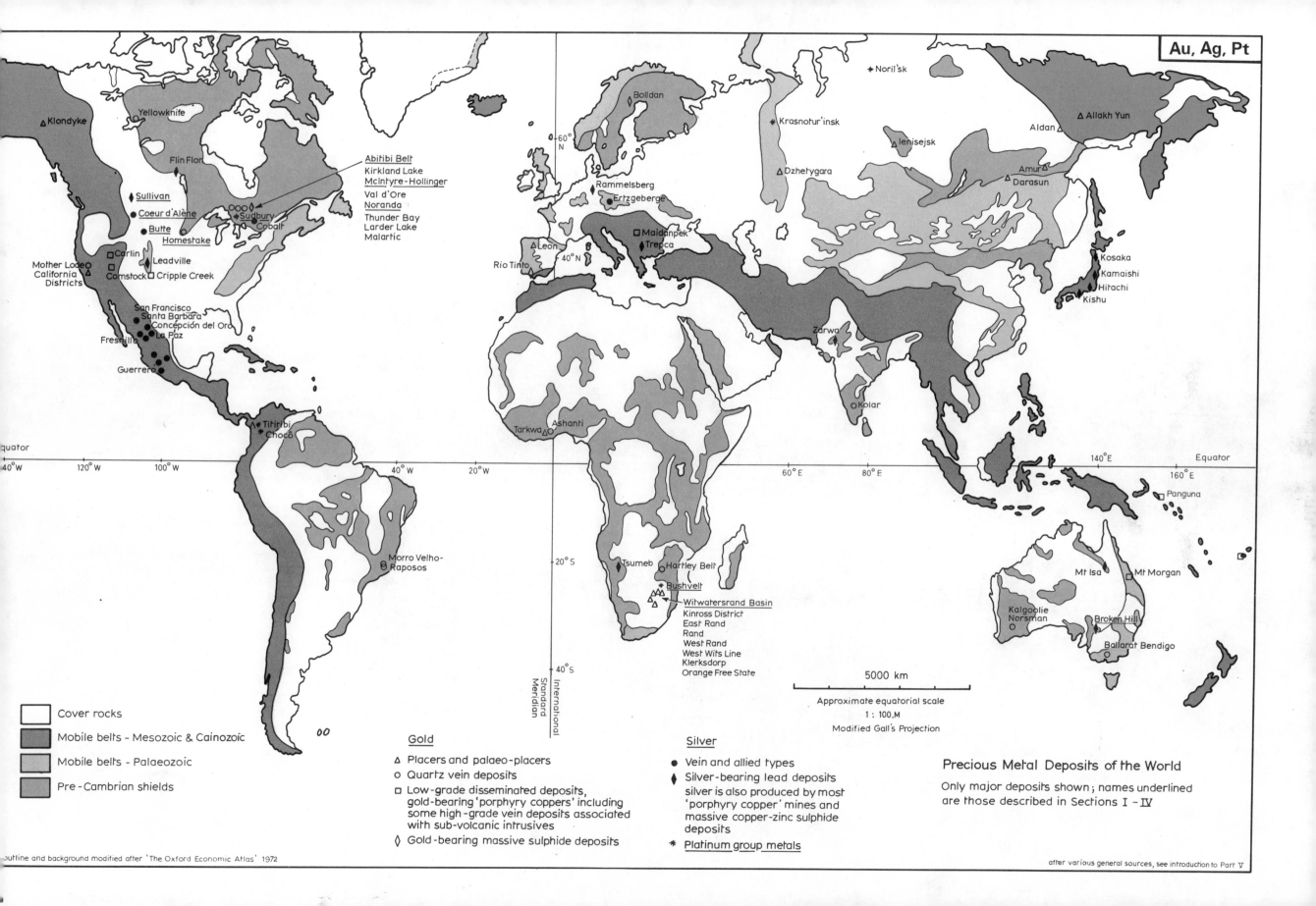

Au, Ag, Pt

* Noril'sk

△ Klondyke
Yellowknife
Bolidan
* Krasnotur'insk
Aldan △
△ Allakh Yun

Flin Flon
△ Ienisejsk

Abitibi Belt
Kirkland Lake
McIntyre-Hollinger
Val d'Ore
Noranda

Sullivan
Coeur d'Alène
Butte
Homestake

Sudbury
Cobalt

Thunder Bay
Larder Lake
Malartic

Rammelsberg
Ertzgeberge
△ Dzhetygara
Amur △
Darasun

□ Carlin
□ Leadville
Mother Lode
California
Districts
△ Comstock □ Cripple Creek

Maidanpek
Trepca
△ Leon
Rio Tinto

Kosaka
Kamaishi
Hitachi
Kishu

San Francisco
Santa Barbara
Concepción del Oro
La Paz
Fresnillo

Zarwa

Guerrero

Ashanti
Tarkwa △
Kolar

△ Titiribi
Chocó

Morro Velho-
Raposos

Tsumeb
Hartley Belt
Bushvelt
Witwatersrand Basin
Kinross District
East Rand
Rand
West Rand
West Wits Line
Klerksdorp
Orange Free State

Panguna

Mt Isa
Mt Morgan

Kalgoolie
Norsman
Broken Hill

Ballarat Bendigo

5000 km

Approximate equatorial scale
1 : 100.M
Modified Gall's Projection

Cover rocks

Mobile belts – Mesozoic & Cainozoic

Mobile belts – Palaeozoic

Pre-Cambrian shields

Gold

△ Placers and palaeo-placers

o Quartz vein deposits

□ Low-grade disseminated deposits,
gold-bearing 'porphyry coppers' including
some high-grade vein deposits associated
with sub-volcanic intrusives

◊ Gold-bearing massive sulphide deposits

Silver

● Vein and allied types

◆ Silver-bearing lead deposits
silver is also produced by most
'porphyry copper' mines and
massive copper-zinc sulphide
deposits

* Platinum group metals

Precious Metal Deposits of the World

Only major deposits shown; names underlined
are those described in Sections I – IV

outline and background modified after 'The Oxford Economic Atlas' 1972

after various general sources, see introduction to Part V

Glossary of Mineral Names

Composition

That given is the simplest emperical composition, arranged as far as possible to indicate any atom-groups that occur at individual lattice sites. Symbols separated by commas are alternative occupants of a lattice site in order of their normal abundance.

Important constituent

Where the mineral is an ore mineral (ore min.) or is used as a source of an element or compound, the average or normal range of weight proportion of the important constituent is given.

Crystal system

Abreviations used are as follows:

Iso – Isometric
Tet – Tetragonal
Orth – Orthorhombic
Hex – Hexagonal
Trig – Trigonal
Mono – Monoclinic
Tric – Triclinic

Density

Given in kg dm^{-3}

Transparency

A transparent (T) mineral is one, the optical properties of which can be measured in a normal thin-section, 30 μm in thickness: an opaque (O) mineral can normally only be examined in reflected light.

Colour

In the case of transparent minerals the normal colour is given; but in most cases there is considerable variation due to impurities, inclusions or other colour centres. In the case of opaque minerals the lustre is given i.e: metallic (Met), sub-metallic (S. Met.), and the colour reflected by the surface of the mineral.

Name	Composition	Important Constituent	Crystal System	Density	Trans-parency	Colour	Use	Remarks
Actinolite	Amphibole $Ca_2(Mg,Fe)_5Si_8O_{22}(OH)_2$	—	Mono	3·2	T	greenish	Fibrous var. as asbestos	
Albite (Ab)	Plagioclase ideally $NaAlSi_3O_8$	—	Tric	2·60	T	white	Rarely used	End member of plagioclase series
Almandine	Garnet $Fe_3Al_2(SiO_4)_3$	—	Iso	4·0	T	deep red	Abrasive for wood and leather	
Alunite	$KAl_3(SO_4)_2(OH)_6$	—	Trig	2·6	T	white	Has been tried as combined source of K, Al and S	Common alteration mineral in sulphide deposits
Ambligonite	$(Li,Na)AlPO_4(F,OH)$	$Li \leqslant 4·2\%$	Tric	3·1	T	white	Li ore mineral	
Amosite	Amphibole similar to anthophyllite	—	Orth	3·2	T	pale brown	Type of asbestos	
Amphibole (group)	$(Ca,Na,Fe,Al,Mg)_7(Si,Al)_8O_{22}(OH)_2$	—	Orth/Mono	variable	T	variable		See individual members of group
Andalusite	Al_2SiO_5	—	Orth	3·3	T	white	Refractory manufacture	Common contact metamorphic mineral
Andesine	Plagioclase Ab_{50} to Ab_{70}	—	Tric	2·66	T	white	Occasionally used as mineral filler and mild abrasive	
Andradite	Garnet $Ca_3Fe_2(SiO_4)_3$	—	Iso	5·78	T/O	Dark brown or green		
Anglesite	$PbSO_4$	Pb: 68%	Orth	6·4	T	white	Supergene ore min. of lead	
Anhydrite	$CaSO_4$	Si: 23·5%	Orth	2·93	T	white	Sometimes used as source of S or in plaster and cement making	
Ankerite	$Ca(Fe,Mg)(CO_3)_2$ can contain Mn	—	Trig	3·0–3·1	T	pale brown		Mineral similar to dolomite found in veins, etc.
Annabergite	$Ni_3(AsO_4)_2.8H_2O$	—	Mono		T	pale green	Found in oxidized zones of Ni deposits	
Anorthite (An)	Plagioclase ideally $CaAl_2Si_2O_8$	—	Tric	2·74	T	white		End member of plagioclase series
Anthophyllite	Amphibole $(Mg,Fe)_7Si_8O_{22}(OH)_2$	—	Orth	3·0–3·2	T	pale brown	Some varieties asbestiform can be used in refractory making in place of talc	
Antigorite	$Mg_3Si_2O_5(OH)_4$	—	Mono	2·6	T	greenish	see serpentine	Platey form of serpentine
Apatite	$Ca_5(PO_4)_3(F,Cl)$ can contain (OH)	P_2O_5: 42%	Hex	3·2	T	white etc.	Main source mineral of phosphate	
Aragonite	$CaCO_3$ can contain Sr	—	Orth	2·94	T	white	Rarely used but can be used in the same way as calcite	Primarily found in skeletal limestones
Argentite	Ag_2S	Ag: 87·1%	Iso	7·3	O	Met. grey	Silver ore min.	
Arsenopyrite	FeAsS can contain Co,Sb,Au	As: 46%	Orth	6·2	O	Met. white	Source of As but often an Au ore mineral	
Atacamite	$Cu_4Cl_2(OH)_6$	Cu: 59·4	Orth	3·7	T	deep green	Cu ore min. from supergene zones	
Augite	Pyroxene $(Ca,Mg,Fe,Al)(Si,Al)_2O_6$	—	Mono	3·2–3·5	T	dark green to black		
Azurite	$Cu_3(CO_3)_2(OH)_2$	Cu: 51%	Mono	3·8	T	deep blue	Cu ore min. from supergene zones	
Baddeleyite	ZrO_2	—		5·5	T	brown	Can be used as source of zirconia for refractions	
Barite	$BaSO_4$	BaO: 60%	Orth	4·6	T	white	Used as such as a heavy medium or as source of Ba chemicals	
Beryl	$Be_3Al_2(SiO_3)_6$	BeO: 14%	Hex	2·8	T	white and others	Ore min. of Be and gem (emerald)	
Biotite	Mica $K(Mg,Fe)_3(AlSi_3)O_{10}(OH,F)_2$	—	Mono	2·7–3·1	T	dark green or brown		
Bismuthinite	Bi_2S_3	Bi: 81%	Orth	6·5	O	Met. grey	Ore min. of Bi but rare	
Boehmite	AlO.OH	Al: 45%	Orth		T	white	Al ore min. constituent of bauxite	
Bornite	Cu_5FeS_4	Cu: 63·3%	Iso	5·4	O	Met. copper red	Cu ore mineral	
Braunite	Mn_7SiO_{12} can contain Fe	Mn: 78%	Tet	4·8	O	S. Met dark brown	Constituent of Mn ores	
Brochantite	$Cu_4SO_4(OH)_6$	Cu: 56%	Orth	3·9	T	green	Cu ore min. from certain supergene zones	
Bronzite	Pyroxene var. of enstatite q.v.							
Bytownite	Plagioclase Ab_{30} to Ab_{10}	—	Tric	2·72	T	white	Rarely used	
Calcite	$CaCO_3$ can contain Mg,Fe,Mn	CaO: 56%	Trig	2·7	T	white	Many industrial and chemical uses incl. lime and cement making	
Carnallite	$KMgCl_3(H_2O)_6$	K: 14%	Orth	1·60	T	white or pink	Source of K, sometimes of Mg and Cl	
Carnotite	$K_2(UO_2)_2(VO_4)_2(H_2O)_3$	U: 40–45%	Orth	4·5	T	yellow	Ore min. of U & V	Supergene mineral
Cassiterite	SnO_2 often with Fe	$Sn \leqslant 78·6\%$	Tet	6·8–7·1	T	variable brown	Main ore min. of Sn.	
Celestite	$SrSO_4$	Sr: 48%	Orth	3·96	T	white	Main source of Sr	
Cerussite	$PbCO_3$	Pb: 78%	Orth	6·55	T	white	Pb ore min. from supergene zones	
Chalcanthite	$CuSO_4.5H_2O$	Cu: 25·4%	Tric	2·3	T	blue	Cu ore min. in certain supergene zones	
Chalcocite	Cu_2S	Cu: 79·8%	Orth/Hex	5·8	O	Met. grey	Ore min. of Cu	Main supergene enrichment mineral of Cu
Chalcopyrite	$CuFeS_2$	Cu: 34·5%	Tet	4·3	O	Met. yellow	Main ore min. of Cu	
Chamosite	$(Fe,Mg,Al)_6(SiAl)_4O_{10}(OH)_8$	$Fe \simeq 30\%$	Orth	4·0	T	Greenish	Ore min. of Fe	
Chlorite (group)	$(Mg,Fe)_5Al(AlSi_3)O_{10}(OH)_8$	—	Mono	2·6–2·9	T	various greens	Constituent of some clays	Common alteration mineral
Chromite	spinel approx. $(Fe,Mg)O.(Cr,Al)_2O_3$	$Cr \leqslant 46\%$	Iso	4·5–4·8	O	S. Met. dark brown	Ore min. of Cr, source of chromates and refractory	
Chrysocolla	approx. $CuSiO_3 2H_2O$	Cu: 36%	Amrph	2·0–2·2	T	pale blue or green	Cu ore min. in some supergene zones	
Chrysotile	$Mg_3Si_2O_5(OH)_4$	—	Mono	2·6	T	pale green or grey	Common asbestos, var. of serpentine	
Cinnabar	HgS	Hg: 86%	Tri	8·1	T	red	Main ore min. of Hg	
Clay mineral group	A variable group of layer structure silicates							See individual members
Clinoenstatite	Pyroxene $MgSiO_3$	—	Mono	3·4	T	greenish-brown		
Clinozoisite	Epidote $Ca_2Al_3(SiO_4)_3(OH)$	—	Mono	3·4	T	grey or greenish		Common alteration mineral
Cobaltite	CoAsS	Co: 36%	Iso	6·3	O	Met. pinkish white	Ore min. of Co but not common	
Coffinite	$USi(O,OH)_4$	U: 72%			O	dull black	Ore min. of U	
Colemanite	$Ca_2B_6O_{11} 5H_2O$	BO_3: 51%	Mono	2·4	T	white	Source of borate	
Collophane	Hydrous variety of apatite						Ill-defined name for the principal material in phosphatic sediments	
Columbite	$(Fe,Mn)(Nb,Ta)_2O_6$	$Nb \leqslant 55\%$	Orth	$\leqslant 5·3$	O	S. Met. black	Ore min. of Nb	Isomorphous with tantalite

Name	Composition	Important Constituent	Crystal System	Density	Transparency	Colour	Use	Remarks
Copper	Cu sometimes with Ag	pure metal	Iso	8·8	O	Met. reddish	Occasional ore min. of Cu	
Cordierite	$(Mg,Fe)_2Al_3(AlSi_5)O_{18}H_2O$	—	Orth	2·7	T	blue and yellow		
Corundum	Al_2O_3	—	Trig	4·0	T	white, brown, blue, red	Abrasive or gem (sapphire or ruby)	
Covellite	CuS	Cu: 66·4%	Hex	4·6	O/T	deep blue	Occasional mineral in Cu ores	
Cummingtonite	Amphibole $(Mg,Fe)_7Si_8O_{22}(OH)_2$	—	Mono	3·2	T	greyish brown		
Cuprite	Cu_2O	Cu: 89%	Iso	6·1	T/O	red	Occasional Cu ore min.	Supergene mineral
Descloizite	$(Zn,Cu)Pb(VO_4)OH$	V: 12%	Orth	5·4 +	T	orange-brown	Occasional ore min. of V	Supergene mineral isomorphous with mottramite
Diamond	C		Iso	3·5	T	white, black	Gem, abrasive	
Diaspore	AlOOH	Al: 45%	Orth	3·4	T	white	Al ore min. – constituent of bauxite	
Digenite	Cu_2S usually with copper deficiency	Cu: 75–79%	Iso	5·5	O	bluish grey	Similar to chalcocite	
Diopside	Pyroxene $Ca(Mg,Fe)Si_2O_6$	—	Mono	3·2–3·4	T	pale green		
Dioptase	$CuSiO_2(OH)_2$	Cu: 41%	Hex	3·3	T	bluish green	Occasional mineral in supergene Cu ores	
Dolomite	$CaMg(CO_3)_2$ can contain Fe	MgO: 21·7%	Tri	2·9	T	white	Source of MgO and some refractory uses	
Enargite	$Cu_3(As,Sb)S_4$	Cu: 48%	Orth	4·4	O.	Met. dark grey	Ore min. of Cu	
Enstatite	Pyroxene $(Mg,Fe)SiO_3$	—	Orth	3·1–3·3	T	green-brown		
Epidote (group)	$(Ca,Fe)_2(Al,Fe,Mn)_3(SiO_4)_3$	—	Orth/Mono	variable	T	shades of green		See individual members
Erythrite	$Co_3(AsO_4)_2(H_2O)_8$	—	Mono	3·0	T	pink	Occasional supergene mineral of Co	
Fayalite	Olivine with Mg:Fe > 1:9	—	Orth	4·3	T	brown-black		
Feldspar (group)	$(K,Ca,Na)(Al,Si)_4O_8$ can contain Ba	—	Mono/Tri	2·5–2·7	T	white		See individual members
Fluorite	CaF_2 often contains rare earths	F: 48%	Iso	3·25	T	white and others	Principal source of F and rare earths – also used as metallurgical flux	
Fosterite	Olivine with Mg:Fe > 9:1	—	Orth	3·2	T	green	Main mineral of rock dunite – used as refractory	
Gahnite	Spinel $ZnO.Al_2O_3$	—	Iso	3·7	T	dark green	Refractory zinc mineral that can give rise to misleading geochemical anomalies	
Galena	PbS can contain Se, Te, Ag	Pb ≤ 87%	Iso	7·6	O	grey	Main ore min. of Pb and Ag	Ag usually as included Ag sulphides
Garnet (group)	$(Ca,Mg,Fe,Mn)_3(Fe,Al,Cr,Ti)(SiO_4)_3$		Iso	3·5–4·3	T	various		See individual members
Garnierite	$(Ni,Mg)_3Si_2O_5(OH)_4$	Ni ≈ 25%	Mono	2·3–2·8	T	green	Ore min. of Ni	
Gibbsite	$Al(OH)_3$	Al: 34·5%	Mono	2·35	T	white	Ore min. of Al – constituent of bauxite	
Glauconite	$K(Fe,Mg,Al)_{4-6}(Si,Al)_8O_{20}(OH)_4$	—		2·3	T	shades of green	Can be used as ion-exchange medium	
Glaucophane	Amphibole $Na_2(Mg,Fe)_3(Al,Fe)_2Si_8O_{22}(OH)_2$	—	Mono	3·1	T	dark blue		
Goethite	FeO.OH	Fe: 62·9%	Orth	4·4	O	black-brown	Constituent of iron ore	Principal mineral in gossans
Gold	Au usually with Ag and Cu	Au: 70–100%	Iso	10–19·3	O	yellow	Main gold ore min.	
Goslarite	$ZnSO_4(H_2O)_7$		Orth	2·1	T	white	Occasional mineral in supergene Zn zones	
Graphite	C		Trig	2·3	O	black	Refractory, lubricant, and conductor	
Greenalite	approx $Fe_{11}Si_8O_{28}(H_2O)_8$	Fe ≈ 40%	Mono	3·5	T	green	Constituent of certain Fe ores	Probably an Fe-rich serpentine
Greenockite	CdS	Cd: 78%	Trig	5·0	T	yellow-orange	Minor source of Cd	
Grossularite	Garnet $Ca_3Al_2(SiO_4)_3$	—	Iso	3·5	T	pale green		The usual garnet in contact replacement deposits
Grunerite	Amphibole $(Fe,Mg)_7Si_8O_{22}(OH)_2$	—	Mono	3·5	T	brown	Constituent of some metamorphosed iron ores	
Gypsum	$CaSO_4(H_2O)_2$	—	Mono	2·3	T	white	Manufacture of plaster	
Halite	NaCl can contain K, Br	—	Iso	2·2	T	white	Common salt and source of Na and Cl	
Halloysite	Clay min. $Al_4Si_4(O,OH)_{10}(OH)_8$	—	Tric	2·6	T	white	Very similar to kaolinite	
Hausmanite	Mn_3O_4 may contain Fe and Zn	Mn ≤ 72%	Tet	4·8	O	S. Met. dark brown	Constituent of Mn ores	
Hedenbergite	Pyroxene $Ca(Fe,Mg)Si_2O_6$	—	Mono	3·7	T	black		
Hematite	Fe_2O_3	Fe: 70%	Trig	5·3	O	Met. black red	Main ore min. of Fe, some industrial uses	
Hemimorphite	$Zn_4Si_2O_7(OH)_2H_2O$	Zn: 37%	Orth	3·45	T	white	Constituent of supergene Zn ores	
Hornblende	Amphibole with complex composition	—	Mono	3·0–3·5	T	dark green		The common amphibole
Hydrozincite	$Zn_5(CO_3)_2(OH)_6$	Zn: 60%	Mono	3·8	T	white	Zn ore min. from supergene zones	
Hypersthene	Pyroxene $(Fe,Mg)SiO_3$	—	Orth	3·4–3·5	T	brownish green		
Idocrase	Approx $Ca_{10}Al_4(Mg,Fe)_2(Si_2O_7)_2(SiO_4)_5(OH)_9$	—	Tet	3·4	T	pale brown and yellow		Common in contact replacement zones
Illite	Clay similar to muscovite mica	—	Orth	variable	T	white-brown	Main constituent of common clay	
Ilmenite	$FeTiO_3$	TiO_2: 53%	Trig	5·0	O	S. Met. black	Source of TiO_2	
Iridosmine	(Os,Ir) Os > 35%	pure metal	Hex	19·3	O	grey	Minor ore min. of Os and Ir	
Iron	Fe often contains other metals e.g.: Ni	pure metal	Iso	7·8	O	Met. grey		Rare but found in some igneous and volcanic rocks
Jarosite	$K(Fe,Al)_3(SO_4)_2(OH)_6$	—		3·2	T	yellow-brown	Constituent of gossans and leached outcrops	
Kaolinite	Clay min. $Al_4Si_4O_{10}(OH)_8$	—	Tric	2·6	T	white	Ceramics and as filler	
Kyanite	Al_2SiO_5	—	Tric	3·7	T	white or blue	Refractory manufacture	
Labradorite	Plagioclase $Ab_{20}-Ab_{50}$	—	Tric	2·67	T	white	Occasionally used as a gem	
Lawsonite	$CaAl_2Si_2O_6(OH)_4$	—	Orth	3·1	T	white, etc		
Lepidocrosite	FeO.OH	Fe: 62·9%	Orth	4·4	O	reddish brown	Constituent of iron ores	Found in gossans with Goethite
Lepidolite	Mica $KLi_2AlSi_4O_{10}(OH,F)_2$	Li: 4%	Mono	2·9	T	lilac or rose	Common ore min. of Li	
Leucite	$KAlSi_2O_6$	K_2O: 18%	Tet	2·5	T	white	Can be used as K_2O source in fertilizers	
Linnaeite	$(Co,Ni,Fe,Cu)_3S_4$	Co ≤ 58%	Iso	4·8–5·0	O	Met. pale grey	Ore min. of Co	Co and Cu-rich variety called 'carrollite'

Name	Composition	Important Constituent	Crystal System	Density	Transparency	Colour	Use	Remarks
Maghemite	Fe_2O_3	Fe: 70%	Iso	5.0	O	S. Met. dark brown	Constituent of some Fe ores	Oxidation product of magnetite
Magnesite	$MgCO_3$ can contain Fe, Ca, Mn	MgO: 47.6%	Tri	3.1	T	white-grey	Source of MgO for refractory use	
Magnetite	Fe_3O_4 can contain Ti, V, Mg	Fe: 72.4%	Iso	5.2	O	Met. black	Ore min. of Fe also used as heavy medium	
Malachite	$Cu_2CO_3(OH)_2$	Cu: 57.3%	Mono	4.0	T	bright green	Cu ore min. from supergene zones	
Manganite	$MnO.OH$	Mn: 62.5%	Orth	4.4	O	S. Met. dark grey	Constituent of Mn ores	
Marcasite	FeS_2	S: 53.4%	Orth	4.9	O	Met. pale yellow	Occasional source of S	
Melanerite	$FeSO_4(H_2O)_7$		Mono	1.9	T	pale green	Occasional mineral in supergene zones	
Mercury	Hg	Hg: 100%	Liquid	13.6	O	Met. white	Used as such	
Metatyuyamunite	$Ca(UO_2)_2(VO_4)_2(H_2O)_{5-7}$		Orth	3.4	T	yellow	Constituent of supergene U ores	
Mica (group)	Layer structure silicates based on Si_4O_{10}	—	Mono	variable	T	various		See individual members
Microcline	Feldspar similar to orthoclase	—	Tri	2.56	T	white	Can be used in same way as orthoclase	
Minnesotaite	$(Fe,Mg)_3Si_4O_{10}(OH)_2$ an Fe-rich talc	Fe: 27%	Mono	3.0	T	Greenish or brownish	Constituent of some iron ores	
Molybdenite	MoS_2 often contains Re	Mo: 60%	Hex	4.8	O	Met. pale bluish grey	Ore min. of Mo and Re	
Monazite	$(La,Ce)PO_4$ contains other rare earths and Th	Th(up to 9%)	Mono	4.9-5.3	T	pale brown	Can be used as source of Th	
Monmorillonite	Clay min. very similar to smectite		Mono	variable	T	white or brown	Absorbent clay	Main constituent of bentonite and fullers earth
Montroseite	$(V_3Fe)O.OH$	V: 36%	Orth		T		Constituent of oxidized V ores	Sometimes found in ores of U
Mottramite	$(Cu,Zn)Pb(VO_4)(OH)$	V: 12%	Orth	6.2	T	orange-brown	Ore min. of V	Supergene mineral isomorphous with decloisite
Muscovite	Mica $KAl_2(AlSi_3)O_{10}(OH,F)_2$	—	Mono	2.76	T	white	Insulator and dialectric in large sheets	
Nepheline	$(Na,K)AlSiO_4$	—	Hex	2.5	T	white	Sometimes used in glass-making	
Niccolite	$NiAs$ can contain Co, Fe or S	Ni: 44%	Hex	7.6	O	Met. pale reddish	Occasional ore min. of Ni	
Nitratine	$NaNO_3$ associated with Iodine	—	Tri	2.29	T	white	Source of nitrate	Principal mineral in the nitrate deposits of Chile
Oligoclase	Plagioclase Ab_{70} to Ab_{90}	—	Tri	2.64	T	white	Rarely used	
Olivine (group)	$(Mg,Fe)_2SiO_4$	—	Orth	3.2-4.7	T	green	As refractory	See end members: fosterite and fayalite
Orthoclase	Feldspar $KAlSi_3O_8$	K_2O:	Mono	2.57	T	white	As flux in ceramics	End member of feldspar group
Orthopyroxene	Generic name for pyroxenes with orthorhombic structures							See individual members
Osmiridium	(Ir,Os) $Os < 35\%$	pure metal	Hex	21.1	O	grey	Minor ore of Ir and Os	
Palygorskite	$Mg_5(Si_4O_{10})_2(OH)_4(H_2O)_4$		Orth	variable but low	T	pale brown or white	Absorbent clay	
Paragonite	Mica Na analogue of sericite	—	Mono	2.7	T	white		Found in alteration zones
Pentlandite	$(Fe,Ni)_8S_9$ to $(Fe,Ni)S$	Ni ⩽ 22%	Iso	5.0	O	Met. bronze yellow	Main ore min. of Ni	Usually occurs as a fine intergrowth with pyrrhotite
Perthite	Feldspar: unmixing intergrowth of potash and plagioclase feldspar							
Petalite	$LiAlSi_4O_{10}$	Li 2%	Mono	2.4	T	white	Li ore mineral	
Phlogopite	Mica $KMg_3(AlSi_3)O_{10}(F,OH)_2$	—	Mono	2.78-2.85	T	white – pale brown		
Pigeonite	Clinopyroxene $(Mg,Fe,Ca)SiO_2$	—	Mono	3.3	T	brownish		
Plagioclase (subgroup of feldspars)	$(Na,Ca)(Al,Si)_3O_8$	—	Tri	2.6-2.77	T	white		See individual members
Platininum	Pt usually with Pd, Rh, Ir, Os, Ru, Fe.	Pt: 45-85%	Iso	21.5	O	Met. grey	Ore min. of Pt metals	
Pollucite	$(Cs,Na)_2Al_2Si_4O_{12}(H_2O)$	Cs ≈ 30%	Iso	2.9	T	white	Rare but has been used as ore min. of Cs	
Polyhalite	$K_2MgCa_2(SO_4)_4(H_2O)_2$	K: 13%	Tric	2.0	T	white or pink	Can be used as source of K	
Powellite	$CaMoO_4$ can contain W		Tet	4.3	T	white	Can be used as Mo ore min. but rare	Often found with its isomorph scheelite
Proustite	Ag_3AsS_3	Ag: 65%	Hex	5.6	O/T	S. Met. red	Ore min. of Ag associated with tennantite	
Psilomelane	$(Ba,Mn'')Mn_4O_8(OH)_2$	Mn ⩽ 52%	Amrph	4	O	S. Met. black	Ore min. of Mn	
Pumpellyite	$Ca_4(Al,Fe,Mg)_6Si_6O_{23}(OH)_3(H_2O)_2$	—	Orth	3.2	T	bluish-green		
Pyrargyrite	Ag_3SbS_3	Ag: 60%	Hex	5.9	O/T	S. Met. purplish red	Ore min. of Ag associated with tetrahedrite	
Pyrite	FeS_2 can contain Co, Ni, Se, Te, Au	S: 53.4%	Iso	5.1	O	Met. pale yellow	Source of S and sometimes minor source of Fe, Co, Au	
Pyrochlore	$(Ca,Na,Ce)(Cb,Ti,Ta)(O,OH,F)_7$ can contain Th	—	Iso	4.2-4.4	T	pale brown	Can be used as ore min. of Cb and Th	
Pyrolusite	MnO_2	Mn: 63%	Orth	4.8	O	Met. grey	Main ore min. of Mn	
Pyromorphite	$Pb_5(PO_4)_3Cl$	Pb: 76%	Hex	7.1	T	yellowish green	Constituent of supergene lead ores	
Pyrope	Garnet $Mg_3Al_2(SiO_4)_3$	—	Iso	3.7	T	deep crimson	Gem garnet	
Pyrophyllite	$Al_2Si_4O_{10}(OH)_2$	—	Mono	variable	T	white	Can be used in place of talc	
Pyrrhotite	Fe_8S_9 to FeS often with Ni	Ni up to 5%	Hex	4.6	O	Met. pinkish brown	Ore min. of Ni usually with pentlandite	
Quartz	SiO_2	Si: 46.7%	Tri	2.65	T	white and various	Source of silicon, silicate and some industrial uses	
Rhodochrosite	$MnCO_3$ can contain Fe, Mg, Ca	Mn: 47.8%	Tri	3.6	T	pale pink	Ore min. of Mn	
Rhodonite	$MnSiO_3$	—	Tri	3.5	T	pink	Can be a prot-ore mineral of Mn deposits	
Riebeckite	Amphibole $Na_2Fe_5Si_8O_{22}(OH)_2$	—	Mono	3.4	T	blue	Fibrous variety is blue asbestos	
Roscoelite	Mica $K(V,Al)_3Si_3O_{10}(OH)_2$	—	Mono	2.7	T	rose pink	Can be prot-ore mineral of supergene V deposits	
Rutile	TiO_2	Ti: 60%	Tet	4.2	T	reddish brown	Ore min. of Ti source of TiO_2 and industrial uses	
Sanidine	Feldspar, same as orthoclase	—	Mono	2.57	T	white		High temp. form of orthoclase
Scapolite	$((Na,Ca)(Si,Al)_4O_8)_3(NaCl,CaCO_3)$	—	Tet	2.6-2.75	T	white		

Name	Composition	Important Constituent	Crystal System	Density	Transparency	Colour	Use	Remarks
Scheelite	$CaWO_4$ can contain Mo	W: 63·8%	Tet	6·1	T	white	Ore min. of W	
Sepiolite	$Mg_3Si_4O_{11}5H_2O$	—	Mono	variable	T	pale colours	A clay-like mineral similar to smectites	
Sericite	Mica similar to muscovite	—	Mono	2·7	T	white	Constituent of some clays	
Serpentine	$Mg_3Si_2O_5(OH)_4$	—	Mono	2·5	T	greenish	Sometimes used as refractory	See also antigorite and chyrsotile
Siderite	$FeCO_3$ can contain Mn, Mg, Ca	Fe: 48·3%	Tri	3·9	T	greenish or brown	Ore min. of Fe	
Sideroplesite	$(Fe,Mg)CO_3$ 5% < Mg < 30%		Tri	3·5	T	pale buff to brown		A magnesium siderite found in some veins etc.
Sillimanite	Al_2SiO_5	—	Orth	3·23	T	white	Refractory manufacture	
Silver	Ag often with Cu, Au or others	pure metal	Iso	11·1	O	Met. white	Constituent of some silver ores	
Skutterudite	$(Co,Ni,Fe)As_3$ to $(Co,Ni,Fe)_2As_5$	Co ≤ 28%	Iso	5·7–6·8	O	Met. pale grey	Occasional ore min. of Co	
Smectite	$(Na,Ca)(Al,Mg)_4Si_8O_{10}nH_2O$		Mono	variable	T	white or brown	Absorbent clay	Generic name for group of reactive clay minerals
Smithsonite	$ZnCO_3$ can contain Cd, Fe, Mn, Ca	Zn ≤ 52%	Trig	4·5	T	white	Zn ore min. from supergene zones	
Sperrylite	$PtAs_2$ usually contains Pd, Rh, Ru, Ir, Os	Pt metals: 57%	Iso	10·6	O	white	Ore min. of Pt metals	
Spessartite	Garnet $Mn_3Al_2(SiO_4)_3$		Iso	4·2	T	pink or brown	Sometimes the prot-ore of Mn deposits	
Sphalerite	$(Zn,Fe,Cd)S$	Zn: 50–67%	Iso	3·9–4·2	O/T	pale to dark brown	Main ore min. of Zn	Principal source of Cd
Sphene	$CaTiSiO_5$	—	Mono	3·54	T/O	brown		
Spinel (group)	$(Mg,Fe,Zn,Mn)O(Al,Cr,Fe)_2O_3$		Iso	wide var.	O/T	various	Of the group magnetite and chromite are used q.v.	
Spodumene	A Li clino-pyroxene $LiAlSi_2O_6$	Li: 3·7%	Mono	3·2	T	white	Li ore min.	
Stannite	Cu_2SnFeS_4	Sn: 27·5%	Tet	4·4	O	Met. grey	Ore min. of Sn	
Staurolite	approx $(Al_2SiO_5)_2(Fe,Mg)(OH)_2$		Orth	3·7	T	shades of brown		Metamorphic min. of very complex composition
Stibnite	Sb_2S_3	Sb: 72%	Orth	4·6	O	Met. grey	Main ore min. of Sb	
Stilpnomelane	Mica Fe rich var. of biotite	Fe ≈ 28%	Mono	3·4	T	dark brown-black	Constituent of certain iron-ores	
Strotianite	$SrCO_3$ can contain Ca	SrO: 70%	Orth	3·7	T	white	Source of Sr chemicals	
Sulphur	S may contain Se & Te	S: 100%	Orth	2·07	T	yellow	Principal source of S	
Sylvite	KCl	K: 52·4%	Iso	2·0	T	white or pink	Principal source of K	
Talc	$Mg_3Si_4O_{10}(OH)_2$	—	Mono	2·8	T	white	Various uses as a filler, lubricant and in ceramic and refractory manufacture	
Tantalite	$(Fe,Mn)(Ta,Nb)_2O_6$	Ta ≤ 70%	Orth	7·3	O	S. Met. black	Ore min. of Ta	Isomorphous with columbite
Tennantite	Cu_3AsS_3 to $Cu_{12}As_4S_{13}$ with Ag	Cu ≤ 52%	Iso	4·4	O	Met. grey	Minor ore min. of Cu and common ore min. of Ag	
Tetrahedrite	Cu_3SbS_3 to $Cu_{12}Sb_4S_{13}$ with Ag	Cu ≤ 46%	Iso	5·1	O	Met. grey	(up to 30%) often associated with galena	
Thorianite	ThO_2 usually contains U		Iso	9·3	O	dull black	Ore min. of Th	Isomorph of uraninite
Thucholite	Complex hydrocarbon with U & Th		Amrph		O	black coal-like	Ore min. of U	
Topaz	$Al_2SiO_4(OH,F)_4$	—	Orth	3·6	T	white or various	Gem	Occurs in certain hydrothermal alteration zones
Tourmaline	$(Na,Ca)(Mg,Fe,Li)_3B_3Al_3((Al,Si)_3O_9)_3(OH,F)_4$	—	Trig	3·0	T	various	Occasionally a gem	Occurs in hydrothermal alteration zones
Tremolite	Amphibole $Ca_2Mg_5Si_8O_{22}(OH)_2$	—	Mono	2·9	T	white	Fibrous var. used as asbestos	
Turquoise	$CuAl_6(PO_4)_4(OH)_85H_2O$	—	Tric	2·7	T	greenish blue	Gem, found in oxidized zones of Cu deposits	
Tyuyamunite	$Ca(UO_2)_2(VO_4)_2(H_2O)_{10}$	U: 48%	Orth	3·3–4·3	T	yellow	Supergene ore min. of uranium	
Uraninite	UO_2 but often oxidized	U ≤ 88%	Iso	up to 9·7	O	S Met. black	Main ore min. of U	
Uvavorite	Garnet $Ca_3Cr_2(SiO_4)_3$	—	Iso	3·42	T	pale green		Found in association with chromite deposits
Vanadinite	$Pb_5(VO_4)_3Cl$	V: 11%	Hex	7·1	T	ruby red	Constituent of supergene Pb ores – can be used as ore min. of V	
Vermiculite	Clay or mica-like mineral similar in composition to phlogopite				T	brownish	Expands on heating and is used to produce absorbent and insulating materials	
Violarite	A linnaeite approx $FeNi_2S_3$	N: 40%	Iso	4·8	O	pale violet	Occasional ore min. of Ni	Main supergene enrichment mineral of Ni
Witherite	$BaCO_3$	BaO: 77%	Orth	4·3	T	white	Occasional source of Ba chemicals but rare	
Wolframite	$(Fe,Mn)WO_4$	W: 60%	Mono	7·1–7·9	O	S. Met. dark brown	Ore min. of W	
Wollastonite	$CaSiO_3$	—	Tric	2·9	T	white	Occasionally as refractory	
Zaratite	$(Ni,Mg)_3CO_3(OH)_4(H_2O)_4$	—	—	2·5	T	green	Constituent of supergene Ni ores	
Zeolites	Generic name for a group of hydrated alkali alumino-silicates commonly found as late stage products in volcanic rocks							
Zinnwaldite	Mica $K(Li,Fe,Al)_3(Si,Al)_4O_{10}(F,OH)_2$	—	Mono	2·7	T	pale violet-brown	Minor mineral in Li deposits and alteration zones	
Zircon	$ZrSiO_4$	ZrO_2: 67%	Tet	4·7	T	white	Refractory or as source of zirconia	
Zoisite	Epidote $Ca_2Al_3(SiO_4)_3(OH)$	—	Orth	3·3	T	pale green		Common alteration mineral

Units of Measurement

The units of measurement used in this book are those of the Système Internationale d'Unités (SI) and those commonly used in the mineral industry that are acceptable in association with the S.I. For the convenience of those readers who are not accustomed to using the S.I., a table of equivalents is shown below.

The S.I. is based on fundamental units, each of which has a precise, and internationally agreed definition.

length	metre (m)
mass	kilogramme (kg)
time	second (s)
electric current	ampere (A)
temperature	kelvin (K)

Multiples of these and other units derived from them, are indicated by the use of prefixes i.e.:

giga	G	10^9
mega	M	10^6
kilo	k	10^3
deci	d	10^{-1}
centi	c	10^{-2}
milli	m	10^{-3}
micro	μ	10^{-6}
nano	n	10^{-9}

(Other multiples are used but not in this book)

Quantity	Unit used	Definition in fundamental units	Equivalent in non-S.I. units (in most cases given to three sig. fig.)
Length	μm	10^{-6}m	sometimes called the 'micron'
	mm	10^{-3}m	= 0·0394 inches
	m	–	= 39·4 inches = 3·28 feet = 1·094 yards
	km	10^3m	= 0·621 mile
Mass	ng	10^{-12}kg	
	μg	10^{-9}kg	
	mg	10^{-6}kg	= 0·0154 grains
	g	10^{-3}kg	= 0·643 pennyweights (dwt) = 0·0322 ounces troy = 0·0353 ounce avoir du poids = 5 carat (precious stones)
	kg	–	= 2·204 pounds = 35·3 ounces avoir du poids = 32·2 troy ounces = 1·65 katis (Malaysian measure of tin)
	t	10^3kg	= 0·984 long ton = 1·102 short ton
	Mt	10^9kg	(colloquially the million tonnes)
	Gt	10^{12}kg	(colloquially the billion tonnes)
Time	s	–	
	d	86400s	= the day
	a	31·6 Ms	= the year
	Ma	–	= the million years (m.y)
	Ga	–	= the billion years (b.y)
Temperature	°C	°K − 273·15	i.e.: degree Celsius = $\frac{5}{9}$ (degree Fahrenheit − 32)
Area	m²	–	= 9·84 square feet = 1·094 square yards
	ha	10^4m²	= 2·47 acres
	km²	10^6m²	= 0·386 square miles
Volume and capacity	dm³		effectively the same as the litre (l) = 0·220 Imperial gallons = 0·264 U.S. gallons = 0·0353 cubic feet
	m³		= 1·308 cubic yards = 35·3 cubic feet

Quantity	Unit used	Definition in fundamental units	Equivalent in non-S.I. units (in most cases given to three sig. fig.)
Density	t/m³ = kg/dm³	10^{-3}kg/m³	effectively the same as relative density (formally known as 'specific gravity')
Pressure	kPa	10^3N/m²	= 0·145 pounds per square inch
Mass proportion (grade of ore etc.)	ng/g = mg/t		= part per billion (p.p.b)
	μg/g = g/t		= part per million (p.p.m) = 0·633 dwt per long ton = 0·583 dwt per short ton = 0·0316 troy ounces per long ton = 0·0291 troy ounces per short ton
	mg/g = kg/t		parts per thousand = 2·24 pounds per long ton = 2·00 pounds per short ton
			% is used for parts per hundred
Mass/volume proportion	mg/m³		= 0·0118 grains per cubic yard
	g/m³		= 0·492 pennyweight per cubic yard = 0·0246 troy ounces per cubic yard
	kg/m³		= 1·69 pounds per cubic yard
Rate of mass flow (mine production rate etc.)	t/d		= 0·984 long tons per day = 1·102 short tons per day
	t/a		= 0·984 long tons per annum = 1·102 short tons per annum

Key to Stratigraphic Names

This key is included for the convenience of readers who are not familiar with the stratigraphic terms that are used in the book.

Table of names for the Phanerozoic Eon

Era	Period	Date Ma	Epoch	Age	Alternative terms used in North America
Cainozoic	Quaternary	Present	Holocene	(Versilian)	
			Pleistocene	Tyrrhenian	
				Milazzian	
				Sicilian	
				Emilian	
				Calabrian	
	Tertiary — Neogene	2·5	Pliocene	Piacenzian	
				Zanclean	
			Miocene	Messinian	
				Tortonian	
				Serravallian	
				Langhian	
				Burdigalian	
				Aquitanian	
	Tertiary — Palaeogene	26	Oligocene	Chattian	
				Rupelian	
			Eocene	Bartonian	
				Lutetian	
				Ypresian	
			Palaeocene	Thanetian	
				Montian	
				Danian	

Era	Period	Date Ma	Epoch	Age	Alternative terms used in North America
Mesozoic	Cretaceous	65	Late	Maastrichtian Campanian Santonian Coniacian Turonian Cenomanian	Gulfian
			Early	Albian Aptian Barremian Hauterivian Valanginian Berriasian	Comanchean
	Jurassic	136	Late	Portlandian Kimmeridgian Oxfordian	
			Middle	Callovian Bathonian Bajocian Aalenian	
			Early	Toarcian Pliensbachian Sinemurian Hettangian	
	Triassic	190	Late	Rhaetian Norian Carnian	
			Middle	Ladinian Anisian	
		225	Early	Scythian	

Era	Period	Date Ma	Epoch	Age	Alternative terms used in North America	
Palaeozoic (Upper Palaeozoic) (Lower Palaeozoic)	Permian		Late	Tatarian Kazanian	Ochoan Guadalupian	
		280	Early	Kungurian Artinskian Sakmarian	Leonardian Wolfcampian	
	Carboniferous		Late	Stephanian Westphalian	Virgilian Missourian Desmoinesian Atokan Morrowan	Pennsylvanian
				Namurian	Springerian Chesterian	
		345	Early	Visean Tournaisian	Meramecian Osagean Kinderhookian	Mississippian
	Devonian		Late	Famennian Frasnian	Chautauquan Senecan	
			Middle	Givetian Couvinian	Erian	
		395	Early	Emsian Siegenian Gedinnian	Ulsterian	
	Silurian		Late	Ludlovian Wenlockian	Cayugan Niagran	
		430	Early	Llandoverian	Medinan	
	Ordovician		Late	Ashgillian Caradocian	Cincinnatian	
		500	Early	Llandeilian Llanvirnian Arenigian Tremadocian	Champlainian Canadian	
	Cambrian		Late	Shidertinian Tuorian	Croixian	
			Middle	Mayan Amgan Lenan	Albertan	
		570	Early	Aldanian	Waucoban	

Table of names for the Pre-Cambrian

Eon	Date Ma	Subdivision	Subdivisions used only in North America
Proterozoic	570	Late	Hadrynian
	880		Helikian
	1640	Early	Aphebian
Archean	2390		

Notes

The Phanerozoic time-scale is based on the correlation of fossil faunas and the radiometric dating of a number of key localities. The names of Eras, Periods and Epochs are universal (except for the separation of the Carboniferous into Mississippian and Pennsylvanian in North America). The names of Ages given are those generally in use throughout Europe with some commonly used alternatives used in North America. Many other Age names are used but they have been avoided in this book as far as possible.

Groups of rocks deposited during a period of time are called by different names than the period of time itself as follows:

Time period	Rock group
Era	Erathem
Period	System
Epoch	Series
Age	Stage

In some older publications, these terms may be found used in a different sense.

The Pre-Cambrian time-scale is based on lithological correlations combined with radiometric ages over wide areas. The scale is necessarily more crude than that for the Phanerozoic and the names used vary from place to place. Only the terms used in this book are included in the table.

Index

Numbers in bold type indicate the principal page references. In the case of minerals please refer also to the Glossary on pages 132–6

141